AF595438

Agroforestry and Sustainable Agricultural Production

Agroforestry and Sustainable Agricultural Production

Editor

Víctor Rolo

MDPI • Basel • Beijing • Wuhan • Barcelona • Belgrade • Manchester • Tokyo • Cluj • Tianjin

Editor
Víctor Rolo
University of Extremadura
Spain

Editorial Office
MDPI
St. Alban-Anlage 66
4052 Basel, Switzerland

This is a reprint of articles from the Special Issue published online in the open access journal *Sustainability* (ISSN 2071-1050) (available at: https://www.mdpi.com/journal/sustainability/special_issues/Agroforestry_SAP).

For citation purposes, cite each article independently as indicated on the article page online and as indicated below:

LastName, A.A.; LastName, B.B.; LastName, C.C. Article Title. *Journal Name* **Year**, *Volume Number*, Page Range.

ISBN 978-3-0365-5577-5 (Hbk)
ISBN 978-3-0365-5578-2 (PDF)

Cover image courtesy of Jose Luis Romero Vivas

Contents

About the Editor

Víctor Rolo

Víctor is a Senior Researcher at the Forest Research Group of the University of Extremadura. He holds a Ph. D in Plant Biology and Production from the University of Extremadura. His past working experience includes postdoctoral positions at the University of Mendel (Czech Republic) and the University of Pretoria (South Africa), where he specialized in adaptive management and forest restoration. His research interests focus on assessing soil—plant components of forest and agroforestry systems to enhance their persistence and productivity in a changing environment. He uses experimental, observational and modelling methods in his research and has published several articles and book chapters.

Preface to "Agroforestry and Sustainable Agricultural Production"

Our ability to feed the future world population with current agriculture production practices has been questioned. Intensification has been the usual option taken by modern agriculture, but it still has a limited potential to meet the increasing food demand without degrading the environment. The consequences of intensification include a loss of biodiversity, a decline in soil fertility, and the collapse of agroecosystem functions. There is a large consensus that innovative practices and more sustainable approaches of farming production are needed, but to what extent they have the potential to fill the yield gap is still unclear.

This book collects original contributions on innovative agroecological practices that seek to maximize production as well as deliver multiple services to society, including biodiversity conservation. This book places a special focus on agroforestry because of its known potential to deliver ecological benefits with a wide range of products and services, while at the same time maximizing resource usage. Overall, this book collects several case studies that address the potential of agroforestry to foster food production while minimizing the negative effects on the environment, thereby empowering local communities and building resilience against future climate scenarios.

Víctor Rolo
Editor

 sustainability

Editorial

Agroforestry for Sustainable Food Production

Víctor Rolo

Forest Research Group, INDEHESA, University of Extremadura, 10600 Plasencia, Spain; rolo@unex.es

Agricultural production is considered to be among the largest drivers of global environmental degradation. Agricultural activities are behind a substantial share of greenhouse gasses emissions, occupy a large amount of the Earth's land surface, consume vast quantities of freshwater, and are responsible for the degradation and fragmentation of forests and the loss of biodiversity. At the same time, feeding an increasing global population in the coming decades will be a global challenge. The lack of suitable arable land in the scenario of climate change is argued as the main factor that will increase the gap between food production and demand. The food production gap is magnified by persistently poor management that has degraded the soil in many areas of the world, limiting the land available for agriculture. In this context, there is an increasing interest in the adoption of practices that maintain the productive capacity in a changing climate and limit the degradation of the environment.

Agroforestry, defined as the deliberate combination of woody vegetation with crops and/or animal systems, has been proposed as a suitable method for agricultural management capable of facing the current environmental challenges. The ecological and economic benefits resulting from the integration of various elements that are part of an agroforestry system can foster the multifunctionality of agricultural lands and limit the various trade-offs associated with food production. Agroforestry has been shown to benefit carbon sequestration, reduce soil erosion, limit negative effects on biodiversity, reduce greenhouse gas emissions and nutrient leaching, buffer extreme weather events for crops, and increase the temporal stability of crop production. Moreover, agroforestry systems increase the provision of sociocultural benefits. Nevertheless, there are several challenges, such as a perceived negative view of trees in agricultural lands, poor definition and policy support, or the lack of know-how to manage complex systems, which prevent the widespread adoption of agroforestry systems.

This Special Issue of Sustainability on Agroforestry and Sustainable Agriculture Production gathers several studies on agroforestry systems from around the world, including a variety of types of agroforestry systems, from traditional wood-pastures to tropical cocoa-based systems, and approaches, from literature reviews to state-of-the-art ecological-economic models. The Special Issue highlights the potential of agroforestry as a promising approach for the creation of multifunctional landscapes able to face contemporary environmental challenges.

The loss of soil quality due to decades of mismanagement is a major concern for food production in many areas of the world. The negative consequences of soil loss are especially relevant in uplands, where high slopes can amplify erosive processes. In these locations, the presence of woody elements has proven to be an effective measure for soil conservation. Hussain et al. [1] provide an example and show how soil loss is substantially reduced in an intercropping system of maize–chili and leucaena (*Leucaena leucocephala*) trees hedgerows, while the productive potential of the land increases, all with a minimum usage of tillage and fertilizers. The increased usage of fertilizers is a common approach to reverting the negative effects of soil degradation on crop yields in agriculture. Xing et al. [2] tackle this challenge in the Loess Plateau in China. They propose a soil quality index to evaluate soil fertility and assess the availability of nutrients for sustainable production in potato

Citation: Rolo, V. Agroforestry for Sustainable Food Production. *Sustainability* **2022**, *14*, 10193. https://doi.org/10.3390/su141610193

Received: 10 August 2022
Accepted: 15 August 2022
Published: 17 August 2022

farmlands. It is well-known that trees have a positive effect on soil fertility, in part due to their ability to uptake nutrients from an extended volume of soil that are incorporated into the soil via litterfall or root decay. Karagatzides, Wilton, and Tsuji [3] provide an example of the positive effect of trees on nutrient availability. By using in situ ion exchange membranes, they found that agroforestry plots had a higher availability of PO_4, Ca, and Zn, which were positively related to crop yield. This pattern is also observed in Mediterranean silvopastoral systems; however, in this system, tree's long-term persistence is jeopardized by the lack of successful regeneration. López-Sánchez et al. [4] report the potential negative effect of high stocking rates on tree regeneration. They argue that only by allowing for the presence of nurse shrubs and through the adequate management of livestock will the sustainability of trees be guaranteed.

The positive effect of the trees of the agroforestry systems on soil's organic carbon content, as well as on the carbon stored in their above- and below-ground biomass, can stimulate the sequestration of atmospheric C in agricultural landscapes. However, the presence of trees can reduce yields if the interaction between vegetation layers is mainly competitive. Ballesteros-Possú, Valencia, and Navia-Estrada [5] assess the potential of various agroforestry settings of cocoa (*Theobroma cacao*) with Melina (*Gmelina arborea*) trees to improve yields and carbon sequestration as compared to traditional systems. They develop a series of allometric equations, measure the soil's organic carbon content, and calculate several indices of economic returns. They conclude that alternative agroforestry settings can improve yields and the carbon-sequestration potential of traditional systems. For sensitive crops to endure climate change, by contrast, the presence of trees can be an opportunity to ameliorate less favorable climatic conditions due to their provision, for instance, of shade. This may be the case with coffee plants that naturally originated from the understory of an African forest, but that are currently mainly cultivated under the sun. The negative effects of shade on yields are a concern, but mixed results and a lack of knowledge of the effect on different coffee varieties are the norm. Ehrenbergerová et al. [6] assessed various yield components of coffee plants growing under the shade and did not find significant differences as compared to those growing in full light. In addition, they reported a positive effect on the soil water content, which can help to build resilient systems in a future drier climate scenario.

Agroforestry can ease the negative effects of the scarcity of arable land for food production because it increases the productivity per unit of land. The lack of arable land is of particular concern in developing countries, where the expansion of agriculture activities is associated with a shear increase in forest degradation and fragmentation. Traditional agroforestry has been practiced for decades in these locations, but they are usually small systems and manage as a subsistence practice in many households; therefore, they do not unlock the full potential of agroforestry. There are still many barriers that limit the wide adoption of agroforestry or that hinder the productive potential in already established agroforestry systems. Achmad et al. [7] examine the available literature about the factors that prevent an increase in the productive capacity of smallholder subsistence agroforestry. The socio-economical background, including the literacy level, financial support, and land tenure, are the main handicaps to increasing the productive potential of small agroforestry systems. They report that these barriers are not overcome by the adoption of technological innovations because of the low literacy level. Octavia et al. [8] performed an analysis on the same lines and concluded that policies should target the mainstream adoption of agroforestry systems. They stress that a successful adoption may be achieved by a careful selection of the species and planting arrangements. The introduction of new species or the alternative usage of local crops can boost the rural development and resilience of local communities. Bas et al. [9] report an example of this and argue that the alternative use of a common crop can enhance the socio-economic resilience and nature conservation of native forests. They found that involving local communities in the process of decision making is key to the sustainable development of rural economies.

In developed countries, the challenges that face traditional agroforestry systems are completely different. Policies that support nature conservation are common, but they may still be insufficient to guarantee the long-term persistence of traditional agroforestry systems. Despite the ecological benefits that this system provides to society, the financial support offered by nature conservation policies may not fully cover the maintenance cost of the system. Nishizawa et al. [10] model the effect of subsidies on farmers' decisions regarding trees and biodiversity conservation in Orchard meadows in Switzerland. The authors highlight that the effectiveness of the payments was highly dependent on the farm type. The integrated model that the authors developed allows them to conclude that policies would be more effective if target specific farm types instead of offering the same solution for all farms.

Overall, this Special Issue of Sustainability collects several case studies that address the potential of agroforestry to foster food production while minimizing the negative effects on the environment, thereby empowering local communities, and building resilience against future climate scenarios.

Funding: This research received no external funding.

Institutional Review Board Statement: Not applicable.

Informed Consent Statement: Not applicable.

Data Availability Statement: Not applicable.

Acknowledgments: V.R. acknowledges the support of the regional government of Extremadura (Spain) through a "Talento" fellowship (TA18022).

Conflicts of Interest: The authors declare no conflict of interest.

References

1. Hussain, K.; Ilyas, A.; Bibi, I.; Hilger, T. Sustainable Soil Loss Management in Tropical Uplands: Impact on Maize-Chili Cropping Systems. *Sustainability* **2021**, *13*, 6477. [CrossRef]
2. Xing, Y.; Wang, N.; Niu, X.; Jiang, W.; Wang, X. Assessment of Potato Farmland Soil Nutrient Based on MDS-SQI Model in the Loess Plateau. *Sustainability* **2021**, *13*, 3957. [CrossRef]
3. Karagatzides, J.D.; Wilton, M.J.; Tsuji, L.J.S. Soil Nutrient Supply in Cultivated Bush Bean–Potato Intercropping Grown in Subarctic Soil Managed with Agroforestry. *Sustainability* **2021**, *13*, 8185. [CrossRef]
4. López-Sánchez, A.; Roig, S.; Dirzo, R.; Perea, R. Effects of Domestic and Wild Ungulate Management on Young Oak Size and Architecture. *Sustainability* **2021**, *13*, 7930. [CrossRef]
5. Ballesteros-Possú, W.; Valencia, J.C.; Navia-Estrada, J.F. Assessment of a Cocoa-Based Agroforestry System in the Southwest of Colombia. *Sustainability* **2022**, *14*, 9447. [CrossRef]
6. Ehrenbergerová, L.; Klimková, M.; Cano, Y.G.; Habrová, H.; Lvončík, S.; Volařík, D.; Khum, W.; Němec, P.; Kim, S.; Jelínek, P.; et al. Does Shade Impact Coffee Yield, Tree Trunk, and Soil Moisture on *Coffea canephora* Plantations in Mondulkiri, Cambodia? *Sustainability* **2021**, *13*, 13823. [CrossRef]
7. Achmad, B.; Sanudin; Siarudin, M.; Widiyanto, A.; Diniyati, D.; Sudomo, A.; Hani, A.; Fauziyah, E.; Suhaendah, E.; Widyaningsih, T.S.; et al. Traditional Subsistence Farming of Smallholder Agroforestry Systems in Indonesia: A Review. *Sustainability* **2022**, *14*, 8631. [CrossRef]
8. Octavia, D.; Suharti, S.; Murniati; Dharmawan, I.W.S.; Nugroho, H.Y.S.H.; Supriyanto, B.; Rohadi, D.; Njurumana, G.N.; Yeny, I.; Hani, A.; et al. Mainstreaming Smart Agroforestry for Social Forestry Implementation to Support Sustainable Development Goals in Indonesia: A Review. *Sustainability* **2022**, *14*, 9313. [CrossRef]
9. Bas, T.G.; Gagnon, J.; Gagnon, P.; Contreras, A. Analysis of Agro Alternatives to Boost Cameroon's Socio-Environmental Resilience, Sustainable Development, and Conservation of Native Forests. *Sustainability* **2022**, *14*, 8507. [CrossRef]
10. Nishizawa, T.; Kay, S.; Schuler, J.; Klein, N.; Herzog, F.; Aurbacher, J.; Zander, P. Ecological–Economic Modelling of Traditional Agroforestry to Promote Farmland Biodiversity with Cost-Effective Payments. *Sustainability* **2022**, *14*, 5615. [CrossRef]

Article

Sustainable Soil Loss Management in Tropical Uplands: Impact on Maize-Chili Cropping Systems

Khalid Hussain [1,2,*], Ayesha Ilyas [1], Irshad Bibi [3] and Thomas Hilger [2,*]

[1] Agro-Climatology Laboratory, Department of Agronomy, University of Agriculture, Faisalabad 38040, Pakistan; ayeshakhalid3053@gmail.com
[2] Institute of Agricultural Sciences in the Tropics (Hans-Ruthenberg-Institute) (490), University of Hohenheim, 70599 Stuttgart, Germany
[3] Institute of Soil and Environmental Sciences, University of Agriculture Faisalabad, Faisalabad 38040, Pakistan; irshad.niazi81@gmail.com
* Correspondence: khalid.hussain@uaf.edu.pk (K.H.); Thomas.hilger@uni-hohenheim.de (T.H.); Tel.: +92-335-742-9406 (K.H.)

Abstract: Intensive land use with inappropriate land management is directly degrading South Asian uplands. A field trial was carried out on the uplands of Western Thailand with a 25% slope to examine the effect of land use management on soil loss for sustainable crop production during two consecutive years (2010–2011). Various cropping systems with soil conservation practices were compared to maize sole cropping (MSC). Results revealed that soil loss was at a minimum in the intercropping system of maize-chili-hedgerows with minimum tillage and fertilization that was 50% to 61% and 60% to 81% less than MSC and the bare soil plot during both years, respectively. Yield advantage was at its maximum, as indicated by the highest land equivalent ratios of 1.28 and 1.21 during 2010 and 2011, respectively, in maize-chili-hedgerows-intercropping with minimum tillage and fertilization. The highest economic returns (5925 and 1058 euros ha^{-1} during 2010 and 2011, respectively) were also obtained from maize-chili-hedgerows-intercropping with minimum tillage and fertilization. Chili fresh fruit yield was maximum in the chili alone plot during both years due to the greater area under cultivation compared with intercropping. Maize-chili-hedgerows with minimum tillage and fertilization reduced soil loss and increased land productivity and net returns, indicating its promising features for sustainable crop production on uplands.

Keywords: land use options; soil conservation; intercropping; hedgerows; minimum tillage

Citation: Hussain, K.; Ilyas, A.; Bibi, I.; Hilger, T. Sustainable Soil Loss Management in Tropical Uplands: Impact on Maize-Chili Cropping Systems. *Sustainability* **2021**, *13*, 6477. https://doi.org/10.3390/su13116477

Academic Editors: Victor Rolo and Sean Clark

Received: 13 April 2021
Accepted: 1 June 2021
Published: 7 June 2021

1. Introduction

A large proportion of agricultural land (around 2 billion ha) in the world is already affected by soil erosion [1], whereas around 10 million hectares of land are destroyed every year due to soil erosion, directly impacting world food production [2].

The land area of Thailand is 514,000 km^2: 41% for agriculture, 31% for forest and 28% unclassified. Most of the small holders cultivate maize for food, while some concentrate on cash crops such as chilies. However, in both cases, cultivation is often carried out on uplands with varying slope degrees (10–40%), which encourages deforestation and ultimately soil erosion [3]. Soil erosion has been a very common problem from decades, and concerns about conserving the soils on uplands are increasing. Soil loss is mainly caused by improper farming methods, low soil cover, extensive tillage and mono-cropping systems, whereas rainfall intensity, slope gradient, soil stability, crop management and conservation practices are considered to be the main factors that directly affect soil erosion [4–8] in Thailand and other parts of Asia. These Thai hillsides have moderate to steep (10–30%) slopes and are dominated by natural bamboo forests.

Heavy rainfalls at the time of crop harvest or just after harvest causes soil erosion, ultimately reducing soil fertility. Land degradation and soil loss due to heavy rainfall and improper cropping has already affected the livelihood of the farmers and led them to adopt alternate income-generating sources rather than farming. Most farmers in the region have already left farming and started off-farm jobs in various mills, factories, and institutions. Knowledge of proper soil use and management to preserve available resources is a big challenge [9] for farming communities living in soil-loss hotspot areas.

Sustainable crop production is most important for regional food security and creating better livelihoods for the farming communities, whereas sustainable crop production on the uplands is not possible without soil conservation practices/conservation tillage. Conservation tillage has vital role in soil conservation, crop production and food security on slopes with tropical weather conditions. Conservation tillage is a noninversion tillage system in which around 30% of crop residues are always kept on the soil surface. Cropping systems with conservation tillage/soil conservation practices directly reduce soil loss, maintain soil fertility and enhance farm productivity on uplands [5–8,10]. In Western Thailand, most of the farmers follow a mono/sole cropping system without any soil conservation practices. The crops (like maize and chili) are grown under sole cropping at a wider distance, encouraging soil loss due to heavy rainfall. Sole cropping offers limited opportunities for sustainable agricultural production, especially under degraded soil configurations with the fragile and unusual nature of tropical weather [11]. Instead of sole/monocropping, intercropping has many advantages, such as yield stability [12], efficient use of above and below ground resources, soil conservation, [13], increasing productivity and land use efficiency [14]. In addition, intercropping systems that are blended with soil conservation practices are more stable and less risky for farmers as these reduce the risk of crop damage [10]. Intercropping with soil conservation methods includes agroforestry systems, grass barriers and contour hedgerows.

Agroforestry is a land use that allows trees and crops and/or livestock production from a single piece of land to achieve ecological, economic, cultural, and environmental benefits [15]. These systems originated from developing countries, where the population pressure is high with limited land resources. They differ from traditional forestry in term of their economic and social benefits along with water and soil sustainability and act as buffers to extreme climatic conditions. Hilger et al. [16] indicated the effectiveness of hedgerows in the reduction of fertile soil loss on uplands. Slogans of proper land use and land use management for maintaining soil fertility are increasing over time in many regions of the world. This not only inspired governments but also led farmers to explore proper soil conservation practices in upland agriculture for maintaining soil productivity and structure. Similarly, various government and non-governmental organizations are also active in Thailand, creating awareness about land degradation issues among farming communities.

Minimum tillage with Jack bean relay cropping was suggested as a soil conservation combo under conservation tillage on uplands, reducing soil loss and improving crop productivity on the uplands of Northeast Thailand [17,18]. In minimum tillage, no tillage is carried out except for on soil where the seed is sowed while Jack bean relay cropping covers the soil during the fallow period and is also left in the field after harvest [17]. No till or minimum till can compact the soil, which can reduce soil loss but also reduce the infiltration rate of rainwater. Minimum tillage coupled with Jack bean relay cropping reduces soil compaction, increases infiltration, and reduces soil loss [18].

Farmers in upland regions are reluctant to adopt conservation practices such as hedgerows, intercropping, and minimum tillage coupled with Jack bean relay cropping, despite many benefits. They perceive that soil conservation practices increase input cost, reduce the land area of crops and increase resource competition. To address farmers' concerns about soil conservation techniques and the benefits of best-suited land use on hillsides, a field experiment was conducted on the uplands of Western Thailand with a specific objective: to explore the role of land use with soil conservation practices like intercropping, hedgerows and minimum tillage coupled with Jack bean relay cropping as conservation tillage for soil loss, water runoff reduction and yield improvement. It was hypothesized that cropping systems with hedgerows and conservation tillage (minimum tillage coupled with Jack bean relay cropping) with proper fertilization would reduce soil loss, as well as enhancing land productivity and net economic returns for small land holding upland farmers in Western Thailand.

2. Materials and Methods

2.1. Study Site

The experiment was performed at Ban Bo Wai Village (13°28′ N, 99°15′ E), Ratchaburi Province, Western Thailand. The soil was loamy-skeletal, siliceous, isohyperthermic Kanhaplic Haplustults with shallow and stony nature, often prone to soil erosion. The region receives rainfall of about 1200 mm annually from May–October each year. The average annual temperature is about 28 °C and 14 MJ m^{-2} d^{-1} solar radiations. Climatic variables were automatically recorded at the research site (Figure 1). Soil was analyzed for physiochemical properties, shown in Table 1. The locality is mostly hilly with moderate to steep slopes mostly covered by maize cultivation. Other crops include cassava (*Manihat esculenta* Crantz) and chili (*Capsicum annuum* L.). The cultivation of the maize crop starts just after the onset of the rainy season in June and ends in late September to mid-October.

Table 1. Main soil physiochemical properties of the study site before planting.

Soil Depth	* Soil Texture	Sand (%)	Silt (%)	Clay (%)	* pH	* SOC ($g{\cdot}kg^{-1}$)	* Total N ($g{\cdot}kg^{-1}$)	Extractable P ($mg{\cdot}kg^{-1}$)	Extractable K ($mg{\cdot}kg^{-1}$)	BD ($g{\cdot}cm^{-3}$)
0–15 cm	Loamy soil	38.8	40.2	21	5.8	13	1.6	12.5	220.6	1.7

* Soil texture was measured by the pipette method, pH as soil: water = 1:1, SOC= soil organic carbon measured by the Walkley-Black method, total nitrogen was measured by the Kjeldahl and steam distillation method, extractable P by the Bray II method, extractable K by 1 N NH_4OAc, and BD = bulk density by core methods.

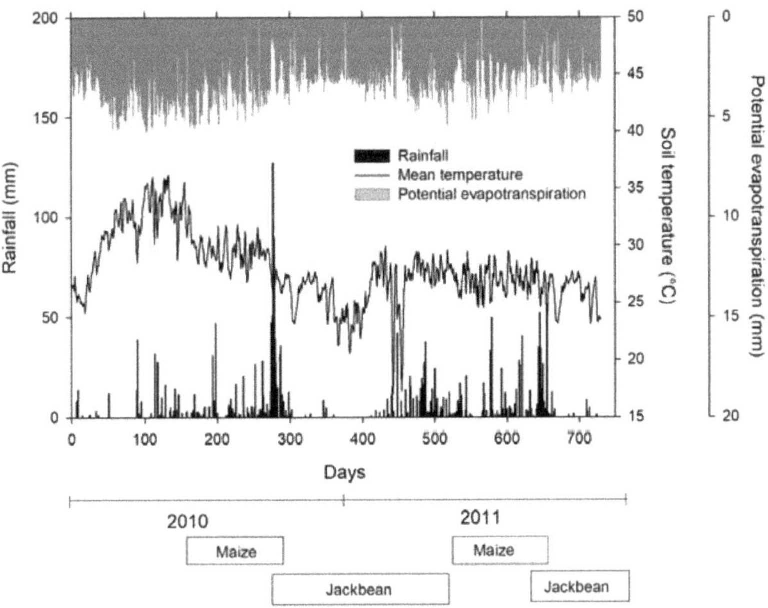

Figure 1. Climatic conditions of the study area during both years of the study.

2.2. Experimental Layout

The study presented here was carried out over two consecutive years of planting (2010 and 2011) on land where treatments were established in 2008–2009 to allow time for soil conservation treatments to express the effect. A randomized complete block design (RCBD) with three replicates was used for execution of the experiment. The plot size was 13 m × 4 m with a 20–25% slope gradient. The treatments were:

T1: Maize sole cropping with tillage and fertilizer application;

T2: Maize intercropped with chilies having fertilizer application and tillage;

T3: Maize intercropped with chilies with minimum tillage, Jack bean (*Canavalia ensiformis* (L.) DC) relay cropping with fertilizer application;

T4: Maize intercropped with chilies and leucaena hedgerows with minimum tillage, Jack bean relay cropping with fertilizer application;

T5: like T3 with no fertilizer application;

T6: like T4 with no fertilizer application.

In addition to these treatments, two bare soil plots were also established along with two sole chili plots. Maize sowing was carried out on 22 June 2010 and 29 June 2011. One-month old chilies were also transplanted on 22 June 2010 and 29 June 2011. Tillage was carried out manually on the tillage plots to a depth of about 20 cm, in which the soil was disturbed fully up to 20 cm depth, whereas minimum tillage was also practiced manually with minimal disturbance of the soil during seeding only (the upper 0–5 cm soil is disturbed at the place where seeds were sown). In minimum tillage treatments, all management practices from planting to harvesting were carried out with minimal soil disturbance. Jack beans were planted as relay cropping on 15 September in both years in the minimum tillage treatments only. Nitrogen was applied to the maize crop in the form of urea in two splits of 31 kg ha^{-1} each, first at 30 days after planting and second at two months after planting. Phosphorus was applied as triple super phosphate at 22 kg ha^{-1} and 36 kg ha^{-1} of K, as potassium chloride was banded at one-month after planting maize. The chilies received nitrogen at 92 kg ha^{-1} at the time of transplanting in an equal amount to top dressing one month after transplanting. T1 had 17 maize rows, whereas every intercropping treatment had eight maize rows (two maize rows followed by two chili rows). Chili rows were six in T2, T3 and T5, while two rows were planted in T4 and T6 (Figure 2).

Maize rows were planted at a 0.75 m distance, while the row distance from maize to chili and the inter-row distance from chili was 1 m, and the distance from maize to hedgerow was 0.25 m. There were 16 maize plants per line in all treatments. Each chili row had four chili plants in the intercropping treatments. In T4 and T6, three hedgerows of 1 m width were planted at the top, middle and bottom end of each plot. The leucaena hedgerows were planted in 2008–2009. Hedges were pruned four times during 2010: one week after maize sowing, 30 and 60 days after maize planting and one month after maize harvest. These were pruned six times during 2011, three times before planting the maize crop in January, May, and June. The remaining prunings were carried out 30, 60 and 105 days after maize planting. Leucaena hedges were always kept to a height of 50 cm during the maize cropping season. Weeding was manually carried out at regular intervals in all plots. Jack beans were planted in all minimum tillage treatments, between all rows, a month before the maize harvest. During dry season, Jack beans were kept in the plots, and their remains were left as mulch on the soil surface. The maize stalks were cut and left as mulch in all treatments. The pruning material of the hedgerow treatment was uniformly dispersed within the respective plots and used as mulch. In 2010, 2.5 and 2.2 kg m^{-2} of leucaena residues were applied at T4 and T6, respectively, and in 2011, around 3.5 and 3.0 kg m^{-2} at T4 and T6, respectively. Each maize row was harvested individually in each treatment during the field experiments of 2010 and 2011. For soil loss and water runoff measurements, a collection tank (150 L) was connected at the base of each treatment (Figure 2). Two bare plots were also established and used as a reference for soil loss without vegetation along with a chili sole cropping plot. All the treatments have a collecting channel at the base of the plot with concrete boundaries established to keep each plot separated and to prevent water coming in and out from the sides of the plot. Measurements of soil loss and water runoff from each treatment were carried out after every rainfall event. The collection tanks were indirectly connected to treatments, with one of the outlets of a divisor box placed between the plot and collecting tank (Figure 2). A divisor box with equal size outlets (8 outlets) was used to minimize the quantity of runoff water going into the collecting tank, in case of heavy rainfall events. After each rainfall event, the volume of runoff water was measured by a meter rod, which was then multiplied by the number of outlets of the divisor to reach the total volume of runoff water. The quantity of lost soil was measured on the basis of suspended and heavier sediment fractions from each plot. The collecting channels developed at the lower end of each plot provided the heavy sediment fractions, which were collected and weighed, whereas suspended fractions were collected from the sediment and water collecting tanks. Thereafter, subsamples were taken and dried to obtain the dry weights of the suspended fractions.

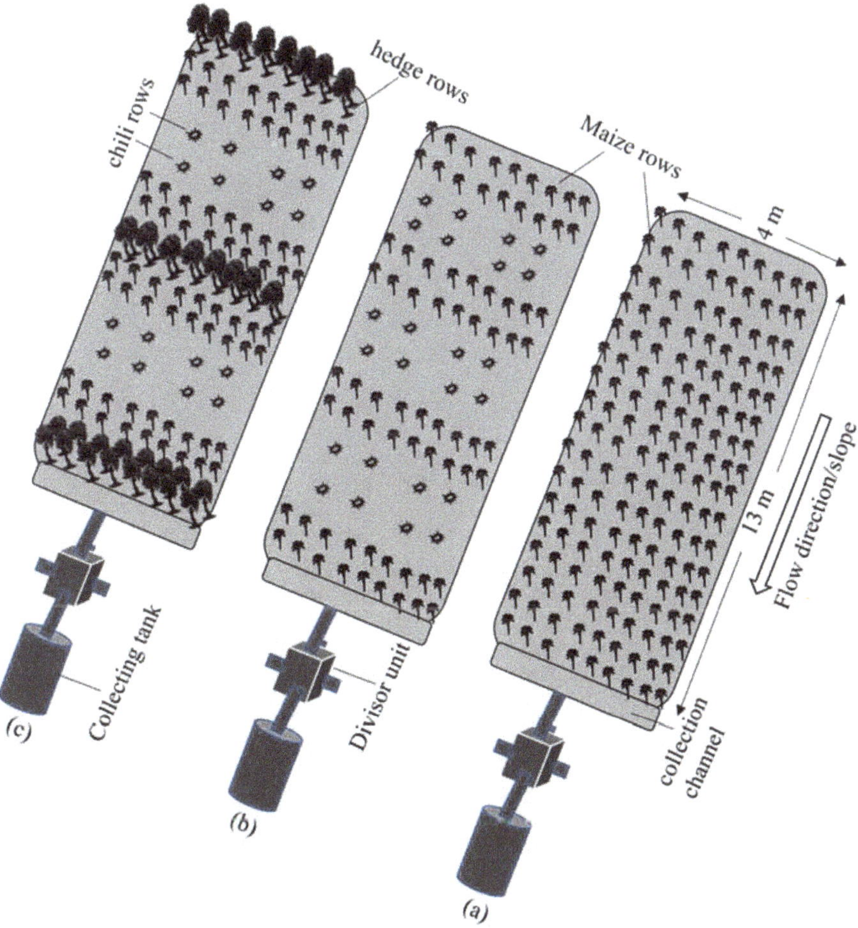

Figure 2. Experimental setup with treatment allocation (**a**) T1: maize mono-cropping (farmers' practice, control,) with tillage and fertilization; (**b**) T2: maize intercropped with chilies with fertilization and tillage; T3: maize intercropped with chilies with mini. tillage, Jack bean relay cropping with fertilizer application; T5: like T3 but without fertilizer application; (**c**) T4: maize intercropped with chilies and leucaena hedgerows with mini. tillage, Jack bean relay cropping with fertilizer application; T6: like T4 but without fertilizer application.

2.3. Productivity Evaluation

Land equivalent ratio (LER) was used to evaluate the productivity of each cropping system. The following formula was used for LER calculations:

$$LER = \frac{MGY_1\ (t\ ha^{-1})}{MGY_s\ (t\ ha^{-1})} + \frac{CFY_1\ (t\ ha^{-1})}{CFY_s\ (t\ ha^{-1})} \tag{1}$$

where MGY_I, MGY_s, CFY_I, and CFY_s are the maize grain yield produced under intercropping and sole cropping, and the yield of chili fruit under intercropping and sole cropping, respectively. Two chili sole crop treatments were additionally established at the test site to provide data on the yield of sole cropped chilies for this assessment. All 17 maize rows were harvested at T1 (control), while all eight maize rows were harvested at both hedge intercrop treatments (T4 and T6). Each line was placed separately. Subsequently, samples from each line were weighed and separated into leaves, stems and grain components. Subsets from each component were dried, and above ground biomass (AGB) was calculated for each treatment. The fresh fruit yield of chilies was taken from time to time, when fruits were established from each plant. Subsequently, the area-corrected yield of maize and chili were calculated for each treatment.

Maize equivalent grain yield was computed as:

$$EY_M = MY_i + [(CY_i * CP)/MP] \tag{2}$$

where EY_M is maize equivalent grain, MY_i is maize grain yield in intercrop, CY_i is chili fruit yield in intercrop, CP is the price of chili fruits, and MP is the price of maize grains.

2.4. Economic Analysis

Economic analysis of the all the treatments studied was carried out as net return/profit to estimate the economic profitability of various land use options:

$$NR\left(EUR\ ha^{-1}\right) = GR\left(EUR\ ha^{-1}\right) - PC\left(EUR\ ha^{-1}\right) \tag{3}$$

where NR is the Net Return, GR is the Gross Return, and PC is Production Cost. The economic analysis was carried out in euros. One EUR was equal to 40 Thai Baht during 2010–2011. The average price of chili fruits in 2010–11 in Thailand was 80 Thai Baht (THB) kg^{-1}, while for maize grain, it was around 10 THB kg^{-1}. The average cost for maize production (sole cropping) was around 850 euros ha^{-1}, 1400 euros ha^{-1} for maize-chili-intercropping with conventional tillage, and 1100 euros ha^{-1} season^{-1} for maize-chili-hedgerows intercropping with minimum tillage. The above-mentioned costs of production are inclusive of all cost from sowing to harvesting.

2.5. Statistical Analysis

Statistical analysis was done in SAS, V-9.2 (SAS Inc., Cary, NC, USA). The RCBD was used in the field study for both years. The bare soil plot was used in soil loss comparisons, while chili sole crop plots were used to compare the chili yield under sole and intercrop conditions. Bivariate techniques were used for analyzing intercropping trials per year [19]. Pairwise comparison of treatments was carried out using Tukey's Honest Significant Difference test at $p = 0.05$.

3. Results

3.1. Soil Loss and Water Runoff

Total soil loss from various land use options was statistically significant (Table 2). A maximum soil loss of 30 t ha^{-1} was observed from the bare plot, followed by T1, T2 and T5 during the 2010 growing season, whereas the minimum quantity of soil (11.6 t ha^{-1}) was lost from fertilized-hedgerow-intercropping-minimum tillage treatment (T4). Hedgerows' inclusion within maize-chili-intercropping with fertilizer application (T4) reduced soil loss by 61% and 53% compared with the bare soil plot and the farmers' practice (maize alone (T1)), respectively. Moreover, hedgerows' inclusion within maize-chili intercropping without fertilizer application (T6) reduced soil loss by 50% and 39% compared with the bare soil plot and the farmers' practice (maize alone (T1)), respectively. Soil loss trends observed during the 2011 growing season were similar to those of the 2010 growing season, but the quantity of soil loss was less in 2011 compared with 2010. Maximum total soil was again lost from the bare soil plot compared with the rest of the treatments during 2011. Minimum soil (3 t ha^{-1}) was lost from the fertilized-hedgerow-intercropping-minimum tillage treatment (T4), followed by the maize-chili intercrop with hedgerows but without fertilizer application treatment (T6) with 3.31 t ha^{-1}. During the 2011 cropping season, T4 reduced 60% of soil loss, followed T6, which reduced soil loss by 55% compared with T1 (farmers' practice).

Table 2. Cumulative soil loss and water runoff from various maize-based cropping systems.

Treatments	Soil Loss (t·ha^{-1})		Water Runoff (m^3·ha^{-1})	
	2010	2011	2010	2011
T1	24.7 b	7.50 b	4091	1227
T2	24.4 b	7.70 b	3955	1268
T3	22.7 b	7.88 b	4431	1429
T4	11.6 c	3.00 c	3980	1123
T5	23.7 b	7.92 b	4004	1111
T6	15.0 bc	3.31 c	3948	1069
Chili sole crop	22.7 b	10.12 b	4392	1533
Bare soil plot	30.0 a	16.10 a	4474	1681
p value	$\leq$0.005	$\leq$0.05	NS	NS

Figures with different small letters are indicating statistically significant differences between the treatments. T1: maize mono-cropping (farmers' practice, control,) with tillage and fertilization; T2: maize intercropped with chilies with fertilization and tillage; T3: maize intercropped with chilies with mini. tillage, Jack bean relay cropping with fertilizer application; T4: maize intercropped with chilies and leucaena hedgerows with mini. tillage, Jack bean relay cropping with fertilizer application; T5: like T3 but without fertilizer application; T6: like T4 but without fertilizer application.

Water runoff measurements also showed statistically significant differences among various land use options during both years (Table 2). During 2010, maximum total water runoff (4474 m^3 ha^{-1}) was observed in the bare soil plot, which was 9.4, 13.1, 0.97, 12.4, 11.7, 13.3 and 1.87% higher compared with T1, T2, T3, T4, T5, T6 and chili sole cropping, respectively. The lowest water runoff during 2010 was observed in both maize-chili-intercropping with hedgerows and with and without fertilizer application treatments (T4 and T6, respectively). In 2011, the total water runoff was similar to that of 2010. Maximum water runoff (1681 m^3 ha^{-1}) was observed in the bare soil plot as observed in 2010. Total water runoff from the bare soil plot was 37, 32.5, 17.6, 49.6, 51.3, 57.2 and 9.7% greater than T1, T2, T3, T4, T5, T6 and chili sole cropping, respectively.

Soil loss measured after each rainfall event during both years from each treatment showed variable trends (Figure 3). All treatments showed variable trends of cumulative soil loss. During early 2010, cumulative soil loss trends were similar with the onset of the rainy season, but later, variability among the treatment increased. Maximum cumulative soil loss during all rainfall events was observed in the bare plot closely followed by the maize alone plot (T1), whereas minimum cumulative soil loss after each rainfall event occurred in the maize-chili intercrop with hedgerow treatment with and without fertilizer application (T4 and T6). Event-based soil loss was maximum at the end of the growing season in all treatments when there were heavy rainfall events. Overall cumulative soil loss in intercrop treatments was lower than in maize sole cropping and the bare soil plot during 2010.

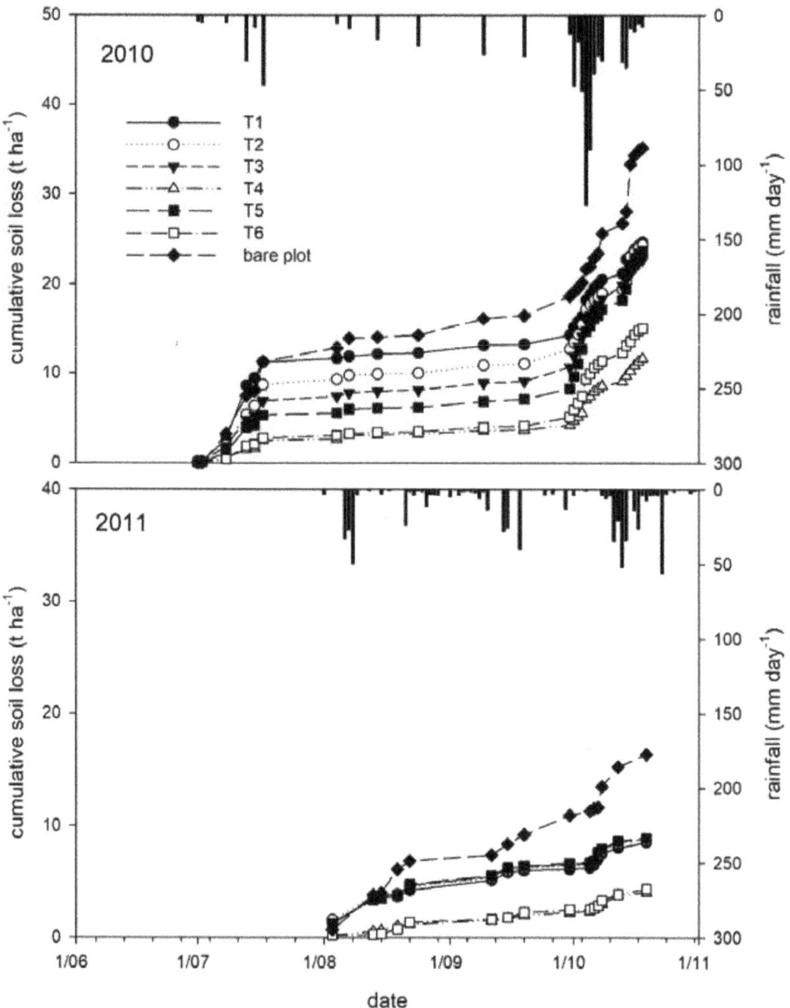

Figure 3. Time series cumulative soil loss from various maize-based cropping systems, T1: Maize monocropping (farmers' practice, control,) having tillage, and fertilization; T2: Maize intercropped with chilies having fertilization and tillage; T3: Maize intercropped with chilies having mini. tillage, Jack bean relay cropping with fertilizer application; T4: Maize intercropped with chilies and leucaena hedgerows having mini. tillage, Jack bean relay cropping with fertilizer application; T5: like T3 but without fertilizer application; T6: like T4 but without fertilizer application.

Cumulative soil loss in all the treatments during 2011 was lower than that of 2010. There were three distinct sets of cumulative soil loss trends in 2011. The highest soil loss occurred in bare plot soil after each rainfall event, followed by maize alone and intercrop treatments with minimum tillage and Jack bean relay cropping (T1, T2, T3, and T5), whereas the lowest cumulative soil loss at each rainfall event was observed in T4 and T6.

3.2. Maize Yield (t ha^{-1})

Maize-based cropping systems significantly affected maize grain, equivalent and biological yield (area corrected) (Table 3). Maize grain yield ranged from 2.45 to 6.86 t ha^{-1} during the 2010 growing season. The maize sole cropping treatment (T1) attained the highest grain yield statistically, compared with the rest of the treatments ($p < 0.001$). Fertilized-hedgerow-intercropping-minimum tillage (T4) yield was highest among all the intercrop treatments, closely followed by T6, which was statistically on par. The lowest grain yield was observed in T5, which was statistically on par with that of T5. During 2011, T1 produced the highest grain yield statistically, compared with the rest of the treatments, but was lower compared with 2010. T4 produced the highest grain yield statistically among intercrop treatments, which was greater than 2010. The lowest grain yield was observed in T5. Maize equivalent grain yield was at its maximum (19.48 t ha^{-1}) under maize-chili intercrop conditions (T2), whereas both intercrop treatments with soil conservation treatments T3 and T4 were statistically on par with equivalent yields of 14.78 t ha^{-1} and 14.77 t ha^{-1}, respectively, during 2010. The lowest maize equivalent yield was 10.94 t ha^{-1}, observed in T6. During 2011, the maximum maize equivalent grain yield (6.91 t ha^{-1}) was observed in T4, which was statistically on par with T2. Minimum maize equivalent grain yield (4.94 t ha^{-1}) was observed in T5.

Table 3. Maize grain, biological and equivalent yield, chili fresh fruit yield and land equivalent ratio obtained from various maize-based cropping systems.

Treatments	Maize Grain Yield (t·ha^{-1})		Maize Biological Yield (t·ha^{-1})		Maize Equivalent Grain Yield (t·ha^{-1})		Chili Fresh Fruit Yield (t·ha^{-1})		Land Equivalent Ratio	
	2010	2011	2010	2011	2010	2011	2010	2011	2010	2011
T1	6.86 a	6.41 a	14.97 a	11.84 a	-	-	-	-	1.00 c	1.00 b
T2	3.06 c	2.97 d	6.84 b	5.46 b	19.48 a	6.54 a	3.28 b	0.71 b	1.23 a	1.17 a
T3	2.57 d	2.69 e	5.55 c	4.97 c	14.78 b	5.79 b	2.44 c	0.62 c	1.08 b	1.03 b
T4	3.33 b	3.78 b	6.71 b	5.51 b	14.77 b	6.91 a	2.29 c	0.62 c	1.28 a	1.21 a
T5	2.45 d	2.33 f	5.03 c	4.13 c	13.06 c	4.94 c	2.12 c	0.52 cd	0.90 d	0.88 c
T6	3.31 b	3.25 c	6.61 b	4.83 c	10.94 d	5.46 b	1.52 d	0.44 d	0.92 d	0.94 c
CSC	-	-	-	-	-	-	6.47 a	1.19 a	1.00 c	1.00 b
p value	≤0.005	≤0.05	≤0.005	≤0.005	≤0.05	≤0.05	≤0.005	≤0.005	≤0.005	≤0.05

Figures with different small letters indicate statistically significant differences between treatments. T1: maize mono-cropping (farmers' practice, control,) with tillage and fertilization; T2: maize intercropped with chilies with fertilization and tillage; T3: maize intercropped with chilies with mini. tillage, Jack bean relay cropping with fertilizer application; T4: maize intercropped with chilies and leucaena hedgerows with mini. tillage, Jack bean relay cropping with fertilizer application; T5: like T3 but without fertilizer application; T6: like T4 but without fertilizer application; CSC: chili sole cropping.

Maize biological yield was statistically highest in T1, closely followed by T2, which was statistically on par with that of T4 during the 2010 growing season. The lowest maize biological yield (5.03 t ha^{-1}) was attained by T5, which was statistically on par with T3. Similar trends of statistically significant biological yield were observed during 2011 but were lower than 2010 in all treatments.

3.3. Chili Yield (t ha^{-1})

Chili fresh fruit yield was statistically significant in both growing seasons (Table 3). During 2010, chili sole cropping produced the highest chili fresh fruit yield statistically, compared with intercropping treatments with and without soil conservation techniques. Among intercrop treatments, T2 produced a higher chili fresh fruit yield, while the lowest chili fresh fruit yield was observed in T6, where chili was intercropped with maize and hedgerows with zero fertilization. During 2011, maximum chili yield was obtained from chili sole cropping as observed during the 2010 growing season. Moreover, chili production under the intercrop condition was at its maximum in T2, where chili was intercropped with maize, minimum tillage and fertilizer application, while minimum chili fresh fruit yield was observed in T6, as observed in 2010. Overall, chili fresh fruit yield was many folds higher during 2010 compared with the 2011 growing season in all treatments due to an insect attack.

3.4. Cropping System Productivity Evaluation and Economic Analysis

Cropping system productivity evaluation was carried out by calculating the land equivalent ratio (Table 4). During 2010, statistical analysis showed that the maximum land equivalent ratio (1.28) was observed in maize-chili-intercropping with leucaena hedges, minimum tillage, Jack bean relay and fertilizer application (T4), which was statistically on par with maize-chili-intercropping (T2). The lowest land equivalent ratio (0.90) was observed in T5, which was statistically on par with maize-chili-intercropping with leucaena hedges, minimum tillage, and Jack bean relay but without fertilizer application (T6). Similar trends of land equivalent ratios were observed during 2011, while the values were slightly lower than in 2010. Statistical analysis showed that the maximum land equivalent ratio (1.21) was observed in maize-chili-intercropping with leucaena hedges, minimum tillage, Jack bean relay and fertilizer application (T4), which was statistically on par with maize-chili-intercropping (T2). The lowest land equivalent ratio (0.88) was observed in T5, which was statistically on par with maize-chili-intercropping with leucaena hedges, minimum tillage, and Jack bean relay but without fertilizer application (T6).

Table 4. Economic analysis of investigated land use treatments during both cropping seasons.

Treatments	Maize Return	Chili Fruit Return	Gross Return	Production Cost	Net Return
2010					
T1	1715	-	1715	850	865
T2	765	6560	7325	1400	5925
T3	643	4880	5523	1100	4423
T4	833	4580	5413	1100	4313
T5	613	4240	4853	1100	3753
T6	828	3040	3868	1100	2768
2011					
T1	1603	-	1603	850	753
T2	743	1420	2163	1400	763
T3	673	1240	1913	1100	813
T4	945	1240	2185	1100	1085
T5	583	1040	1623	1100	523
T6	813	880	1693	1100	593

The cost and net return values present in the table are Euros/hectare (1 Euro = 40 Baht). T1: maize mono-cropping (farmers' practice, control,) with tillage and fertilization; T2: maize intercropped with chilies with fertilization and tillage; T3: maize intercropped with chilies with mini. tillage, Jack bean relay cropping with fertilizer application; T4: maize intercropped with chilies and leucaena hedgerows with mini. tillage, Jack bean relay cropping with fertilizer application; T5: like T3 but without fertilizer application; T6: like T4 but without fertilizer application.

The crops present in the intercropping condition provided greater economic returns compared with farmers' practice (Table 4). During 2010, maximum net return (EUR 5925) was obtained from T2, followed by T3, T4, T5 and T6, respectively, while the lowest net return (865 EUR) was obtained from T1, where maize was planted as sole cropping, whereas, during 2011, maize-chili-intercropping with leucaena hedgerows and fertilizer application (T4) produced the highest net return value (EUR 1085) followed by T1. Minimum net return was obtained from T5. Economic analysis of the land use options also indicated that the net return values were higher during 2010 compared with 2011.

4. Discussion

4.1. Effect of Various Land Use Options on Soil Loss and Water Runoff Dynamics

Soil health is essential for sustainability productivity. In Thailand, no strict laws are available to restrict farmers using uplands without thinking about soil health. Soil degradation has accelerated over the last decade [20]. Maize sole cropping (T1) reduced soil loss by around 18–50% compared with the bare soil plot during 2010–2011, respectively. This reduction in soil loss was due to soil covered by maize plants. Soil loss between

T1 and intercrop treatments with and without conservation tillage were non-significant. This means that intercropping and minimum tillage effects were like maize sole cropping and tillage. Maize was planted in rows 75 cm apart, which provided enough soil cover to reduce soil loss compared with intercropping with conservation tillage. In intercropping treatments, maize rows were planted at 75 cm apart, but chili rows were planted 1 m apart, which favored soil loss even under conservation tillage.

On the other hand, intercropping treatments with hedgerows and conservation tillage practices (T4) reduced soil loss by many folds (61.33% and 53%) compared with the bare soil plot and farmers' practice (T1) during both growing seasons. Hedgerows reduced soil loss in two ways: first, Leucaena pruning provided the soil with additional cover, which reduced the direct abrasion of splash raindrops; and second, it enhanced the organic matter of the soil. Minimum tillage directly reduces soil loss compared with conventional tillage [21]. In maize alone (T1), the soil was tilled, which may have also reduced the soil's capacity to conserve moisture to some extent and may have facilitated soil loss, while minimum tillage associated with Jack bean relay cropping and subsequent mulching with leucaena hedgerow pruning in the soil conservation treatment may have improved soil structure, which reduced soil loss [18]. Better moisture conservation by hedgerows as they slowed down water runoff additionally facilitated water infiltration and ultimately reduced soil loss [5,17].

Cropping systems with soil conservation practices and conservation tillage may create sustainability in production, plays an important role in increasing land use efficiency on a long-term basis, and has increased interest of the farmers due to its potential benefits of increasing yields and reduction of soil erosion risks. Pansak et al. [17] indicated that intercropping systems with leucaena hedgerows along with conservation tillage (minimum tillage and Jack bean relay cropping) not only reduced soil loss but also water runoff, except during the hedgerows' establishment phase.

4.2. Effect of Land Use Options on Crop Productivity

Low maize yield in all the treatments during 2011 was due to reduction in soil fertility due to fertile soil loss over time compared with 2010 (Table 2). Most cultivation on uplands is carried out on freshly cleared forests, and it has been observed that, over time, the fertility of land decreases because of land mismanagement (heavy tillage/conventional tillage), high loss of fertile topsoil, low fertilizer inputs and intensive land use.

Soil analysis data of the field experiment showed reductions for organic matter and extractable P from 2010 to 2011 (2010: 10.6 mg kg^{-1} extractable P, 1.97% soil organic matter vs. 2011: 9.5 mg kg^{-1} extractable P, 1.76% soil organic matter at a soil depth of 0–45 cm) [10]. This argument was further supported by the grain nitrogen concentration values of the same experiment, which were lower in 2011 compared with 2010 [6]. Grain nitrogen concentration is directly related to nitrogen availability and its utilization. If there is low availability of nitrogen, then the uptake of plants is low, and low concentrations are found in the grains.

Maize grain and biological yield were highest in maize alone/farmers' practice land use due to the optimum area under the maize crop. Whereas land use options with intercropping and soil conservation practices occupied some space, which reduced the area under the maize crop but was compensated with the yield of the intercrop, which increased land use efficiency. Among intercropping land use options (T2–T6), maximum maize grain yield and biological yield were observed in T4, where maize was intercropped with chili along with soil conservation practices (hedgerows and minimum tillage) during both years, closely followed by T6 and T2. Soil conservation methods with minimum tillage or no tillage can reduce fertile soil loss [21], which ensures sustainable crop production. The lower values of the equivalent maize grain yield during 2011 were due to lower chili fresh fruit yield because chili plants were infested by cercospora leaf spot at around 15–20 days after transplanting, which later created defoliation of chili plants [5,7,8]; similarly, it also reduced the net economic return.

The intercropping of maize with chili in soil conservation practices was economically and environmentally viable on uplands for crop production, indicated by the high values of maize equivalent grain yield under intercrop land use options [22]. Moreover, conservation agricultural practices always have a positive impact on system productivity [23]. Land use options with intercropping in soil conservation practices would provide a sustainable solution to soil loss management on uplands with economic benefits. The soil conservation measures improved soil fertility, lowered soil loss and also provided higher organic inputs from leucaena prunings and Jack bean harvest residues, and additional N through biological N fixation by *Leucaena leucocephala* hedgerows (28 kg ha^{-1} y^{-1} N added to soil as biological fixed nitrogen).

Crop productivity was good even under non-fertilized land use treatments with and without hedgerows, probably due to minimum tillage associated with legume relay cropping, which enhanced the organic matter of the plots by biological N fixation and residue incorporation [24]. Positive effects of minimum tillage in combination with mulching and growing a relay cover crop (legumes) on crop yield have been documented in several studies [25,26]. Moreover, chili fresh fruit yield compensated maize production in intercrop and soil conservation practices (with hedgerows) and fertilizer application. Zuazo et al. [27] and Quinkenstein et al. [28] testified that grass barriers and hedgerows are quite effective for soil conservation and sustainable agricultural production on slopes.

Farmers' choice of cultivation on the uplands of Western Thailand depended on two main points: the first focus is household food security, and the second focus is the market demand of a specific crop. Most smallholders grow maize as domestic food, while some also concentrate on growing cash crops like chilies due to their high market value. Wider space between chili plants and rows (1 m^2) makes the soil prone to soil loss if chili is grown alone as sole crop. In this study, we successfully practiced intercropping with soil conservation practice to provide an option for farmers to grow both food for their families in terms of maize grain and earn cash by selling green chilies without compromising through loss of land resources. Maximum chili yield was obtained in the intercrop treatment with soil conservation practice is another indication of sustainable soil productivity in soil conservation land use treatments. [5], while carrying out productivity analysis of the same treatments indicated that all intercrop treatments with fertilizer application showed that 3 to 21% more land area would be required for a sole cropping system to reach the yield of an intercropping system [25,29]. Higher land equivalent ratios are an indication of the yield advantage of intercropping over sole stands due to the judicious use of available environmental and land resources for plant growth [11].

4.3. Crop Productivity Evaluation and Economic Return

The land equivalent ratios (LER) were higher than 'one' in fertilized intercropping systems during both years (Table 3). This is an indication of yield advantage over sole cropping conditions. This was attributed to the judicious utilization of water, light and nutrients for plant growth [11]. Land equivalent ratios of T2 to T4 were 1.08 to 1.28 in 2010 and 1.03 to 1.21 in 2011, indicating a better LUE of intercrops than those of sole crops. This means 8 to 28% and 3 to 21% extra land is required by a sole-cropped system to attain yield equal to an intercropping system [30,31]. This is a clear indication of better land resource utilization in intercropping systems compared with sole/monocropping. Higher values of LER in maize-chili-intercropping with leucaena hedgerows, minimum tillage, Jack bean relay cropping and proper fertilizer application (T4) would be convincing for upland farmers to adopt these types of cropping systems to increase land productivity along with soil conservation.

Modern agriculture around the globe is focused on economics. Sustainable production and economic profitability gains are more important when the land holding is small. The intercropping systems with soil conservation techniques showed greater net returns compared with sole cropping conditions due to better market incentives for chili as a cash crop (Table 3). Earlier studies carried out in various environments also mentioned the superiority of intercropping in raising farm income and soil fertility restoration compared with the mono/sole cropping of component crops [14], because chili and maize grains are an important part of daily cuisines in Thailand. Chili is consumed in its fresh and dried forms and has the highest market value due to its high consumption, which increases the net return of soil-conservation-based intercropping land uses. The net return was less during 2011 due to the infestation of chili plants by cercospora leaf spot around three weeks after transplanting, which led to the defoliation of chili plants and ultimately reduced chili yield [5].

5. Conclusions

The role of cropping systems in conservation practices is very crucial in reducing soil loss and improving sustainable crop production on uplands. Fertilized-hedgerow-intercropping with minimum tillage reduced fertile soil loss, which ensured sustainable crop production. Minimum tillage and Jack bean relay cropping proved to be the best conservation option on the tropical uplands for reducing soil loss, increasing land utilization and economic returns. They acted as a buffer against soil loss and water runoff to save the fertile topsoil and optimize the soil moisture conditions for sustainable crop production on the uplands. Experiments emphasizing the nuances of land use with soil conservation options are useful for yield maximization and livelihood uplift for upland farming communities in Southeast Asia. Long-term experiments on crop production with soil conservation on uplands should be initiated locally at a government level to address farmers' concerns that are visualized while practicing soil conservation techniques during cropping. This will ultimately enhance confidence in farmers to use soil conservation techniques for sustainable crop production on uplands.

Author Contributions: K.H. and T.H. planned the experiments, K.H. conducted experiments, K.H., A.I., I.B. and T.H. interpreted the results and made the writeup. All authors have read and agreed to the published version of the manuscript.

Funding: The funding for the study was provided by K.U. Leuven under project OT/07/045.

Institutional Review Board Statement: Not applicable.

Informed Consent Statement: Not applicable.

Data Availability Statement: The data presented in this study are available on request from the corresponding author.

Acknowledgments: We would like to thank K.U. Leuven for funding field research under project OT/07/045, the University of Agriculture, Faisalabad for supporting first author stay in Germany and the Forschungszentrum Jülich GmbH for providing the equipment. We also extend gratitude's to Thanuchai Kongkaew (who passed away on 16 February 2012) for supporting field experiments and student's facilitation during field campaigns in Thailand.

Conflicts of Interest: The authors confirm that there is no conflict of interest with the networks, organizations and data centers referred in the manuscript.

References

1. Hillel, D.; Hatfield, J.L. *Encyclopedia of Soils in the Environment*; Elsevier: Amsterdam, The Netherlands, 2005; Volume 3.
2. Pimentel, D.; Burgess, M. Soil Erosion Threatens Food Production. *Agriculture* **2013**, *3*, 443–463. [CrossRef]
3. Forsyth, T. Sustainable Livelihood Approaches and Soil Erosion Risks: Who Is to Judge? *Int. J. Soc. Econ.* **2007**, *34*, 88–102. [CrossRef]
4. Bahadur, K.C.K. Mapping Soil Erosion Susceptibility Using Remote Sensing and GIS: A Case of the Upper Nam Wa Watershed, Nan Province, Thailand. *Environ. Geol.* **2009**, *57*, 695–705. [CrossRef]
5. Hussain, K.; Wongleecharoen, C.; Hilger, T.; Vanderborght, J.; Garré, S.; Onsamrarn, W.; Sparke, M.-A.; Diels, J.; Kongkaew, T.; Cadisch, G. Combining δ 13 C Measurements and ERT Imaging: Improving Our Understanding of Competition at the Crop-Soil-Hedge Interface. *Plant Soil* **2015**, *393*, 1–20. [CrossRef]
6. Hussain, K.; Wongleecharoen, C.; Hilger, T.; Ahmad, A.; Kongkaew, T.; Cadisch, G. Modelling Resource Competition and Its Mitigation at the Crop-Soil-Hedge Interface Using WaNuLCAS. *Agroforest. Syst.* **2016**, *90*, 1025–1044. [CrossRef]

7. Hussain, K.; Ilyas, A.; Wajid, A.; Ahmad, A.; Mahmood, N.; Hilger, T.; Kongkaew, T. Alley Cropping Simulation: An Opportunity for Sustainable Crop Production on Tropical Uplands. *Pak. J. Agric. Sci.* **2019**, *56*, 109–112. [CrossRef]
8. Hussain, K.; Ilyas, A.; Wongleecharoen, C.; Hilger, T.; Wajid, A.; Ahmad, A.; Cadisch, G. Sustainable Land Use Options for Optimum Resources Use in Maize Based Cropping System on Uplands of Western Thailand. *Agroforest. Syst.* **2020**, *94*, 2289–2300. [CrossRef]
9. Frene, J.P.; Gabbarini, L.A.; Wall, L.G. Soil Physiology Discriminates between No-till Agricultural Soils with Different Crop Systems on Winter Season. *Soil Use Manag.* **2020**, *36*, 571–580. [CrossRef]
10. Hussain, K. Measuring and Modelling Resource Use Competition at the Crop-Soil-Hedge Interface on a Hillside in Western Thailand. Ph.D. Thesis, University of Hohenheim, Stuttgart, Germany, 2015.
11. Banik, P.; Sasmal, T.; Ghosal, P.K.; Bagchi, D.K. Evaluation of Mustard (Brassica Compestris Var. Toria) and Legume Intercropping under 1:1 and 2:1 Row-replacement Series Systems. *J. Agron. Crop Sci.* **2000**, *185*, 9–14. [CrossRef]
12. Lithourgidis, A.S.; Vlachostergios, D.N.; Dordas, C.A.; Damalas, C.A. Dry Matter Yield, Nitrogen Content, and Competition in Pea–Cereal Intercropping Systems. *Eur. J. Agron.* **2011**, *34*, 287–294. [CrossRef]
13. Javanmard, A.; Nasab, A.D.M.; Javanshir, A.; Moghaddam, M.; Janmohammadi, H. Forage Yield and Quality in Intercropping of Maize with Different Legumes as Double-Cropped. *J. Food Agric. Environ.* **2009**, *7*, 163–166.
14. Maitra, S.; Hossain, A.; Brestic, M.; Skalicky, M.; Ondrisik, P.; Gitari, H.; Brahmachari, K.; Shankar, T.; Bhadra, P.; Palai, J.B.; et al. Intercropping—A Low Input Agricultural Strategy for Food and Environmental Security. *Agronomy* **2021**, *11*, 343. [CrossRef]
15. Sollen-Norrlin, M.; Ghaley, B.B.; **Rintoul**, N.L.J. Agroforestry Benefits and Challenges for Adoption in Europe and Beyond. *Sustainability* **2020**, *12*, 7001. [CrossRef]
16. Hilger, T.; Keil, A.; Lippe, M.; Panomtaranichagul, M.; Saint-Macary, C.; Zeller, M.; Pansak, W.; Dinh, T.V.; Cadisch, G. Soil Conservation on Sloping Land: Technical Options and Adoption Constraints. In *Sustainable Land Use and Rural Development in Southeast Asia: Innovations and Policies for Mountainous Areas*; Springer: Berlin/Heidelberg, Germany, 2013; pp. 229–279.
17. Pansak, W.; Hilger, T.H.; Dercon, G.; Kongkaew, T.; Cadisch, G. Changes in the Relationship between Soil Erosion and N Loss Pathways after Establishing Soil Conservation Systems in Uplands of Northeast Thailand. *Agric. Ecosyst. Environ.* **2008**, *128*, 167–176. [CrossRef]
18. Pansak, W.; Hilger, T.; Lusiana, B.; Kongkaew, T.; Marohn, C.; Cadisch, G. Assessing Soil Conservation Strategies for Upland Cropping in Northeast Thailand with the WaNuLCAS Model. *Agroforest. Syst.* **2010**, *79*, 123–144. [CrossRef]
19. Mutsaers, H.J.W. *A Field Guide for On-Farm Experimentation*; IITA: Ibadan, Nigeria, 1997.
20. Paleari, S. Is the European Union Protecting Soil? A Critical Analysis of Community Environmental Policy and Law. *Land Use Policy* **2017**, *64*, 163–173. [CrossRef]
21. Chowaniak, M.; Głąb, T.; Klima, K.; Niemiec, M.; Zaleski, T.; Zuzek, D. Effect of Tillage and Crop Management on Runoff, Soil Erosion and Organic Carbon Loss. *Soil Use Manag.* **2020**, *36*, 581–593. [CrossRef]
22. Agegnehu, G.; Ghizaw, A.; Sinebo, W. Yield Performance and Land-Use Efficiency of Barley and Faba Bean Mixed Cropping in Ethiopian Highlands. *Eur. J. Agron.* **2006**, *25*, 202–207. [CrossRef]
23. Kumar, V.; Gathala, M.K.; Saharawat, Y.S.; Parihar, C.M.; Kumar, R.; Kumar, R.; Jat, M.L.; Jat, A.S.; Mahala, D.M.; Kumar, L.; et al. Impact of Tillage and Crop Establishment Methods on Crop Yields, Profitability and Soil Physical Properties in Rice–Wheat System of Indo-Gangetic Plains of India. *Soil Use Manag.* **2019**, *35*, 303–313. [CrossRef]
24. Thomas, G.A.; Dalal, R.C.; Standley, J. No-till Effects on Organic Matter, PH, Cation Exchange Capacity and Nutrient Distribution in a Luvisol in the Semi-Arid Subtropics. *Soil Till. Res.* **2007**, *94*, 295–304. [CrossRef]
25. Sogbedji, J.M.; van Es, H.M.; Agbeko, K.L. Cover Cropping and Nutrient Management Strategies for Maize Production in Western Africa. *Agron. J.* **2006**, *98*, 883–889. [CrossRef]
26. Shafi, M.; Bakht, J.; Jan, M.T.; Shah, Z. Soil C and N Dynamics and Maize (*Zea May* L.) Yield as Affected by Cropping Systems and Residue Management in North-Western Pakistan. *Soil Tillage Res.* **2007**, *94*, 520–529. [CrossRef]
27. Zuazo, V.H.D.; Martínez, J.R.F.; Pleguezuelo, C.R.R.; Raya, A.M.; Rodríguez, B.C. Soil-Erosion and Runoff Prevention by Plant Covers in a Mountainous Area (SE Spain): Implications for Sustainable Agriculture. *Environmentalist* **2006**, *26*, 309–319. [CrossRef]
28. Quinkenstein, A.; Woellecke, J.; Böhm, C.; Grünewald, H.; Freese, D.; Schneider, B.U.; Hüttl, R.F. Ecological Benefits of the Alley Cropping Agroforestry System in Sensitive Regions of Europe. *Environ. Sci. Policy* **2009**, *12*, 1112–1121. [CrossRef]
29. Midya, A.; Bhattacharjee, K.; Ghose, S.S.; Banik, P. Deferred Seeding of Blackgram (*Phaseolus Mungo* L.) in Rice (*Oryza Sativa* L.) Field on Yield Advantages and Smothering of Weeds. *J. Agron. Crop Sci.* **2005**, *191*, 195–201. [CrossRef]
30. Dhima, K.V.; Lithourgidis, A.S.; Vasilakoglou, I.B.; Dordas, C.A. Competition Indices of Common Vetch and Cereal Intercrops in Two Seeding Ratio. *Field Crops Res.* **2007**, *100*, 249–256. [CrossRef]
31. Bedoussac, L.; Justes, E. Dynamic Analysis of Competition and Complementarity for Light and N Use to Understand the Yield and the Protein Content of a Durum Wheat-Winter Pea Intercrop. *Plant Soil* **2010**, *330*, 37–54. [CrossRef]

Article

Assessment of Potato Farmland Soil Nutrient Based on MDS-SQI Model in the Loess Plateau

Yingying Xing [1], Ning Wang [1], Xiaoli Niu [2], Wenting Jiang [1] and Xiukang Wang [1,*]

1 College of Life Sciences, Yan'an University, Yan'an 716000, China; xingyingying@yau.edu.cn (Y.X.); wangwangning0@gmail.com (N.W.); jiangwenting@yau.edu.cn (W.J.)
2 College of Agricultural Equipment Engineering, Henan University of Science and Technology, Luoyang 471000, China; niuxiaoli88@haust.edu.cn
* Correspondence: wangxiukang@yau.edu.cn

Abstract: Soil nutrients are essential nutrients provided by soil for plant growth. Most researchers focus on the coupling effect of nutrients with potato yield and quality. There are few studies on the evaluation of soil nutrients in potato fields. The purpose of this study is to investigate the soil nutrients of potato farmland and the soil vertical nutrient distributions, and then to provide a theoretical and experimental basis for the fertilizer management practices for potatoes in Loess Plateau. Eight physical and chemical soil indexes were selected in the study area, and 810 farmland soil samples from the potato agriculture product areas were analyzed in Northern Shaanxi. The paper established the minimum data set (MDS) for the quality diagnosis of the cultivated layer for farmland by principal component analysis (PCA), respectively, and furthermore, analyzed the soil nutrient characteristics of the cultivated layer adopted soil quality index (SQI). The results showed that the MDS on soil quality diagnosis of the cultivated layer for farmland soil included such indicators as the soil organic matter content, soil available potassium content, and soil available phosphorus content. The comprehensive index value of the soil quality was between 0.064 and 0.302. The SPSS average clustering process used to classify SQI was divided into three grades: class I (36.2%) was defined as suitable soil fertility (SQI < 0.122), class II (55.6%) was defined as moderate soil fertility (0.122 < SQI < 0.18), and class III (8.2%) was defined as poor soil fertility (SQI > 0.186). The comprehensive quality of the potato farmland soils was generally low. The proportion of soil nutrients in the SQI composition ranged from large to small as the soil available potassium content = soil available phosphorus content > soil organic matter content, which became the limiting factor of the soil organic matter content in this area. This study revolves around the 0 to 60 cm soil layer; the soil fertility decreased gradually with the soil depth, and had significant differences between the respective soil layers. In order to improve the soil nutrient accumulation and potato yield in potato farmland in northern Shaanxi, it is suggested to increase the fertilization depth (20 to 40 cm) and further study the ratio of nitrogen, phosphorus, and potassium fertilizer.

Keywords: potato; minimum data sets; soil nutrient; soil fertility index

Citation: Xing, Y.; Wang, N.; Niu, X.; Jiang, W.; Wang, X. Assessment of Potato Farmland Soil Nutrient Based on MDS-SQI Model in the Loess Plateau. *Sustainability* **2021**, *13*, 3957. https://doi.org/10.3390/su13073957

Academic Editor: Victor Rolo

Received: 19 February 2021
Accepted: 25 March 2021
Published: 2 April 2021

1. Introduction

Soil is an important natural resource, and soil nutrients are key to contributing to food security, human health, and sustainable development. Soil plays a special role in crop growth and reproduction, such as nutrient storage, nutrient transformation and circulation, rainwater conservation, biological support, stability, and as a buffer of environmental changes [1–3]. The core of soil science research is soil fertility, and soil nutrient content is an important indicator of soil fertility [4]. Soil nutrients have the characteristics of flow and migration in soil. Therefore, the analysis of the soil nutrient content and spatial distribution, the classification of soil nutrient levels and the comprehensive evaluation of soil nutrients are the most effective means to develop a scientific and reasonable fertilization program [5,6]. It is also the main measure to reduce the excessive application of fertilizer, improve the utilization efficiency of fertilizer and protect the sustainable development of the environment.

Soil nutrients play an important role in the growth of the plant and any imbalance in soil fertility affects the crop productivity as well as posing severe environmental hazards [7]. In the agricultural ecosystem, human regulation plays a dominant role, and the main source of nutrients is fertilizer application. According to statistics from 2000 to 2015, it is estimated that the quality of arable land in China is poor, and low- and medium-yield farmland occupies 72% of the total arable land area [8]. The soil degradation in the black soil region of northeast China was severe, and agricultural production capacity was lost in some areas [9]. Many researchers have carried out soil nutrient evaluation in order to ensure the soil water supply and fertilizer capacity and prevent blind fertilization [10–12]. Soil nutrient assessment is essential to the basic investigation and research of efficient agriculture [13]. Precision agriculture requires a combination of crop yield components and national policies to reasonably regulate the use and dosage of fertilizers input into the farmland ecosystem, which is conducive to meeting the production conditions of marginal effect and achieving the maximum utilization rate of fertilizers [14].

Soil nutrient assessment refers to the monitoring and evaluation of soil properties, soil functions and soil conditions. Soil quality is usually assessed through the measurement of selected soil properties. The evaluation of soil quality is difficult due to the heterogeneity of soil, and the physical, chemical, and biological properties of soil vary greatly in different regions. Thus, to comprehensively assess soil quality, a soil quality index (SQI) that integrated soil properties into an overall index was established and has been used widely. Many researchers have selected multiple soil evaluation indexes such as pH value, salinity, and clay weight [15,16]. Some researchers selected multiple soil evaluation indexes such as pH value, C/N, and soil nematode to establish the minimum data set of soil evaluation [17,18]. Some researchers used soil material indexes such as porosity and three-dimensional aggregate characteristics to evaluate farmland soil quality [19,20]. Some researchers selected soil quality evaluation indexes from three aspects of soil physics, chemistry, and microorganism, and established the microbial community and enzyme activity evaluation system [21,22]. Therefore, it is particularly important to select appropriate evaluation indexes scientifically, and soil indexes that mainly affect crop growth should be considered simultaneously. In addition, soil quality and soil fertility assessments have also been carried out because their selection indexes are similar to the easily confused methods of evaluation, which leads to the tendency to evaluate soil fertility in soil quality assessment.

The limiting factors of soil nutrient vary with land use modes, land types, ecosystems, locations, and soil parent materials [23,24]. Therefore, the selection of appropriate indexes is particularly important for the results of soil assessment. The assessment of soil quality could lead to new methods and practices that could be applied to more sustainable development. The establishment of a minimum data set (MDS) is a convenient process for selecting soil quality indicators and evaluation [25]. The MDS can reduce data redundancy by selecting the most appropriate metric among preselected metrics. In addition, the weight of selected indexes can be generated during the establishment of MDS, which reduces the subjective influence of human factors on soil quality and is conducive to subsequent soil quality evaluation [26]. Many researchers have conducted soil quality assessments based on MDS [27–29]. The MDS of soil quality indicators has been used to evaluate the relationship between the indicators and their effects on soil properties and crops [30]. There are many methods to construct the minimum data set, such as the grey correlation method, principal component analysis (PCA) method, and correlation coefficient method, etc. The PCA method has strong objectivity and can ensure the minimum loss of original data information to reflect the impact of indicators on soil quality [31]. The four indicators extracted from the MDS were significantly correlated with the fertility indicators established by all indicators [32]. The MDS can well reflect the regional differences and is suitable for soil nutrient evaluation [25].

The Loess Plateau is one of the most advantageous potato producing areas in China. In recent years, about 50% of the potato yield in northern Shaanxi reached the middle level, and the ratio of high potato yield was less than 6% [33]. The land resources in the Loess Plateau were affected by unreasonable fertilization, which resulted in soil quality degradation and threatened the local ecological security [34]. Research on the soil quality of potato fields in the Loess Plateau has been relatively rare in recent years. There are two problems in previous studies. One is that many studies only use statistical data on the sample scale, and the results are not extended to the whole region. Second, most of the studies focus on regional ecosystems and do not evaluate specific land use types.

The purpose of this study was to evaluate the soil quality of potato cropland on the Loess Plateau, based on soil nutrient indicators from 2017 to 2018. This study attempts to use principal component analysis and correlation analysis to determine the MDS of soil nutrient evaluation in this region, and the fuzzy mathematics membership function method to establish a soil nutrient evaluation index. The results are useful for local potato field scientific fertilization and soil protection.

2. Materials and Methods

2.1. Site Description and Soil Sampling

The experiments were carried out in 10 counties (36°28′–39°37′ N, 108°21′–109°55′ E, altitude of 762 to 1340 m above sea level) in the Loess Plateau region of China from 2017 to 2018 (Figure 1). The region is a typical hilly landform of the Loess Plateau and belongs to an arid and semi-arid continental monsoon climate. The climate is mild and semiarid, with an annual average temperature of 9 °C, a monthly average maximum temperature of 22 °C (July), and a monthly average minimum temperature of −4.6 °C (January). The average annual sunshine duration of 2200 h exceeds 158 days without frost and the mean annual radiation is 490 KJ/cm^2. From 1990 to 2018, the annual average precipitation in this region was 480 mm. The soil type is loess soil, which is classified as Inceptisol according to the USDA soil classification and Cambisol according to the World Reference Database System [21]. The average sand (2.00 to 0.02 mm grain size), silt (0.02 to 0.002 mm), and clay (<0.002 mm) contents in the 0- to 80-cm soil profile was measured with a laser particle size analyzer (Dandong Haoyu Technology Co., Ltd., Dandong, Liaoning, China), and the values were 62%, 25%, and 13%, respectively.

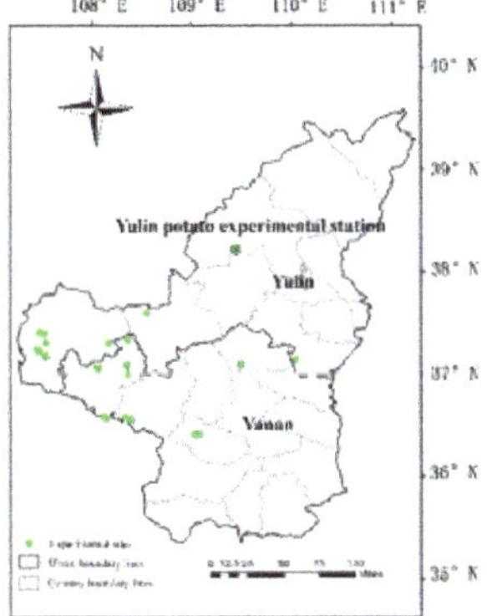

Figure 1. The map depicts the collection area of soil and plant samples on the Loess Plateau of China. These circles indicate the location of the experiment site. The thick line represents the city boundary and the thin line represents the county boundary.

2.2. Sample Collection and Determination Methods

Soil samples should be collected from farmland or test plots with an area greater than 600 m^3. Soil cores were collected using an auger with a 25 mm inner diameter. Five sample points were randomly collected from the same test plot at 0- to 60-cm soil depths. The soil temperature ranged from 12.3 to 17.4 °C when we collected the soil samples. The soil samples were air-dried, homogenized, and hand ground to pass through a 2-mm sieve for the determination of soil physical and chemical index. The soil depth from 0 to 60 cm was determined as S, 0 to 20 cm as S1, 20 to 40 cm as S2, and 40 to 60 cm as S3.

The soil organic matter content was determined by the potassium dichromate volumetric method (external heating method) [33]. The soil available potassium was determined by the flame photometry method with NH_4OAc extraction [33]. The soil available phosphorus was determined by the molybdenum antimony anti-spectrophotometry method with $NaHCO_3$ extraction [33]. The soil acidity (pH) was measured in an aqueous soil extract in de-ionized water (1:2.5 soil:water) [33]. For the soil electrical conductivity 1:5 soil-to-water ratio solution, 50 g of soil was mixed with 250 g of distilled water, and repeated three times, shaking manually four times every 30 min for 1 min and standing for 4 h to achieve balance. Once the 1:5 soil–water solutions reached equilibrium, the soil's electrical conductivity was measured by inserting an Accumet 50 m (Fisher Scientific, Hampton, NH, USA) into the solution [35]. The content of soil alkali-hydrolyzable nitrogen was determined by conductance titration; 0.5 g soil samples were distilled with 2 mol L^{-1} NaOH for 5 h, then treated with 10mol L^{-1} NaOH for 7 min, and then treated with 40 g L^{-1} boric acid for NH_3, which was absorbed and released by direct steam [36]. The soil water content was determined using the gravimetric method. The soil nitrate N content was measured using a spectrophotometer (UV-2600, Shanghai Hengping Scientific Instrument Co. Ltd., Shanghai, China). First, 0.5 g of fresh soil was transferred to a 100-mL Erlenmeyer flask, and then 50 mL of a 2-mol L^{-1} potassium chloride solution was added to the Erlenmeyer flask. Next, the mixture was oscillated for 30 min using a shock machine until reaching uniformity. Finally, the solution was filtered, and then 5 mL of the solution was measured using a spectrophotometer at a wavelength of 210 nm [10].

2.3. *Establishment of Minimum Data Set MDS*

What comes first is to screen soil nutrient indicators, in order to avoid the influence of redundant data and complex multiple correlations of indicators on the constructed soil quality assessment index. Principal component analysis (PCA) was used to reduce dimensionality and representative soil measurement indexes were selected to establish MDS. When performing the principal component analysis, the extract principal components had eigenvalues greater than 1. The index with the same principal component load $\geqq 0.5$ is divided into one group. If the load of a soil parameter in two principal components is higher than 0.5 at the same time, the parameter should be merged into the group with a lower correlation with other parameters.

Second, to solve the vector normal, the vector normal of the evaluation index was calculated. The vector normal is the length of the vector normal mode of the index in a multi-dimensional space composed of components. The longer the length, the greater the comprehensive load of the indicator in all principal components, and the greater its ability to interpret comprehensive information. The vector normal is calculated as follows:

$$N_{ik} = \sqrt{\sum_{i=1}^{k} (u_{ik}^2 \cdot \lambda_k)} \quad (1)$$

In the formula, N_{ik} is the comprehensive load of the i-th index on the first k principal components of the eigenvalue; u_{ik} is the load of the i-th index on the k-th principal component; λ_k is the eigenvalue of the k-th principal component.

2.4. Establishment of Soil Nutrient Evaluation Index

To calculate the soil quality index (SQL) for different data sets [37–39], the calculation formula is as follows:

$$SQL = \sum_{i=1}^{k} W_i \cdot N_i \quad (2)$$

In the formula, SQI is the soil nutrient evaluation index; W_i is the weight of the i-th index; N_i is the membership of the i-th index. The weight calculation formula is as follows:

$$W_i = \frac{M_i}{\sum M_i} \quad (3)$$

$$M_i = \frac{\sum_{i=1}^{k}\left(\frac{u_{ik}}{\sqrt{\lambda_k}} \cdot \theta_{ik}\right)}{\sum_{i=1}^{k} \theta_{ik}} \quad (4)$$

In the formula, W_i is the weight of the i-th index; θ_{ik} is the variance percentage of the i-th index on the k-th principal component. The membership degree rising function f(x) calculation formula is as follows:

$$f(x) = \begin{cases} 1.0\ x > x_2 \\ \frac{0.9(x-x_1)}{(x_2+x_1)} + 0.1\ x_1 < x \le x_2 \\ 0.1\ x < x_1 \end{cases} \quad (5)$$

In the formula, x_1 is the minimum value of the soil index threshold; x_2 is the maximum value of the soil index threshold.

2.5. Data Processing Methods

SPSS 22.0 statistical software and Sigma Plot 14.0 were used for statistical analysis and data plotting, respectively. Tukey's multiple comparison test was used to detect differences among treatments at the 0.05 significance level. In addition, the relationships among all the parameters (soil nutrient content) were calculated using a bivariate correlation analysis (Pearson correlation coefficient and double-tailed significance test).

3. Results and Analysis

3.1. Soil Physical and Chemical Properties

The soil organic matter content in soil layer S1 was significantly different from that in soil layer S2 and S3, but there was no significant difference between S2 and S3 (Figure 2A). The organic matter content in the soil layer S1 was up to 8.4 g kg^{-1} on average, which was 30% and 40.7% higher than that in higher than that in S2 and S3, respectively. The available potassium content in soil layer S1 was significantly different from that in soil layer S2 and S3, but there was no significant difference between S2 and S3 (Figure 2B). The available potassium content in soil layer S1 was the highest at 69.1 mg kg^{-1} on average, which was 19.7% higher than that in S2 and 26% higher than that in S3. The soil available phosphorus content in soil layer S1 was significantly different from that in soil layer S2 and S3, but there was no significant difference between S2 and S3 (Figure 2C). As the depth of the soil layer decreased, the available phosphorus content showed a decreasing trend. The soil available phosphorus content in soil layer S1 was the highest at 16.9 mg kg^{-1} on average, which was 49.3% and 58.7% higher than that in S2 and S3, respectively. The soil pH in S1 soil layer is significantly different from S2 and S3 (Figure 2D). As the depth of the soil layer decreased, the pH value increased. The average soil pH value of the S3 soil layer was 8.6, which was 1.9% and 0.3% higher than that in S1 and S2, respectively.

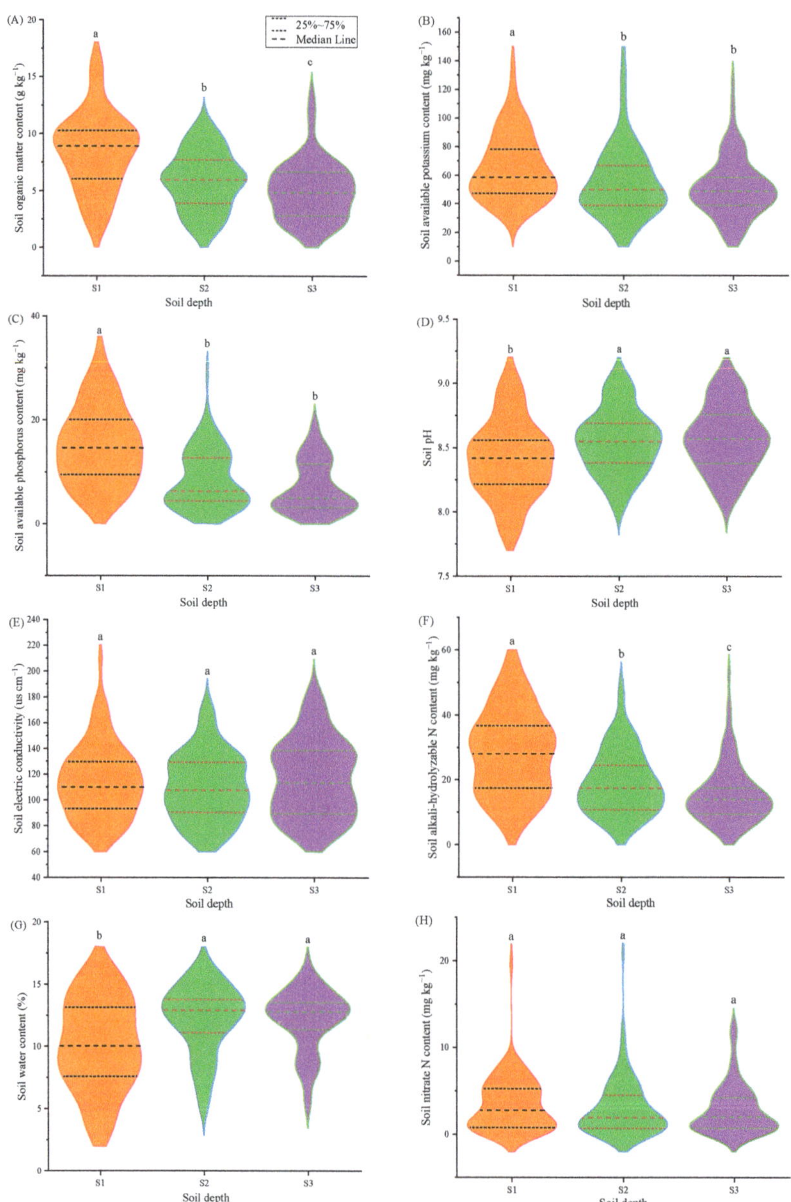

Figure 2. The relationship between soil depth and soil organic matter content (**A**), soil available potassium content (**B**), soil available phosphorus content (**C**), soil pH (**D**), soil electric conductivity (**E**), soil alkaline hydrolyzable N content (**F**), soil water content (**G**), and soil nitrate N content (**H**). 0 to 20 cm was determined as S1, 20 to 40 cm was determined as S2, and 40 to 60 cm was determined as S3. The different letters indicate significant differences at $p < 0.05$.

There is no significant difference in the electrical conductivity among soil layers (Figure 2E). The soil alkaline hydrolyzable N content in soil layer S1 was significantly different from that in soil layer S2 and S3, and the difference between S2 and S3 was significant (Figure 2F). The soil alkali-hydrolyzable N content decreased with the decrease of soil depth. The soil alkali-hydrolyzable N content in soil layer S1 was the highest at 28.5 mg kg^{-1} on average, which was 32% and 45.2% higher than that in S2 and S3, respectively. The soil water content in soil layer S1 was significantly different from that in soil layer S2 and S3 (Figure 2G). The soil water content of soil layer S2 (12.18%) and S3 (12.14%) was higher on average, which was 14.4% and 19.1% higher than that of S1. There was no significant difference in the soil nitrate nitrogen content between the soil layers (Figure 2H).

3.2. The Establishment of the Minimum Data Set MDS

Based on the results of principal component analysis (Table 1), the eigenvalues of the first three principal components are greater than 1, and the cumulative contribution rate of the three principal components is 64%.

Table 1. Eigenvalue and contribution rate in principal components analysis.

Component	Initial Eigenvalues			Extract the Sum of Squares and Load It		
	Total	Variance (%)	Accumulation (%)	Total	Variance (%)	Accumulation (%)
1	2.386	29.825	29.825	2.386	29.825	29.825
2	1.591	19.89	49.715	1.591	19.89	49.715
3	1.142	14.269	63.984	1.142	14.269	63.984

According to the principal component analysis load matrix (Table 2) and the screening criteria, the final soil organic matter content, soil pH value, soil electric conductivity, and soil alkaline hydrolyzable N content entered PC1, thesoil available potassium content and soil nitrate N content entered PC2, and soil available phosphorus content and soil water content entered PC3.

Table 2. Principal component loading matrix and calculated vector normal.

Indicator	Principal Component			Vector Normal
	PC1	PC2	PC3	
SOM	0.854	0.067	−0.069	1.324
K	0.362	0.673	−0.161	1.031
P	0.480	−0.021	0.685	1.042
pH	−0.556	0.492	0.226	1.087
EC	0.532	−0.292	−0.015	0.901
SAN	0.811	0.118	0.096	1.266
SWC	0.191	0.546	−0.602	0.988
SNN	−0.088	0.703	0.468	1.027

Note: SOM, soil organic matter content; K, soil available potassium content; P, soil available phosphorus content; EC, soil electric conductivity; SAN, soil alkaline hydrolyzable N content; SWC, soil water content; SNN, soil nitrate N content.

The vector normal of each indicator is shown in Table 2, and selects the indicator whose vector normal is in the range of 10% of the highest total score in each group. The indicators entering MDS are the soil organic matter content, soil available potassium content, and soil available phosphorus content. Correlation tests between soil indicators were used to screen MDS indicators. Highly correlated indicators with the highest vector normal entered the smallest data set. Low correlations entered the same group of indicators into the MDS. It can be known from the Table 3 that the MDS index has a significant correlation with each soil nutrient index ($p < 0.05$), which satisfies the comprehensive evaluation of nutrients instead of all soil indicators. Therefore, the indicators that finally entered MDS were the soil organic matter content, soil available potassium content and soil available phosphorus content.

Table 3. Correlation coefficient between matrix among soil fertility indicators.

Indicators	SOM	K	P	pH	EC	SAN	SWC	SNN
SOM	1							
K	0.322 **	1						
P	0.278 **	0.069	1					
pH	−0.355 **	0.034	−0.178 **	1				
EC	0.403 **	−0.06	0.112	−0.210 **	1			
SAN	0.601 **	0.254 **	0.346 **	−0.312 **	0.292 **	1		
SWC	0.158 *	0.300 **	−0.139 *	0.014	−0.018	0.142 *	1	
SNN	−0.051	0.213 **	0.111	0.365 **	−0.111	0.039	0.112	1

Note: ** Significant correlation at level 0.01 (both sides); * Significant correlation at level 0.05 (both sides); SOM, soil organic matter content; K, soil available potassium content; P, soil available phosphorus content; EC, soil electric conductivity; SAN, soil alkaline hydrolyzable N content; SWC, soil water content; SNN, soil nitrate N content.

3.3. Comprehensive Evaluation of Nutrients

According to the results of the principal component analysis, the weight of the soil nutrient index of MDS was calculated, and the results were shown in Table 4. The weights of soil indexes mainly included soil organic matter (0.176), soil available potassium (0.164), and soil available phosphorus (0.192). The soil organic matter, available potassium and available phosphorus in this survey all had a positive effect on the growth of potato, which belonged to the ascending membership function. The turning point value of the function curve and the determination of index membership are shown in Table 4.

Table 4. Communality and weight of soil quality indicators in the MDS.

MDS	Common Factor Variance	Weightiness	Turning Point		Membership Function
			X_1	X_2	
SOM	0.260	0.176	6	40	Distribution curve of upper ring type
K	0.242	0.164	20	200	
P	0.283	0.192	3	40	

Note: SOM, soil organic matter content; K, soil available potassium content; P, soil available phosphorus content.

The soil quality evaluation index (SQI) is shown in Table 5. The SQI of this study ranged from 0.064 to 0.302, which was classified as Grade I (SQI < 0.122), Grade II (0.122 < SQI < 0.186), and Grade III (SQI > 0.186). In this study, the soil nutrient evaluation index showed moderate variation, among which SQI was 8.2% of Grade III, 55.6% of Grade II, and 36.2% of Grade I. As the depth of the soil layer decreases, the SQI shows a downward trend, and the SQI varies significantly between layers. Grade I of SQI in S1 is relatively high, while Grade II is the lowest. The proportion of Grade II in SQI of S2 layer and S3 layer is relatively high, while that of Grade III is relatively low, showing that the proportion of Grade I and II tends to be stable, and the proportion of Grade III tends to 0.

Table 5. Classification of scores for comprehensive evaluation of soil nutrients.

Soil depth	SQI Rangeability	SQI Mean	SQI Standard Deviation	SQI Coefficient of Variation	The Proportion of Different Soil Fertility		
					I	II	III
					SQI < 0.122	0.122 < SQI < 0.186	SQI > 0.186
S	0.064–0.302	0.125	0.043	34.8%	0.362	0.556	0.082
S1	0.083–0.281	0.160 a	0.045	27.9%	0.605	0.173	0.222
S2	0.069–0.302	0.113 b	0.034	30.0%	0.259	0.716	0.025
S3	0.064–0.161	0.101 c	0.025	24.8%	0.222	0.778	0.000

Note: Significant differences in lowercase letters ($p < 0.05$).

3.4. Evaluation of Nutrients in Different Soil Layers

The cumulative value of the soil nutrient quality evaluation index based on the composition of organic matter, available potassium and available phosphorus is shown in Figure 3. The soil nutrient quality evaluation index is a large proportion of medium-speed potassium and fast-effect phosphorus, and the minimum of organic matter is the main problem of soil nutrient in the area.

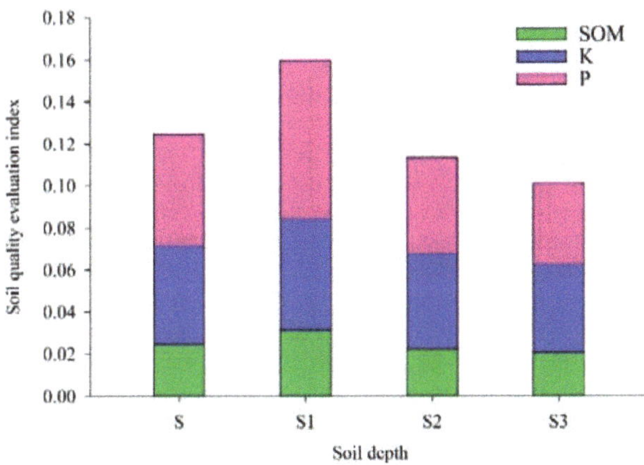

Figure 3. Cumulative histogram of soil quality comprehensive index.

With the increase of soil depth, the proportion of organic matter, available potassium and available phosphorus in the evaluation index all decreased. The proportion of soil available phosphorus in layer S1 is the largest, and the proportion of soil available potassium and phosphorus in layer S2 and layer S3 is similar.

4. Discussion

Many researchers use quantitative evaluation and mathematical models to study soil nutrients [40–42]. In this study, principal component analysis was used to screen eight soil indicators, and finally soil organic matter, available potassium and available phosphorus and nitrogen were determined as MDS. The cumulative contribution rate of soil indicators in the MDS constructed in this study was 63.98%, and correlation analysis showed that each soil index was significantly correlated with the construction of MDS soil indicators. Soil organic matter, soil available potassium, soil available phosphorus and soil nitrate nitrogen play an important role in constructing the MDS of soil nutrient evaluation [43]. Different factors are selected to participate in the evaluation of soil quality, leading to slightly different key factors in the final evaluation. The soil quality evaluation is sufficient to study specific land functions and land use types.

The SQI range of this survey is 0.064 to 0.302, which is divided into three levels, namely Grade I (SQI < 0.122), Grade II (0.122 < SQI < 0.186), and Grade III (SQI > 0.186). The soil nutrient quality evaluation index of 91.8% in the Loess Plateau was in grade I and grade II, and the soil nutrient evaluation index was poor overall. Some studies have also shown that the Loess Plateau belongs to low quality cultivated land, with large topographic fluctuation, a lack of irrigation conditions, poor soil quality and low soil nutrient content [44,45]. The scope of this study was 0 to 60 cm soil layer, and the SQI showed a downward trend as the soil layer decreased, and there were significant differences in the SQI among different layers. This is similar to the distribution of soil nutrients in farmland by many researchers [46,47]. The proportion of SQI of topsoil S1 in each grade ranged from large to small as Grade I > Grade II > Grade III; SQI varies with the change of region and the range of change is large in this study.

Continuous fertilizer application resulted in the enrichment of soil nutrients in the topsoil, but the soil nutrients were lost due to ammonia volatilization, runoff and leaching, etc., resulting in a great change in the content of soil nutrients in the topsoil [48–50]. The SQI of middle soil S2 and lower soil S3 was higher in Grade II and lower in Grade III. With the advance of the growth period, soil nutrients slowly migrate downward, and the amount of nutrients taken away by crops in the middle and lower soils gradually increases, resulting in the decrease of nutrient content in the middle and lower soils. The same study showed that the soil nutrient content in the 0 to 40 cm soil layer of sloping farmland on the Loess Plateau was the highest, and the nutrient content decreased with the increase of profile depth [51].

In the composition of soil nutrient quality index in this study, soil nitrate nitrogen and soil available potassium accounted for a large proportion, while soil organic matter accounted for the smallest proportion, indicating that soil organic matter became the limiting factor of soil nutrient quality in this region. In the past 20 years, soil nutrient observation data in the Loess Plateau region showed that organic matter content was still in a deficient state [52]. Soil organic matter plays an important role in providing nutrients and improving soil fertility. The results showed that the long-term application of organic fertilizer significantly increased the content of soil organic carbon, and the combined application of nitrogen and phosphate fertilizer increased the content of soil available nutrients [53]. Therefore, it is suggested to increase the input of organic matter in this area for a long time to improve and enhance soil fertility.

The reasons for the poor soil nutrient quality are poor soil physical and chemical properties, low input or large nutrient loss. This area is an ecologically fragile area, and concentrated rainfall leads to the coexistence of soil and water loss and nutrient loss. A large number of investigations showed that the traditional fertilization situation in this area was the insufficient application of organic fertilizer, excessive and insufficient input of nitrogen and phosphorus fertilizer, and less applications of potassium fertilizer. This study shows that the soil nutrients in this region are in a state of scarcity, so fertilization measures should be formulated to enhance the soil nutrient content in this region and prevent serious soil degradation caused by planting. The results show that the adjustment of the fertilization depth in Loess Plateau is beneficial to improve the utilization efficiency of fertilizer [54]. In order to improve the effective accumulation of soil nutrients in potato fields, it is suggested to apply deep fertilizer, increase the input of organic fertilizer and reasonable N-P-K application in this area.

5. Conclusions

In this study, the MDS evaluation indexes of soil nutrients in northern Shaanxi were soil organic matter, soil available potassium and soil available phosphorus. The SQI value range was 0.0638 to 0.469, and the overall soil nutrient evaluation index is poor. The proportion of soil nutrient indicators in the SQI was in the order of soil available K = soil available P > soil organic matter, and organic matter has become the limiting factor for soil nutrients in the region. In order to ensure the soil nutrient accumulation and sustainable agricultural development of potato farmland in the Loess Plateau, it is suggested to increase the fertilization depth, especially the potato root layer depth (20 to 40 cm), and further study is needed to determine the optimal ratio of nitrogen, phosphorus and potassium fertilizer.

Author Contributions: Y.X., N.W. and X.N. wrote the paper, X.W. designed the study idea and applied the model to calculated, Y.X., N.W., X.N. and W.J. collected and arranged the whole database, Y.X. and X.W. analyzed and discussed. All the authors revised the manuscript. All authors have read and agreed to the published version of the manuscript.

Funding: This research was funded the National Natural Science Foundation (51669034, 51809224) and Key Scientific Research Program of Education Department of Shaanxi Province (18JS117). The APC was funded by the National Natural Science Foundation (51809224).

Institutional Review Board Statement: Not applicable.

Informed Consent Statement: Not applicable.

Data Availability Statement: The data presented in this study are available on request from the corresponding author.

Acknowledgments: We thank Jia Yun, Huihui Zhang, Tao Guo and Tianmei Li for providing us with experimental field plots and helping us to collect the data.

Conflicts of Interest: The authors declare no conflict of interest.

References

1. Cowie, A.L.; Penman, T.D.; Gorissen, L.; Winslow, M.D.; Lehmann, J.; Tyrrell, T.D.; Twomlow, S.; Wilkes, A.; Lal, R.; Jones, J.W.; et al. Towards sustainable land management in the drylands: Scientific connections in monitoring and assessing dryland degradation, climate change and biodiversity. *Land Degrad. Dev.* **2011**, *22*, 248–260. [CrossRef]
2. Mucina, L.; Wardell-Johnson, G.W. Landscape age and soil fertility, climatic stability, and fire regime predictability: Beyond the OCBIL framework. *Plant Soil* **2011**, *341*, 1–23. [CrossRef]
3. Panigrahi, P.; Srivastava, A.K.; Panda, D.K.; Huchche, A.D. Rainwater, soil and nutrients conservation for improving productivity of citrus orchards in a drought prone region. *Agric. Water Manag.* **2017**, *185*, 65–77. [CrossRef]
4. Yageta, Y.; Osbahr, H.; Morimoto, Y.; Clark, J. Comparing farmers' qualitative evaluation of soil fertility with quantitative soil fertility indicators in Kitui County, Kenya. *Geoderma* **2019**, *344*, 153–163. [CrossRef]
5. Hengl, T.; Leenaars, J.G.B.; Shepherd, K.D.; Walsh, M.G.; Heuvelink, G.B.M.; Mamo, T.; Tilahun, H.; Berkhout, E.; Cooper, M.; Fegraus, E.; et al. Soil nutrient maps of Sub-Saharan Africa: Assessment of soil nutrient content at 250 m spatial resolution using machine learning. *Nutr. Cycl. Agroecosyst.* **2017**, *109*, 77–102. [CrossRef]
6. Zhao, Z.; Liu, G.; Liu, Q.; Huang, C.; Li, H.; Wu, C. Distribution characteristics and seasonal variation of soil nutrients in the Mun River Basin, Thailand. *Int. J. Environ. Res. Public Health* **2018**, *15*, 1818. [CrossRef]
7. Wang, X.; Fan, J.; Xing, Y.; Xu, G.; Wang, H.; Deng, J.; Wang, Y.; Zhang, F.; Li, P.; Li, Z. Chapter three-The effects of mulch and nitrogen fertilizer on the soil environment of crop plants. In *Advances Agronomy*; Sparks, D.L., Ed.; Academic Press: Cambridge, MA, USA, 2019; Volume 153, pp. 121–173.
8. Wang, X. Consumption of cropland with high yield and low irrigation water intensity by urban expansion in China during 2000–2015. *Ecol. Indic.* **2020**, *117*, 106644. [CrossRef]
9. Gong, H.; Meng, D.; Li, X.; Zhu, F. Soil degradation and food security coupled with global climate change in northeastern China. *Chin. Geogr. Sci.* **2013**, *23*, 562–573. [CrossRef]
10. Wang, X.; Guo, T.; Wang, Y.; Xing, Y.; Wang, Y.; He, X. Exploring the optimization of water and fertilizer management practices for potato production in the sandy loam soils of Northwest China based on PCA. *Agric. Water Manag.* **2020**, *237*, 106180. [CrossRef]
11. Wang, H.; Wang, X.; Bi, L.; Wang, Y.; Fan, J.; Zhang, F.; Hou, X.; Cheng, M.; Hu, W.; Wu, L.; et al. Multi-objective optimization of water and fertilizer management for potato production in sandy areas of northern China based on TOPSIS. *Field Crops Res.* **2019**, *240*, 55–68. [CrossRef]
12. Grant, C.A.; Peterson, G.A.; Campbell, C.A. Nutrient considerations for diversified cropping systems in the Northern Great Plains. *Agron. J.* **2002**, *94*, 186–198. [CrossRef]
13. Techen, A.-K.; Helming, K.; Brüggemann, N.; Veldkamp, E.; Reinhold-Hurek, B.; Lorenz, M.; Bartke, S.; Heinrich, U.; Amelung, W.; Augustin, K.; et al. Chapter Four—Soil research challenges in response to emerging agricultural soil management practices. In *Advances in Agronomy*; Sparks, D.L., Ed.; Academic Press: Cambridge, MA, USA, 2020; Volume 161, pp. 179–240.
14. Van Evert, F.K.; Gaitán-Cremaschi, D.; Fountas, S.; Kempenaar, C. Can precision agriculture increase the profitability and sustainability of the production of potatoes and olives? *Sustainability* **2017**, *9*, 1863. [CrossRef]
15. Wang, F.; Yang, S.; Wei, Y.; Shi, Q.; Ding, J. Characterizing soil salinity at multiple depth using electromagnetic induction and remote sensing data with random forests: A case study in Tarim River Basin of southern Xinjiang, China. *Sci. Total Environ.* **2021**, *754*, 142030. [CrossRef] [PubMed]
16. Wu, C.; Liu, G.; Huang, C.; Liu, Q. Soil quality assessment in Yellow River Delta: Establishing a minimum data set and fuzzy logic model. *Geoderma* **2019**, *334*, 82–89. [CrossRef]
17. Obriot, F.; Stauffer, M.; Goubard, Y.; Cheviron, N.; Peres, G.; Eden, M.; Revallier, A.; Vieublé-Gonod, L.; Houot, S. Multi-criteria indices to evaluate the effects of repeated organic amendment applications on soil and crop quality. *Agric. Ecosyst. Environ.* **2016**, *232*, 165–178. [CrossRef]
18. Paz-Ferreiro, J.; Fu, S. Biological indices for soil quality evaluation: Perspectives and limitations. *Land Degrad. Dev.* **2016**, *27*, 14–25. [CrossRef]
19. Feng, Y.; Wang, J.; Liu, T.; Bai, Z.; Reading, L. Using computed tomography images to characterize the effects of soil compaction resulting from large machinery on three-dimensional pore characteristics in an opencast coal mine dump. *J. Soils Sediments* **2019**, *19*, 1467–1478. [CrossRef]
20. Behrends Kraemer, F.; Soria, M.A.; Castiglioni, M.G.; Duval, M.; Galantini, J.; Morrás, H. Morpho-structural evaluation of various soils subjected to different use intensity under no-tillage. *Soil Tillage Res.* **2017**, *169*, 124–137. [CrossRef]

21. Qu, Q.; Xu, H.; Xue, S.; Liu, G. Stratification ratio of rhizosphere soil microbial index as an indicator of soil microbial activity over conversion of cropland to forest. *Catena* **2020**, *195*, 104761. [CrossRef]
22. Zungu, N.S.; Egbewale, S.O.; Olaniran, A.O.; Pérez-Fernández, M.; Magadlela, A. Soil nutrition, microbial composition and associated soil enzyme activities in KwaZulu-Natal grasslands and savannah ecosystems soils. *Appl. Soil Ecol.* **2020**, *155*, 103663. [CrossRef]
23. Lisetskii, F.; Stolba, V.F.; Marinina, O. Indicators of agricultural soil genesis under varying conditions of land use, Steppe Crimea. *Geoderma* **2015**, *239–240*, 304–316. [CrossRef]
24. Carter, M.R. Soil quality for sustainable land management. *Agron. J.* **2002**, *94*, 38–47. [CrossRef]
25. Li, X.; Li, H.; Yang, L.; Ren, Y. Assessment of soil quality of croplands in the corn belt of Northeast China. *Sustainability* **2018**, *10*, 248. [CrossRef]
26. Mahajan, G.; Das, B.; Morajkar, S.; Desai, A.; Murgaokar, D.; Kulkarni, R.; Sale, R.; Patel, K. Soil quality assessment of coastal salt-affected acid soils of India. *Environ. Sci. Pollut. Res.* **2020**, *27*, 26221–26238. [CrossRef]
27. Choudhury, B.U.; Mandal, S. Indexing soil properties through constructing minimum datasets for soil quality assessment of surface and profile soils of intermontane valley (Barak, North East India). *Ecol. Indic.* **2021**, *123*, 107369. [CrossRef]
28. Wang, D.; Bai, J.; Wang, W.; Zhang, G.; Cui, B.; Liu, X.; Li, X. Comprehensive assessment of soil quality for different wetlands in a Chinese delta. *Land Degrad. Dev.* **2018**, *29*, 3783–3794. [CrossRef]
29. Karaca, S.; Dengiz, O.; Demirağ Turan, İ.; Özkan, B.; Dedeoğlu, M.; Gülser, F.; Sargin, B.; Demirkaya, S.; Ay, A. An assessment of pasture soils quality based on multi-indicator weighting approaches in semi-arid ecosystem. *Ecol. Indic.* **2021**, *121*, 107001. [CrossRef]
30. Armenise, E.; Redmile-Gordon, M.; Stellacci, A.; Ciccarese, A.; Rubino, P. Developing a soil quality index to compare soil fitness for agricultural use under different managements in the Mediterranean environment. *Soil Tillage Res.* **2013**, *130*, 91–98. [CrossRef]
31. Yuan, P.; Wang, J.; Li, C.; Xiao, Q.; Liu, Q.; Sun, Z.; Wang, J.; Cao, C. Soil quality indicators of integrated rice-crayfish farming in the Jianghan Plain, China using a minimum data set. *Soil Tillage Res.* **2020**, *204*, 104732. [CrossRef]
32. Andrews, S.S.; Karlen, D.L.; Cambardella, C.A. The soil management assessment framework. *Soil Sci. Soc. Am. J.* **2004**, *68*, 1945–1962. [CrossRef]
33. Wang, N.; Xing, Y.; Wang, X. Exploring options for improving potato productivity through reducing crop yield gap in Loess Plateau of China based on grey correlation analysis. *Sustainability* **2019**, *11*, 5621. [CrossRef]
34. Li, W.; Wang, D.; Liu, S.; Zhu, Y.; Yan, Z. Reclamation of cultivated land reserves in Northeast China: Indigenous ecological insecurity underlying national food security. *Int. J. Environ. Res. Public Health* **2020**, *17*, 1211. [CrossRef]
35. Sukweenadhi, J.; Balusamy, S.R.; Kim, Y.J.; Lee, C.H.; Kim, Y.J.; Koh, S.C.; Yang, D.C. A growth-promoting bacteria, Paenibacillus yonginensis DCY84^T enhanced salt stress tolerance by activating defense-related systems in *Panax ginseng*. *Front. Plant Sci.* **2018**, *9*, 813. [CrossRef]
36. Bremner, J. Determination of nitrogen in soil by the Kjeldahl method. *J. Agric. Sci.* **1960**, *55*, 11–33. [CrossRef]
37. Raiesi, F. A minimum data set and soil quality index to quantify the effect of land use conversion on soil quality and degradation in native rangelands of upland arid and semiarid regions. *Ecol. Indic.* **2017**, *75*, 307–320. [CrossRef]
38. Vasu, D.; Tiwari, G.; Sahoo, S.; Dash, B.; Jangir, A.; Sharma, R.P.; Naitam, R.; Tiwary, P.; Karthikeyan, K.; Chandran, P. A minimum data set of soil morphological properties for quantifying soil quality in coastal agroecosystems. *Catena* **2020**, *198*, 105042. [CrossRef]
39. Yu, P.; Liu, S.; Zhang, L.; Li, Q.; Zhou, D. Selecting the minimum data set and quantitative soil quality indexing of alkaline soils under different land uses in northeastern China. *Sci. Total Environ.* **2018**, *616*, 564–571. [CrossRef] [PubMed]
40. Kwiatkowska-Malina, J. Qualitative and quantitative soil organic matter estimation for sustainable soil management. *J. Soils Sediments* **2018**, *18*, 2801–2812. [CrossRef]
41. Abdel-Fattah, M.K.; Mohamed, E.S.; Wagdi, E.M.; Shahin, S.A.; Aldosari, A.A.; Lasaponara, R.; Alnaimy, M.A. Quantitative evaluation of soil quality using Principal Component Analysis: The case study of El-Fayoum depression Egypt. *Sustainability* **2021**, *13*, 1824. [CrossRef]
42. Wang, Z.; Wang, G.; Zhang, Y.; Wang, R. Quantification of the effect of soil erosion factors on soil nutrients at a small watershed in the Loess Plateau, Northwest China. *J. Soils Sediments* **2020**, *20*, 745–755. [CrossRef]
43. Li, H.; Zhu, N.; Wang, S.; Gao, M.; Xia, L.; Kerr, P.G.; Wu, Y. Dual benefits of long-term ecological agricultural engineering: Mitigation of nutrient losses and improvement of soil quality. *S. Total Environ.* **2020**, *721*, 137848. [CrossRef] [PubMed]
44. Ma, J.; Chen, Y.; Wang, H.; Wang, H.; Wu, J.; Su, C.; Xu, C. Newly created farmland should be artificially ameliorated to sustain agricultural production on the Loess Plateau. *Land Degrad. Dev.* **2020**, *31*, 2565–2576. [CrossRef]
45. Gong, J.; Chen, L.; Fu, B.; Huang, Y.; Huang, Z.; Peng, H. Effect of land use on soil nutrients in the loess hilly area of the Loess Plateau, China. *Land Degrad. Dev.* **2006**, *17*, 453–465. [CrossRef]
46. D'Hose, T.; Cougnon, M.; De Vliegher, A.; Vandecasteele, B.; Viaene, N.; Cornelis, W.; Van Bockstaele, E.; Reheul, D. The positive relationship between soil quality and crop production: A case study on the effect of farm compost application. *Appl. Soil Ecol.* **2014**, *75*, 189–198. [CrossRef]
47. Jokela, W.; Posner, J.; Hedtcke, J.; Balser, T.; Read, H. Midwest Cropping System Effects on Soil Properties and on a Soil Quality Index. *Agron. J.* **2011**, *103*, 1552–1562. [CrossRef]

48. Rotz, C.A.; Taube, F.; Russelle, M.P.; Oenema, J.; Sanderson, M.A.; Wachendorf, M. Whole-farm perspectives of nutrient flows in grassland agriculture. *Crop Sci.* **2005**, *45*, 2139–2159. [CrossRef]
49. Carmo, M.; García-Ruiz, R.; Ferreira, M.I.; Domingos, T. The N-P-K soil nutrient balance of Portuguese cropland in the 1950s: The transition from organic to chemical fertilization. *Sci. Rep.* **2017**, *7*, 1–14. [CrossRef] [PubMed]
50. Smith, K.A. Changing views of nitrous oxide emissions from agricultural soil: Key controlling processes and assessment at different spatial scales. *Eur. J. Soil Sci.* **2017**, *68*, 137–155. [CrossRef]
51. Tian, H.; Qiao, J.; Zhu, Y.; Jia, X.; Shao, M.A. Vertical distribution of soil available phosphorus and soil available potassium in the critical zone on the Loess Plateau, China. *Sci. Rep.* **2021**, *11*, 1–10. [CrossRef]
52. Wang, X.; Wang, G.; Tao Guo, T.; Xing, Y.; Mo, F.; Wang, H.; Fan, J.; Zhang, F. Effects of plastic mulch and nitrogen fertilizer on the soil microbial community, enzymatic activity, and yield performance in a dryland maize cropping system. *Eur. J. Soil Sci.* **2021**, *72*, 400–412. [CrossRef]
53. Manna, M.; Swarup, A.; Wanjari, R.; Ravankar, H.; Mishra, B.; Saha, M.; Singh, Y.; Sahi, D.; Sarap, P. Long-term effect of fertilizer and manure application on soil organic carbon storage, soil quality and yield sustainability under sub-humid and semi-arid tropical India. *Field Crops Res.* **2005**, *93*, 264–280. [CrossRef]
54. Wang, X.; Wang, N.; Xing, Y.; Yun, J.; Zhang, H. Effects of plastic mulching and basal nitrogen application depth on nitrogen use efficiency and yield in maize. *Front. Plant Sci.* **2018**, *9*, 1446. [CrossRef] [PubMed]

Article

Soil Nutrient Supply in Cultivated Bush Bean–Potato Intercropping Grown in Subarctic Soil Managed with Agroforestry

Jim D. Karagatzides [1,*], Meaghan J. Wilton [2] and Leonard J. S. Tsuji [2]

1 Engineering and Environmental Technologies, Georgian College, Barrie, ON L4M 3X9, Canada
2 Department of Physical and Environmental Sciences, University of Toronto, Toronto, ON M1C 1A4, Canada; m.wilton@utoronto.ca (M.J.W.); leonard.tsuji@utoronto.ca (L.J.S.T.)
* Correspondence: jim.karagatzides@georgiancollege.ca

Abstract: To address food insecurity in northern Canada, some isolated communities started gardening initiatives to reduce dependencies on expensive foods flown in to communities. From 2012–2014, soils in northern Ontario James Bay lowlands were cultivated with bush beans and potatoes, grown in sole and intercropping configurations, in an open field and an agroforestry system enclosed by willow trees. The objective of this study was to compare the supply rates of 15 plant-available nutrients in these soils using in situ ion exchange membranes. After three years of cultivation, the agroforestry site had significantly greater supply of PO_4, Ca, and Zn and these nutrients had positive correlations with yield. By contrast, the open site had significantly greater supply of Mg, SO_4, and B; these nutrients, and Al, had negative correlations with yield. Whilst there were no differences between sole and intercropping configurations, significantly greater supply of NO_3, Ca, Cu, Fe, and Zn occurred early in the growing season, compared to significantly greater supply of K, SO_4, B, and Al later in the season. Significantly greater yields have been harvested in the agroforestry site and it is suspected that the presence of a willow shelterbelt improves the microclimate and plant-available PO_4, Ca, and Zn.

Keywords: nutrient supply rate; subarctic agriculture; agroforestry; bush bean; potato; northern agriculture management; ion exchange resins; PRS probes

Citation: Karagatzides, J.D.; Wilton, M.J.; Tsuji, L.J.S. Soil Nutrient Supply in Cultivated Bush Bean–Potato Intercropping Grown in Subarctic Soil Managed with Agroforestry. *Sustainability* **2021**, *13*, 8185. https://doi.org/10.3390/su13158185

Academic Editor: Victor Rolo

Received: 28 June 2021
Accepted: 16 July 2021
Published: 22 July 2021

1. Introduction

Agriculture is expanding northwards as global warming increases air temperatures and lengthens growing seasons in the higher latitudes [1,2]. The forms of northern agriculture occurring include intensification within existing areas, and extension to new or historically cultivated areas [1,3,4]. In some subarctic communities of Canada, agriculture is being introduced as a climate change adaptation strategy to improve food sufficiency and mitigate current food insecurities [3–5]. There is also an increasing international interest in northern food production to abate anticipated global food insecurities [1,2]. Both assessment and development of adaptive northern agricultural practices are currently lacking yet are imperative for high-latitude food production initiatives to become sustainable strategies [1,2]. There are concerns that northern agriculture growth, without adapted practices, would negatively impact environmental health, including the deterioration of biodiversity, increased carbon losses, and adverse effects on water and nutrient cycles [1,2]. These concerns are compelling as agriculture expands into the boreal ecoregion—which consists of vast forests and wetlands.

A survey on climate change and northern expansion of agriculture found that investigating nutrient cycles was voted as highly relevant by participating researchers [1]. Having a greater understanding of nutrient cycles and dynamics for agriculture in the boreal deciphers the impacts of land conversion and supports the creation of policy and best management strategies [1,2,5]. Recently, Kedir et al. [5] assessed nutrient conditions within sandy, acidic boreal soils under agricultural management across Newfoundland and Labrador, Canada, and determined N and P inefficiencies. This nutrient assessment gives guidance for creating a regional nutrient management plan for other regions with similar soil characteristics but is not transferable to all soil types within the boreal, such as the contrasting soil pH and texture of our study in the James Bay Lowlands region of the Canadian Boreal.

The present study assesses the bioavailability of macro- and micronutrients across a growing season in calcareous silty loam soil, cultivated with bush bean (*Phaseolus vulgaris* L.) and potato (*Solanum tuberosum* L.) under open and agroforestry (tree-based) management practices. Agroforestry is the practice of combining woody perennials (e.g., trees, shrubs) with herbaceous crops (plants or pastures) or livestock, in spatial and temporal arrangements [6,7]. There are many forms and designs of agroforestry to meet specific ecological, environmental, economic, or cultural needs [6,7]. Agroforestry research that addresses food security and environmental issues has been more pronounced in (sub)tropical systems than temperate systems [6,7]. In comparison, agroforestry research in boreal settings is minimal. Peer-reviewed studies on boreal agroforestry have examined silviculture—an agroforestry practice that controls forest growth and structure to produce value-added products—such as for timber, biofuel [8,9], and non-timber forest products [10]. In our study site, agroforestry was initially utilized in the form of a shelterbelt to protect bean and potato crops from cold wind [11]. Shelterbelts comprise row arrangements where the distance between rows varies depending on the agrosystem and its purpose. In temperate climate systems, shelterbelts protect crops, pasture, and livestock by minimizing wind, erosion, nutrient run-off, and snowdrifts, and regulating temperature and moisture [6,7]. To our knowledge, utilizing shelterbelts to support the production of fruits and vegetables in the subarctic climate has been undertaken only in the Fort Albany First Nation.

Potato and bush bean were chosen for our study as the far-north remote community wanted to grow fruits and vegetables locally under ambient conditions and be less reliant on imported produce [12,13]. Moreover, potato and bush bean have practical and nutritional purposes. Historical records indicate that potato was briefly cultivated in the region. Additionally, potato is considered a world crop staple, and it grows well in cool and wet climates [14,15]. Bush bean was selected because it is fast-growing, can withstand cooler temperatures, and is a legume that permits nitrogen fixation [13,16]. Both potato and bean are crops used within intercropping systems—the practice of cultivating two or more crops simultaneously on the same field [17]. In some regions, potato is intercropped to mitigate the high risks associated with this high-valued crop [17]. Intercropping with bean is frequently applied to reduce nitrogen fertilization requirements when used in conjunction with non-N-fixing companion crops [16]. The combination of potato–bean as an intercrop is not frequently used [17]. Nevertheless, Sharaiha and Battikh [17] and Gitari et al. [18] found that the bean–potato combination improved the equivalent yield in semi-arid and subtropical climates. At our subarctic study location, Barbeau et al. [13] found bean–potato intercrops increased yields significantly when implemented with agroforestry (as opposed to an open field).

Both intercropping and agroforestry practices have features that enhance agroecosystem services (i.e., enhancing biodiversity, microclimates, and soil formation) and resource efficiencies (i.e., supporting natural pest management and nutrient cycling) [7,16,19]. The services provided from intercropping and agroforestry that improved yields at the Fort Albany First Nation site are not known. Previous studies used soil-extractable nutrients to assess soil fertility [3,11] and foliar macronutrient concentrations of bush bean and potatoes [20]. However, Spiegelaar et al. [11] expressed that there may be a discrepancy between the amount of nutrients extracted from the soil and the actual nutrient uptake by crops due to the uncommon subarctic environment conditions for cultivating soils.

In this study, we examine if agroforestry and intercropping management practices provide soil nutrient advantages compared to cultivating in an open field and sole cropping. Secondly, we test for correlations between soil nutrient supply rate and yield and plant nutrition. This study provides information that contributes to developing knowledge on northern agroecosystems and insights into different management practices in a subarctic climate.

2. Materials and Methods

2.1. Location

The research location was in the boreal community of Fort Albany First Nation (FAFN), which resides in Mushkeguwuk Cree territory on the west coast of James Bay, Ontario (52°30′64″ N; 81°72′77″ W). The region's terrain is described as flat, 14 m above sea level, and dominated by muskeg or peat moss situated on limestone bedrock. The region is in a subarctic climate (Köppen climate classification Dfc) with mean annual daily temperatures of −0.5 °C, and mean total annual rainfall of 503 mm and snowfall of 227 cm (1981–2010) for Moosonee, ON [21], the nearest weather station 125 km southeast. The frost-free growing period during this study in 2014 was 94 days from 21 June to 22 Sept (Julian day of year (DOY) 172–265).

Two sites were constructed for agriculture in 2012—an agroforestry site (A), and an open field site (O; Figure 1). Historically, the sites were boreal forest that were cleared in the 1930s and drained for agriculture production until the 1970s. The fields became disused and progressed naturally as grasslands [11]. In the Agroforestry section, willows (*Salix* spp.) had naturally grown for 40 years in thicket rows along manually made drainage ditches. The two willow thicket rows utilized for the Agroforestry site ran parallel from southwest to northeast and were approximately 5 m in height and 6 m wide with a 15 ± 3% optical porosity [22]. The cultivated study area was within the 14 m space that separated the two thickets. The Open site was located 200 m southwest of the Agroforestry site, 58 m from the nearest tree, and surrounded by grassland. Additional details on the location history and site description are in Wilton et al. [20].

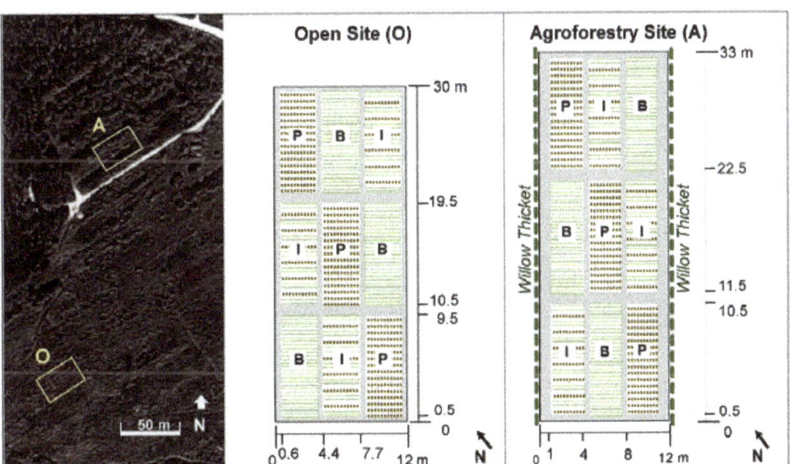

Figure 1. Location of Open (O) and Agroforestry (A) sites where PRS probes were used in Fort Albany First Nation, Ontario (left, map modified from Google Earth, 2017, DigitalGlobe, Westminster, CO, USA). Plot schematics are management practices for sole cropping of bush bean (B, green lines) or potato (P, brown dots) and bean–potato intercropping (I). Figure modified from Wilton et al. [20].

The soils at both the Agroforestry and Open site had a silty loam texture consisting of 18% sand, 66% silt, and 16% clay [20]. The physical characteristics of soil (0–20 cm depth) at each site were similar for pH (A = 7.7, O = 7.6), bulk density (A and O = 0.66 g cm^{-3}), C/N ratios (A = 15, O = 16), and cation exchange capacities (A and B = 41 cmol + kg^{-1}) [20].

The seasonal mean soil temperature and moisture content at 10 cm depth were 17.5 °C with a standard deviation of 2.8 °C and 0.34 ± 0.052 $m^3\ m^{-3}$ in the Agroforestry site, and 15.4 ± 2.8 °C and 0.37 ± 0.023 $m^3\ m^{-3}$ in the Open site. No fertilizer amendments had been added to the sites since being established. Though supplementing the soil with nutrients is essential for agriculture production, there is value in studying subarctic soils and management practices before the inclusion of agricultural inputs [6].

2.2. Site Preparations

A Latin square was used for the experimental design to account for variation due to tree effects (i.e., edge and shading effects). The Open site dimensions were 30 m × 12 m compared to 33 m × 12 m for the Agroforestry site, which also had an additional 1 m wide border from the willow treeline (Figure 1). Each site had 3 rows and 3 columns (totaling 9 sub plots per site), and therefore the subplot dimensions were 9 m × 3.3 m in the Open site, and 10 m × 3 m in the Agroforestry site.

Sites were rototilled (Troy-Bilt Rear Tine Tiller 208 CC, OH, USA) at 15 cm depth from 16 June to 19 June of 2014 (DOY 167–170). Within each plot, there were three crop management treatments: sole cropped bean (Provider Var.), sole cropped potato (AC Chaleur Var.), and row-intercropped bean and potato. Each treatment was assigned a subplot randomly in each of the three plots in each site. Potatoes were planted in rows on June 24 (DOY 175) at 10 cm depth. Sole cropped potato had a row seeding density of 3.5 tubers m^{-1} with an inter-row spacing of 60 cm. Bush bean was sown in rows at a planting depth of 3 cm on 26 June (DOY 177). The seeding density was 30 seeds m^{-1} with a 20 cm inter-row spacing in the sole cropped bean treatment. The row ratio was 1 row potato to 2 rows of beans for the intercropping treatment. Both potatoes and beans in the intercropping subplots maintained seeding row densities: bean–bean inter-row was 20 cm, and the bean–potato inter-row was 60 cm. Wilton et al. [20] describe the leaf content of N, P, K, Ca, and Mg for these crops.

2.3. Ion Exchange Resin Probes

Plant root simulator (PRS) probes (Western Ag Technologies, Saskatoon, SK, Canada) were used to quantify the plant-available nutrient supply rate for soil ions in each plot for each site [23]. The advantage of using PRS probes is that they act as an analog for root uptake of bioavailable nutrients. They can also be used on a broad range of soils within different environments [24], and they are practical tools to utilize at remote research sites. The PRS probes consist of a two-sided anion or cation exchange membrane with a surface area of 176 cm^2. Positively charged anion probes adsorbed NO_3^-–N, phosphorus in the form of $H_2PO_4^-$, and SO_4^{2-}, while negatively charged cation probes adsorbed NH_4^+–N, K, Ca, Mg, B, Al, Cd, Cu, Fe, Mn, Pb, and Zn. Within a one-meter border of each plot, three pairs of cationic and anionic PRS probes were vertically inserted randomly into the soil to 10 cm depth on 16 June to 24 June (DOY 167–175), 31 July to 8 August (DOY 212–220), 31 August to 7 September (DOY 243–250), and 26 September to 3 October (DOY 269–276). Probes were left in the soil for 7 or 8 days as determined by consultants at Western Ag using soil-extractable nutrient results found for the same study plots [3].

After removal, the probes were rinsed with reverse osmosis-treated water to remove soil particles, refrigerated at 4 °C, and shipped to Western Ag Technology, Saskatoon, Canada for analysis. Nitrogen compounds were determined colorimetrically using an automated flow injection system, and all other nutrients measured using inductively coupled plasma spectrometry. Results are expressed as the weight of nutrient adsorbed to the surface area of the ion exchange membrane (adjusted to 10 cm^2) over the burial duration (8-day burials adjusted to 7 days) as dividing the time into smaller units is invalid since the adsorption was unlikely to be linear over time.

2.4. Statistical Analysis

Statistical analyses were completed using SPSS 26 [25]. Each nutrient extracted from the PRS probes was assessed in a 3-way analysis of variance (anova) for the main effects of site (Open, Agroforestry), crop (bush bean sole cropping, potato sole cropping, intercropping), month (June, August, September, October), and all possible 2-way and 3-way interactions. The Student–Newman–Keuls (SNK) post hoc test was used to distinguish significant differences among treatments and months. Levene's test on the variances using the median revealed that variances were homogeneous for all nutrients ($p > 0.2$). Linear correlations were obtained for leaf tissue chemistry for each crop on 10 August 2014 [20] against the nutrient supply rate for June and August. Linear correlations were also obtained for total annual crop yield and total nutrient supply rate across the growing season (the sum of PRS data for the four time periods).

3. Results

3.1. Spatiotemporal Soil Nutrient Supply Rate in Agroforestry and Open Sites

Although there was substantial variation for the nutrient supply rate (e.g., Open site: NO_3 in Sept and B in Oct; Agroforestry site: Mn and Cu in June, Fe in August)—likely because of the small sample size (n = 3) of PRS probes sampling a heterogeneous soil environment across a growing season—many significant differences were observed. There were significant differences between Agroforestry and Open sites for six of 13 nutrients and significant seasonal differences for nine of 13 nutrients, as well as three significant interaction effects between site and month main effects (Table 1). However, there were no significant differences among the three crop treatments nor for any interaction effect with crop. Additionally, the supply rate of NH_4–N, Pb, and Cd was excluded from this analysis because >70% of samples collected for each nutrient were below detection limits (DL). For NH_4–N (DL = 2 ppm) and Pb (DL = 0.2 ppm), only samples collected in June were above the detection limit. For Cd, only two of 72 samples collected across the four sampling periods had detectable levels of Cd (DL = 0.2 ppm).

Nitrate supply rates were similar between sites, but a significant seasonal difference was found for June, with rates 75% greater than at other times of the growing season (Figure 2). In contrast to N, the supply rate of soil PO_4–P did not reveal seasonal differences, but the Agroforestry site had a significantly greater (nearly double) supply rate of PO_4–P compared to the Open site.

Soil K was one of only two nutrients (B being the other) where the supply rate was significantly greater later in the season, with a 50% increase in K measured in October compared to June (Figure 2). There was also a significant site × month interaction effect: compared to the Agroforestry site, the Open site had a greater supply rate of soil K in June but a significantly smaller supply rate of K in August and October. Similar to K, B had a significantly greater supply rate at the end of the growing season—nearly double the supply rate in October compared to earlier sampling periods. The significantly greater supply rate of B in soil during October was influenced by the Open site and lead to the significant site × month interaction effect for soil B ($p = 0.001$). Regardless of crop type, the Open site had, on average, a 52% greater supply rate of B than that measured in the Agroforestry site.

Table 1. Results from three-way ANOVA (*F*- and *p*-values) for soil nutrient supply rates ($n = 3$ cationic and anionic PRS probes per site × crop × month) in cultivated sites in the subarctic community of Fort Albany First Nation (James Bay Ontario, Canada) for 2014. Sites were cultivated from 2012–2014 in either an open field or an agroforestry field enclosed by a willow thicket. The crop treatment was either sole planting (bush bean or potato) or a bean–potato intercrop. Bold font indicates significant difference $\alpha = 0.05$.

	Site	*p*	Crop	*p*	Month	*p*	Site × Crop	*p*	Site × Month	*p*	Crop × Month	*p*	Site × Crop × Month	*p*
NO_3	2.19	0.146	0.93	0.402	5.25	**0.003**	0.26	0.774	2.656	0.059	0.295	0.937	0.19	0.978
PO_4	31.23	**1.1×10^{-6}**	0.11	0.898	1.01	0.395	0.02	0.980	1.41	0.251	1.74	0.133	1.13	0.359
K	0.85	0.361	0.74	0.485	3.57	**0.021**	0.87	0.426	3.82	0.016	1.24	0.301	0.82	0.557
Ca	9.59	**0.003**	0.32	0.730	4.58	**0.007**	0.60	0.556	0.60	0.615	0.70	0.648	2.08	0.073
Mg	61.22	**4.1×10^{-10}**	1.18	0.316	0.88	0.456	1.854	0.168	0.91	0.445	0.69	0.659	1.37	0.247
SO_4	301.76	**2.4×10^{-22}**	0.38	0.683	4.89	**0.005**	0.36	0.699	6.29	**0.001**	0.47	0.829	0.099	0.996
B	4.90	0.032	0.59	0.561	3.32	**0.027**	1.06	0.354	6.02	**0.001**	0.19	0.979	0.46	0.837
Al	3.44	0.070	1.61	0.211	4.34	**0.009**	0.45	0.639	0.57	0.637	1.39	0.240	0.69	0.659
Cu	1.22	0.274	0.28	0.756	5.32	**0.003**	0.53	0.593	2.08	0.116	0.25	0.956	0.59	0.737
Fe	1.68	0.201	0.38	0.685	4.28	**0.009**	0.81	0.450	0.89	0.454	0.53	0.780	0.43	0.855
Mn	2.81	0.100	0.94	0.398	1.79	0.162	1.29	0.285	0.78	0.511	0.66	0.678	0.88	0.517
Zn	28.08	**2.9×10^{-6}**	1.74	0.187	43.50	**9.9×10^{-14}**	0.26	0.773	2.75	0.053	0.78	0.594	0.98	0.449

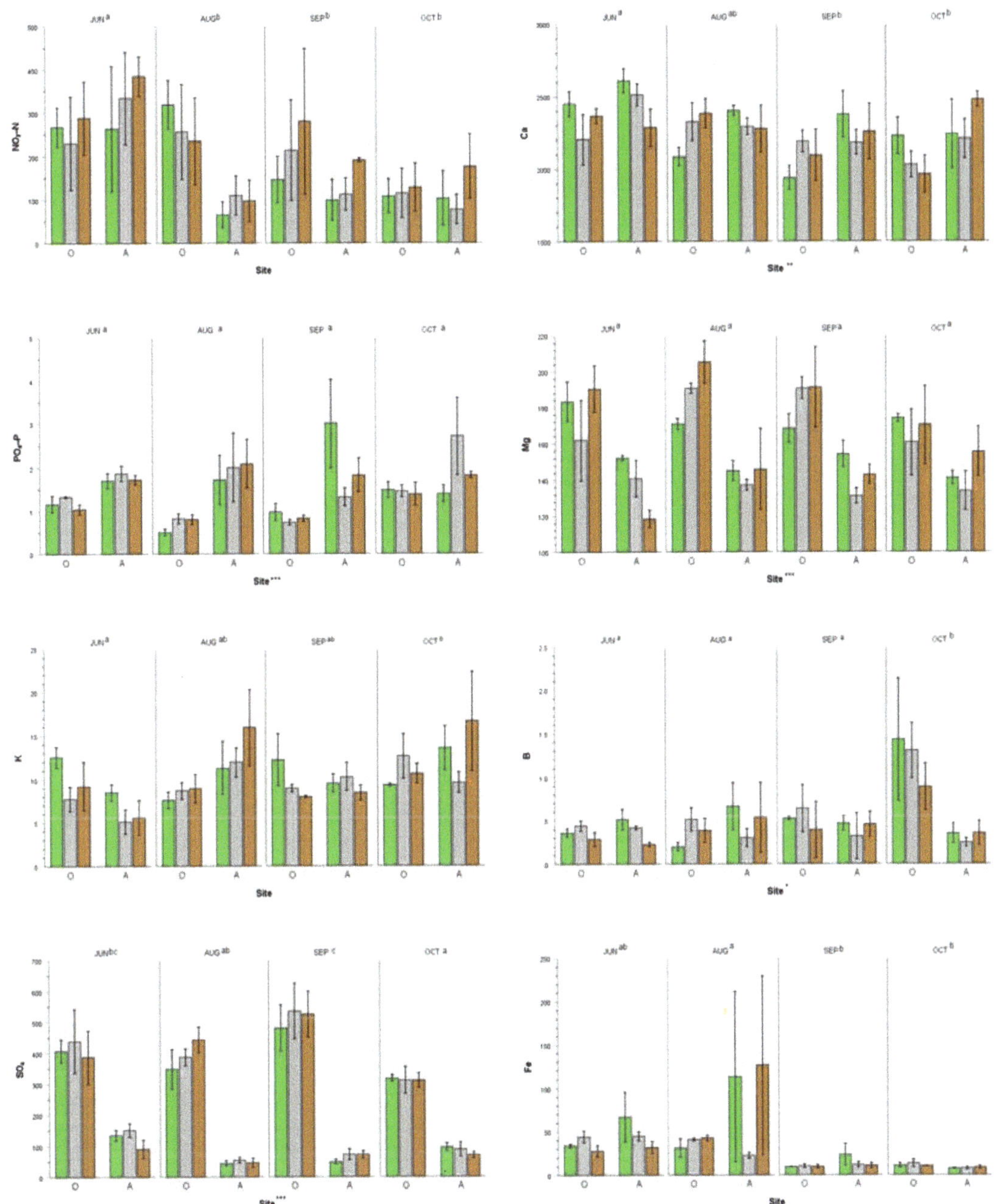

Figure 2. *Cont.*

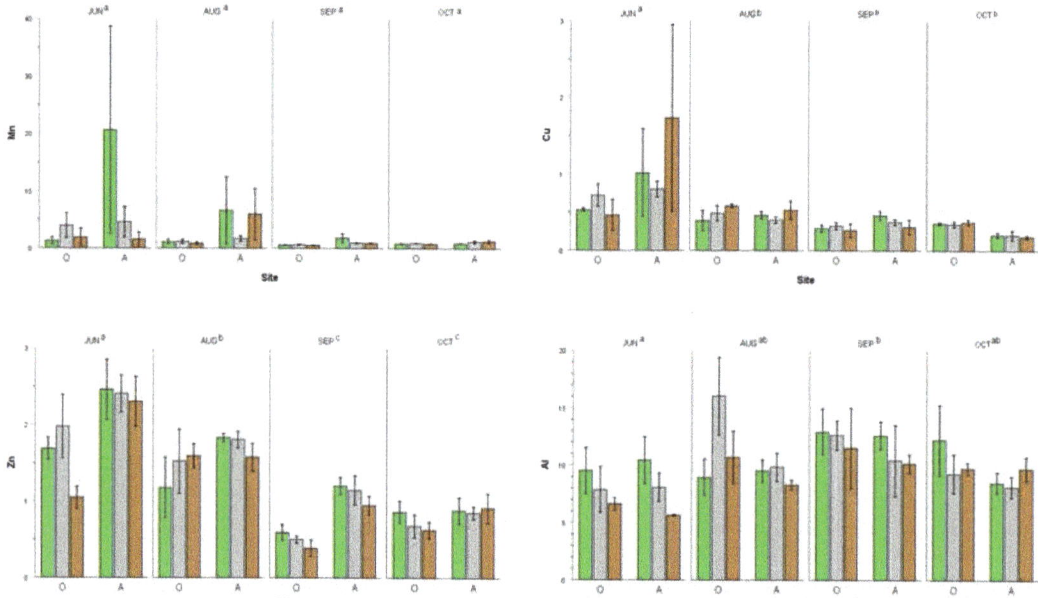

Figure 2. Comparisons of mean (±1 SE, n = 3) soil nutrient supply rate (μg 10 cm^{-3} $week^{-1}$) in Open and Agroforestry sites in Fort Albany (James Bay, Ontario). Supply rate was measured in soil planted with bush bean (green), potato (brown), and intercropped (gray), during the 2014 growing season. Within each figure panel, significant differences between sites are shown with asterisks (* $p = 0.05$, ** $p = 0.01$, *** $p = 0.001$), and lowercase letters for temporal differences whereby months with different letters are significantly different ($p < 0.05$, SNK post hoc test). There were no significant differences among the three types of crop plantings.

Contrasting patterns in the spatial and temporal supply rate were found between soil Mg and Ca (Figure 2). A small but significant difference in soil Ca supply rate between sites (Agroforestry had ~7% greater Ca supply rate than soil in the Open site) but the opposite trend was found for soil Mg with ~25% greater supply rate in soil in the Open site. In addition, whilst there were no significant seasonal differences for soil Mg, there was a 10% greater supply rate of soil Ca in June compared to September and October.

The supply rate of SO_4 represents one of the largest differences between sites as soil in the Open site had five times greater supply rate of SO_4 compared to the Agroforestry site (Figure 2). Sulfate was also the only soil nutrient with significantly greater supply rate measured in September, which on average was 33% greater than in August and 50% greater than the SO_4 supply rate measured in October. There was also a significant site × month interaction effect for SO_4, likely the result of the lower supply rate in October in the Open site, compared to the other sampling periods, which produced the smallest difference between sites in October.

For the five metals (Fe, Mn, Zn, Cu, Al), there were no significant differences between sites, apart from Zn where the supply rate in the Agroforestry soil was ~44% greater than in the Open site when averaged across the growing season (Figure 2). There were, however, significant seasonal differences for four of the five metals, the exception being Mn (Table 1). The supply rate of Fe was nearly five times greater in August than in September and October, driven largely by the greater supply rate of Fe in Agroforestry site soil than in the Open site. Similar to Fe, the supply rate of Zn was two times greater in August compared to later in the growing season, and June had an even greater supply rate of Zn in soil (25% greater in June than in August). The supply rate of Cu in soil also was significantly greater in June—2- to 3-fold greater in June than in other sampling periods—and, like Fe, driven by the substantially greater increase in the Agroforestry site. In contrast to the other metals, the supply rate of Al was significantly lower in June compared to September, with an average increase of ~45% from June to September.

3.2. The Relationship between Soil Nutrient Supply Rate and Leaf Tissue Chemistry

There were only two significant correlations for each of bush bean and potato between soil nutrient supply rate in June or August and leaf tissue chemistry (Table 2). Significant positive correlations with leaf nutrient concentration in bush bean were found for soil supply rate of PO_4 in June ($r = 0.60$, $p = 0.04$) and for Zn in August ($r = 0.76$, $p = 0.004$). Likewise for potato, a significant positive correlation was found for the supply rate of PO_4 and potato leaf P concentration, though in this case the relationship was significant with the supply rate of both June ($r = 0.71$, $p = 0.01$) and August ($r = 0.69$, $p = 0.01$; Table 2). In addition, there was a significant negative correlation between the supply rate of Ca in soil and potato leaf ($r = -0.58$, $p = 0.05$).

Table 2. Correlations (two-tailed, $n = 12$) of soil nutrient supply rate with leaf chemistry and yield in Fort Albany (James Bay, Ontario) during the 2014 growing season. Correlation with yield is for the sum of the soil supply rate measurements in June, August, September, and October. Bold font indicates significant differences at $\alpha = 0.05$.

Bush Bean	June PRS + Leaves		August PRS + Leaves		Yield	
Soil	*r*	*P*	*r*	*p*	*r*	*p*
NO_3–N	0.57	0.06	0.28	0.38	0.05	0.88
PO_4–P	**0.60**	**0.04**	0.25	0.42	0.56	0.06
K	0.45	0.14	0.14	0.66	−0.16	0.61
Ca	−0.27	0.40	0.41	0.19	0.72	0.01
Mg	−0.20	0.53	−0.37	0.23	−0.28	0.38
B	−0.05	0.44	0.16	0.31	−0.20	0.54
Fe	−0.22	0.50	−0.01	0.99	−0.22	0.50
Mn	−0.21	0.52	−0.24	0.46	−0.15	0.64
Zn	0.52	0.08	**0.76**	**0.004**	**0.70**	**0.01**
Cu	−0.15	0.65	−0.34	0.28	−0.30	0.35
Al	–	–	–	–	0.16	0.63
SO_4	–	–	–	–	−0.54	0.07
Potato	**June PRS + Leaves**		**August PRS + Leaves**		**Yield**	
Soil	*r*	*p*	*r*	*P*	*r*	*p*
NO_3–N	0.50	0.10	0.43	0.16	−0.05	0.88
PO_4–P	**0.71**	**0.01**	**0.69**	**0.01**	**0.72**	**0.01**
K	0.06	0.84	0.13	0.69	0.26	0.42
Ca	−0.03	0.92	**−0.58**	**0.05**	0.36	0.25
Mg	0.46	0.14	0.34	0.28	**−0.61**	**0.04**
B	−0.10	0.75	0.06	0.87	−0.54	0.07
Fe	−0.08	0.80	−0.24	0.44	0.46	0.13
Mn	−0.30	0.34	−0.19	0.55	0.42	0.18
Zn	−0.12	0.72	0.25	0.44	0.51	0.09
Cu	−0.06	0.85	0.26	0.42	**0.60**	**0.04**
Al	–	–	–	–	**−0.81**	**0.002**
SO_4	–	–	–	–	**−0.79**	**0.002**

3.3. The Relationship between Soil Nutrient Supply Rate and Yield

Significant positive correlations between the June + August rate of soil nutrient supply with bush bean yield were found for Ca ($r = 0.72$, $p = 0.01$) and Zn ($r = 0.70$, $p = 0.01$; Table 2). More correlations between soil nutrient supply rates and yield were found for potato than for bean (Table 2). Significant positive correlations were found with PO_4 ($r = 0.72$, $p = 0.01$) and Cu ($r = 0.60$, $p = 0.04$), as well as significant negative correlations with the supply rate of Mg ($r = -0.61$, $p = 0.04$), SO_4 ($r = -0.79$, $p = 0.002$), and Al ($r = -0.81$, $p = 0.002$).

4. Discussion

The objective of this study was to measure spatiotemporal supply rates for 15 nutrients in open and agroforestry (treed shelterbelt) sites in a northern Ontario community, as a potential factor leading to the greater yield of bush bean and potato previously found in this agroforestry garden [3,15]. The Agroforestry site had significantly greater nutrient supply rates for PO_4–P, Ca, and Zn, and these nutrients had positive correlations with crop yield. By contrast, the Open site had a significantly greater supply of Mg, SO_4, B, and Al, and these nutrients had negative correlations with crop yield. Whilst there were no differences among crop treatments, temporal differences in nutrient supply rates were revealed: a significantly greater supply of NO_3–N, Ca, Cu, Fe, and Zn occurred early in the growing season, compared to a significantly greater supply of K, SO_4, B, and Al later in the growing season. In addition, four significant correlations were found between nutrient supply rate and tissue nutrition. Positive correlations were found between P and Zn for bean. While a positive correlation was also found for soil and tissue P in potato, there was a negative correlation between soil and leaf tissue Ca for potato. Below, we describe the implications of these spatiotemporal differences in soil nutrition supply rates for the differences observed in crop yield.

4.1. Agroforestry Sites Had Greater Supply Rates of Phosphorus, but It Is Still Inadequate for Maximum Crop Yield

There was a greater supply rate of PO_4–P in soils in the Agroforestry site and significant positive correlations ($p \leq 0.05$) of PO_4–P supply with potato yield and for tissue concentrations of P in both crops. Wilton et al. [20] also determined that leaf P in the Agroforestry site was significantly greater than in the Open site, but leaf N:P ratios indicated deficient amounts of P in the foliage of bean and potato, suggesting that the greater supply of P in the agroforestry plots may still be inadequate for maximum yield. Indeed, the amount of PO_4–P adsorbed on the PRS probes was considerably lower than the median value of 6 μg P/10 cm^2/week found in the Western Ag agriculture soil database [23].

Small P adsorption is not unexpected in subarctic wetland regions, such as the James Bay lowlands, where cool soil temperatures slow microbial activity, reducing P mineralization and availability [26–28]. At our research location, soil temperature at 10 cm depth indicated that the Agroforestry site was +2.1 °C warmer than the Open site for the period between sowing and harvest [20]. Other agroforestry studies set in temperate climates also observed increased soil temperatures and other microclimate-enhancing effects [6,29–32]. Under subarctic climates, the increase in soil temperature may have a greater impact on nutrient cycling than that observed in temperate climates. The effect of increased soil temperatures on microbial processes in northern Sweden determined that adding leaf litter to the subarctic soil and increasing the soil temperature by 1 °C increased soil bacterial growth rate and net mineralization of P [33]. Similarly, increasing soil temperatures by 0.9–2 °C doubled nematode populations, and increased microbial and fungal biomass, enhancing N and P mineralization and the nutrient availability to plants [34]. Utilizing agroforestry practices to improve soil fertility by enhancing soil microbial communities is promising [6]; however, research on the subject is limited [35], focused on tropical climates [36,37], and often contradictory or inconclusive [31,38–40].

Phosphorus limitations are common in calcareous soil [41] where the alkaline pH encourages P to be fixed to Ca, preventing the availability of HPO_4^{2-} compounds [27,41–43]. Additionally, large soil organic matter content and CEC, as seen at our study location, promote binding of P to clay fractions and organic complexes [44,45]. Though the soil had a high potential for fixing P, the root characteristics of willows may have facilitated improved P soil availability in the Agroforestry site. Willows adapt to a wide range of environmental conditions. For example, when grown in soils with high concentrations of Ca, willows secrete organic acids from extensive fibrous root systems [46] and thereby solubilize P precipitates fixed to Ca [42,43].

Another feature of willow (and many other tree species) to acquire P is the ability to explore large volumes of soil. Phosphorus and other immobile nutrients are located through the tree's expansive root networks and symbiotic relationships with ectomycorrhiza and vesicular–arbuscular endomycorrhizae [42,43,47]. Phosphorus acquired by willow can become available to nearby crops through leaf litter decomposition, tree root–crop root interactions, or from sloughed off tissue during the growth and decay of tree roots [6,43,48]. Rytter [49] calculated that the fine root biomass of willow grown in Sweden had annual growth of 900–7200 kg ha^{-1} and decay of 500–6800 kg ha^{-1} in a 0–50 cm deep soil profile, while Phillips et al. [50] monitored the root performance of willow poles (3 m unrooted stem cuttings) in a temperate climate and found that roots extended to 9 m within a 9-month period. Considering the willow trees were mature at our Agroforestry site, and the two thickets were 14 m apart, the tree root coverage likely extended into all cultivated plots within the treed site.

The correlations between P supply rates revealed a significant positive association ($p = 0.008$) for potato yield, suggesting that even a small additional supply of PO_4–P improves crop yield. We addressed this issue at Fort Albany by testing the effect of local compost on bush bean growth in a pot experiment. Wilton et al. [51] found that increasing amounts of this local compost significantly increased bean yield—both with the production of more beans per plant and larger individual bean mass.

Although our study does not reveal causal factors directly improving P in the agroforestry sites, it is likely that the willow thicket provides ecological services that improved the availability of P for bean and potato through tree root mechanisms and an enhanced microclimate. Theoretically, the organic matter derived from the willows provides internal recycling of nutrients to the microbial community [6,43].

4.2. Spatiotemporal Dynamics of Macronutrient Supply Rates in Subarctic Cultivated Soil

Nearly all the N collected on the PRS probes was in the form of NO_3–N as most of our samples had NH_4–N levels below the detection limit of 2 ppm. Nitrate levels were in the typical range of 10–600 μg/10 cm^2/week for agricultural soils using the Western Ag PRS probes [23], particularly in June when the NO_3–N supply rate in the agroforestry site was >3 times the rate later in the growing season. The availability of soil N was reflected in the N concentrations in bean and potato leaf tissue that were generally above the critical level of 3% N [20].

The trend of smaller soil NO_3–N supply with time has been observed in other PRS probes studies [41,52]. They note that the seasonal decline in NO_3–N was influenced by increased root biomass, crop uptake, and loss through soil-emitted N. However, warmer temperatures can influence N loss through nitrate leaching [53,54]. Therefore, northern agriculture practices should consider mitigating NO_3^- entering waterways even when fertilizing is minimal. Applying agroforestry practices have been shown to reduce NO_3^- leaching [6,54–57]. Specifically, willows encompass a high filtering capacity for N and their root zone promotes denitrification [55]. It appears that utilizing willow thickets in the James Bay lowlands has great potential for managing potentially excess NO_3^- from agriculture initiatives, as these trees are fast-growing, adapt to the climate shifts, and can be a source of bioenergy [47,58–60].

In sharp contrast to NO_3–N, the supply rate of K was significantly greater later in the growing season. However, all soil K supply rates estimated in this study from different sites, times of year, and crop management methods were substantially less than the 40–370 ug/10 cm^2/week rates found for agricultural soils using the Western Ag PRS probes [23]. Wilton et al. [20] also found K deficiency in leaf tissue for bean and potato. The small supply of K in June may indicate that environmental conditions earlier in the summer either do not produce sufficient K for plant growth or that the supply of K is below optimal crop requirements throughout the growing season.

Throughout the growing season, both sites had supply rates for Ca and Mg that were within the range reported for agricultural soils of 250–2700 μg Ca/10 cm^2/week and

126–262 μg Mg/10 cm^2/week, occasionally approaching the maximum levels observed of 2700 μg Ca/10 cm^2/week in agricultural soils [23]. However, there were spatial differences between these nutrients. The supply rate of Ca was significantly greater in the Agroforestry site (similar to PO_4–P and Zn). In comparison, the Mg supply rate was significantly greater in the Open site (along with SO_4^-).

The predominant source of Ca and Mg at Fort Albany is from the underlying limestone bedrock [61]; having these two elements in excess can cause crop deficiencies in P and K [27,43,62]. Our findings from the adsorbed soil nutrients and results from exchangeable nutrients [3,11] confirmed that the soils have P and K deficiencies for agricultural purposes. Agroforestry management with willows may have mitigated P growth limitations, but it did not improve K availability. Both sites had historic drainage of peatland and are prone to annual spring flooding, which influences the loss of exchangeable K through leaching [27]. It is suspected that crops had to rely on K fixed to clay and organic complexes since the sites were never supplemented for nutrient losses [27,63,64]. Limited soil K may explain why foliar tissues for both crops had Ca and Mg concentrations exceeding critical thresholds for adequate growth [20]. In general, cation competition between K, Ca, and Mg can occur during plant uptake [27,44,65]. Hence, it is possible that potatoes unintentionally obtained Ca and Mg while acquiring K for tuber development [66].

Sulfur is important for crop production with many functions in the plant, including synthesizing proteins, enzymes, vitamins, and chlorophyll, and is a crucial component for nodule development in legumes [67]. Though sulfur deficiency is a common issue in agriculture production [67,68], the supply rates of SO_4^- indicate that this nutrient was abundantly available in the soils at our research location. The median SO_4^- soil supply rates adsorbed on PRS probes for agricultural soils is 40 10 cm^{-1} $week^{-1}$ [23]. The Agroforestry site had adsorption rates that ranged from 30-221 10 cm^{-1} $week^{-1}$, while the PRS probes in the Open site had adsorbed a significantly greater amount of SO_4^-, ranging from 249–737 10 cm^{-1} $week^{-1}$. The source of the surplus SO_4^- is not confirmed; gypsum may exist deeper within the bedrock formation [69], or SO_4^- could be derived from the mineralization of organic matter in the cultivated soils or sourced from the surrounding peatland and wetlands [68]. The surplus of S can be beneficial in calcareous soils to improve the solubility and availability of Zn, Cu, Mn, P, and N and lower soil pH [67,70–72]. Conversely, S toxicity is not common but can occur [73]. In our study, the supply rate of S had a significant negative correlation with potato yield. The contrast in SO_4^- supply rates between the two sites suggests that agroforestry may aid in regulating S in the soil—this warrants further study when developing agriculture practices for the James Bay lowlands and other northern regions.

4.3. *Spatiotemporal Dynamics of Micronutrient Supply Rates in Subarctic Cultivated Soil*

There were also significant spatiotemporal differences among micronutrients at the James Bay study site. For example, we found a decreasing supply rate of Zn from June through October. Zn is least available in alkaline soil, and cold soils inhibit the uptake of Zn [74] (pp. 300–301) and both conditions occur in Fort Albany soil. However, we also found greater supply rates of Zn in the Agroforestry site where the average soil temperature was warmer than in the Open site [20]. Compared to other agricultural soils, the supply rate of Zn is adequate [23]. Zinc also had a significant positive correlation with the yield of bush bean ($p = 0.01$), a pattern consistent with PO_4–P and Ca, the other nutrients with greater supply rates in the Agroforestry site. This suggests that the greater supply rate of Zn in the Agroforestry site may contribute to the increased yield.

Boron is an essential nutrient required in small amounts and can be toxic at slightly larger amounts [74] (p. 219). Though the supply rate of B is adequate compared to other agricultural soils [23], we found significantly greater B supply later in the growing season and a greater supply rate of B in the Open site—where previous work measured lower crop productivity [3,13]. Collectively, these results suggest that B levels in the Open site may be approaching toxic concentrations. As found for B, the greatest supply of Al occurred later in the growing season and in the Open site. Compared to agricultural soils, the supply rate for Al was small [23], but Al still had a negative correlation with potato yield.

Similar to Zn, the supply rate of Fe and Cu were significantly greater in June and August compared to later in the growing season. Compared to agricultural soils, the supply of Fe was high [23], which may create P shortages through P fixation in acidic soils but not in soil with pH > 7 [45], the pH measured in both sites of this study. Lastly, compared to agricultural soils, the supply of Cu was adequate, as was the case for Mn [23]. Manganese was the sole nutrient without significant differences between sites, crop treatments, or across the months sampled, nor any significant interaction effects.

5. Conclusions

Agriculture is expanding to boreal communities adapting to a warming climate as a promising approach to replace food importation and to meet food security needs. Baseline information on subarctic land cultivation is essential to determine northern agriculture practices that are suitable for effective crop production and to mitigate potential environment and health risks. This study examined plant available nutrients with two types of low-input management technologies—willow agroforestry and bean–potato intercropping. Previous research found greater crop yield in agroforestry sites than in open sites [3,13] and our present study suggests that it is associated with greater soil supply of PO_4–P, Ca, and Zn, and a concomitant smaller supply rate of Mg, SO_4, B, and Al. Intercropping bean with potato did not improve the availability of soil nutrients. Moreover, PO_4–P and Zn had significant positive correlations with bean and potato yield, with the additional positive correlation between soil Ca and bean yield. Conversely, soils in the open site had significant negative correlations between the supply of Mg, SO_4, B, and Al with potato yield (though not for bean). However, despite the greater supply of P in the agroforestry site, P (and K) deficiencies persist in cultivated soil in the James Bay lowlands of northern Ontario. In order to maintain crop production, additional nutrient supplementation appears to be crucial which can involve local composting initiatives [51]. Agroforestry in the form of a willow shelterbelt is a promising practice to utilize in the boreal region for many potential purposes, including protecting crops from wind, enhancing the microclimate, promoting biodiversity, and supporting wildlife habitats. Additionally, as observed in this study on calcareous silty loam soils, willow shelterbelts increased P bioavailability and potentially regulated and reduced N and S run-off.

Author Contributions: Conceptualization, methodology, interpretation, and writing: J.D.K., M.J.W., and L.J.S.T.; field investigation: M.J.W.; statistical analysis: J.D.K.; funding acquisition, L.J.S.T. All authors have read and agreed to the published version of the manuscript.

Funding: Funding was provided by the Canadian Institutes of Health Research (FRN 169169), Georgian College Research Fund, and Western Ag Innovations Inc.

Data Availability Statement: Data are stored with Western Ag https://www.westernag.ca/ (accessed on 30 April 2021).

Acknowledgments: We thank the Fort Albany First Nation Chief and Council and community members for hosting this project. We greatly appreciate the guidance and expertise Eric Bremer provided with respect to the PRS probes.

Conflicts of Interest: The authors declare no conflict of interest. The funders had no role in the design of the study; in the collection, analyses, or interpretation of data; in the writing of the manuscript, or in the decision to publish the results.

References

1. Altdorff, D.; Borchard, N.; Young, E.H.; Galagedara, L.; Sorvali, J.; Quideau, S.; Unc, A. Agriculture in boreal and Arctic regions requires an integrated global approach for research and policy. *Agron. Sustain. Dev.* **2021**, *41*, 23. [CrossRef]
2. Hannah, L.; Roehrdanz, P.R.; Krishna Bahadur, K.C.; Fraser, E.D.G.; Donatti, C.I.; Saenz, L.; Wright, T.M.; Hijmans, R.J.; Mulligan, M.; Berg, A.; et al. The environmental consequences of climate-driven agricultural frontiers. *PLoS ONE* **2020**, *15*, e0228305. [CrossRef]
3. Barbeau, C.D.; Wilton, M.J.; Oelbermann, M.; Karagatzides, J.D.; Tsuji, L.J.S. Local food production in a subarctic Indigenous community: The use of willow (*Salix* spp.) windbreaks to increase the yield of intercropped potatoes (*Solanum tuberosum*) and bush beans (*Phaseolus vulgaris*). *Int. J. Agric. Sust.* **2017**, *16*, 29–39. [CrossRef]
4. Chen, A.; Natcher, D. Greening Canada's Arctic food system: Local food procurement strategies for combating food insecurity. *Can. Food Stud.* **2019**, *6*, 140–154. [CrossRef]
5. Kedir, A.J.; Zhang, M.; Unc, A. Understanding soil fertility status in Newfoundland from standard farm soil tests. *Can. J. Soil Sci.* **2021**, 1–15. [CrossRef]
6. Smith, J.; Pearce, B.D.; Wolfe, M.S. Reconciling productivity with protection of the environment: Is temperate agroforestry the answer? *Renew. Agric. Food Syst.* **2013**, *28*, 80–92. [CrossRef]
7. Bhardwaj, D.R.; Navale, M.R.; Sharma, S. Agroforestry practices in temperate regions of the world. In *Agroforestry*; Dagar, J., Tewari, V., Eds.; Springer: Singapore, 2017; pp. 163–187. [CrossRef]
8. Hynynen, J.; Niemistö, P.; Viherä-Aarnio, A.; Brunner, A.; Hein, S.; Velling, P. Silviculture of birch (*Betula pendula* Roth and *Betula pubescens* Ehrh.) in northern Europe. *Forestry* **2010**, *83*, 103–119. [CrossRef]
9. Dubois, H.; Verkasalo, E.; Claessens, H. Potential of birch (*Betula pendula* Roth and *B.* pubescens Ehrh.) for forestry and forest-based industry sector within the changing climatic and socio-economic context of Western Europe. *Forests* **2020**, *11*, 336. [CrossRef]
10. Pohjanmies, T.; Jašková, A.; Hotanen, J.P.; Manninen, O.; Salemaa, M.; Tolvanen, A.; Merilä, P. Abundance and diversity of edible wild plants in managed boreal forests. *For. Ecol. Manag.* **2021**, *491*, 119151. [CrossRef]
11. Spiegelaar, N.F.; Tsuji, L.J.S.; Oelbermann, M. The potential use of agroforestry community gardens as a sustainable import-substitution strategy for enhancing food security in subarctic Ontario, Canada. *Sustainability* **2013**, *5*, 4057–4075. [CrossRef]
12. Skinner, K.; Hanning, R.M.; Desjardins, E.; Tsuji, L.J. Giving voice to food insecurity in a remote indigenous community in subarctic Ontario, Canada: Traditional ways, ways to cope, ways forward. *BMC Public Health* **2013**, *13*, 427. [CrossRef]
13. Barbeau, C.D.; Oelbermann, M.; Karagatzides, J.D.; Tsuji, L.J.S. Sustainable agriculture and climate change: Producing potatoes (*Solanum tuberosum* L.) and bush beans (*Phaseolus vulgaris* L.) for improved food security and resilience in a Canadian subarctic First Nations community. *Sustainability* **2015**, *7*, 5664–5681. [CrossRef]
14. Birch, P.R.; Bryan, G.; Fenton, B.; Gilroy, E.M.; Hein, I.; Jones, J.T.; Prashar, A.; Taylor, M.A.; Torrance, L.; Toth, I.K. Crops that feed the world 8: Potato: Are the trends of increased global production sustainable? *Food Secur.* **2012**, *4*, 477–508. Available online: https://link.springer.com/article/10.1007/s12571-012-0220-1 (accessed on 21 May 2021). [CrossRef]
15. Spiegelaar, N.F.; Tsuji, L.J. Impact of Euro-Canadian agrarian practices: In search of sustainable import-substitution strategies to enhance food security in subarctic Ontario, Canada. *Rural Remote Health* **2013**, *13*, 2211. [CrossRef] [PubMed]
16. Rodriguez, C.; Carlsson, G.; Englund, J.E.; Flöhr, A.; Pelzer, E.; Jeuffroy, M.H.; Makowski, D.; Jensen, E.S. Grain legume-cereal intercropping enhances the use of soil-derived and biologically fixed nitrogen in temperate agroecosystems. A meta-analysis. *Eur. J. Agron.* **2020**, *118*, 126077. [CrossRef]
17. Gitari, H.I.; Gachene, C.K.; Karanja, N.N.; Kamau, S.; Nyawade, S.; Sharma, K.; Schulte-Geldermann, E. Optimizing yield and economic returns of rain-fed potato (*Solanum tuberosum* L.) through water conservation under potato-legume intercropping systems. *Agric. Water Manag.* **2018**, *208*, 59–66. [CrossRef]
18. Sharaiha, R.K.; Hadidi, N.A. Micro-environmental effects on potato and bean yields grown under intercropping system. *Agron. Ser.* **2008**, *51*, 209–219. Available online: http://www.uaiasi.ro/revagrois/PDF/2008_1_209.pdf (accessed on 21 May 2021).
19. Techen, A.K.; Helming, K.; Brüggemann, N.; Veldkamp, E.; Reinhold-Hurek, B.; Lorenz, M.; Vogel, H.J. Soil research challenges in response to emerging agricultural soil management practices. *Adv. Agron.* **2020**, *161*, 179–240. [CrossRef]
20. Wilton, M.J.; Karagatzides, J.D.; Tsuji, L.J. Nutrient concentrations of bush bean (*Phaseolus vulgaris* L.) and potato (*Solanum tuberosum* L.) cultivated in subarctic soils managed with intercropping and willow (*Salix* spp.) agroforestry. *Sustainability* **2017**, *9*, 2294. [CrossRef]
21. Environment Canada. Available online: https://climate.weather.gc.ca/climate_normals/ (accessed on 26 April 2021).
22. Kenney, W.A. A method for estimating windbreak porosity using digitized photographic silhouettes. *Agric. For. Meteorol.* **1987**, *39*, 91–94. [CrossRef]
23. Western Ag Technologies. Available online: https://www.westernag.ca/ (accessed on 30 April 2021).
24. Mohamed, M.A.; Stamatakis, A.; Keramidas, V. Anionic resin–extractable phosphorus as an index of phosphorus availability in calcareous soils of Crete amended or not amended with pig manure compost. *Commun. Soil Sci. Plant Anal.* **2013**, *44*, 50–61. [CrossRef]
25. *SPSS 26*; SPSS Inc.: Chicago, IL, USA, 2017.
26. Van Meeteren, M.J.M.; Tietema, A.; Westerveld, J.W. Regulation of microbial carbon, nitrogen, and phosphorus transformations by temperature and moisture during decomposition of *Calluna vulgaris* litter. *Biol. Fert. Soils* **2007**, *44*, 103–112. [CrossRef]

27. Rydin, H.; Jeglum, J.K.; Hooijer, A. Nutrients, light and temperature. In *The Biology of Peatlands*, 2nd ed.; Rydin, H., Jeglum, J.K., Eds.; Oxford University Press: New York, NY, USA, 2013; pp. 176–198. [CrossRef]
28. Sun, D.; Bi, Q.; Li, K.; Dai, P.; Yu, Y.; Zhou, W.; Lv, T.; Liu, X.; Zhu, J.; Zhang, Q.; et al. Significance of temperature and water availability for soil phosphorus transformation and microbial community composition as affected by fertilizer sources. *Biol. Fertil. Soils* **2018**, *54*, 229–241. [CrossRef]
29. Foereid, B.; Bro, R.; Mogensen, V.O.; Porter, J.R. Effects of windbreak strips of willow coppice—Modelling and field experiment on barley in Denmark. *Agric. Ecosyst. Environ.* **2002**, *93*, 25–32. [CrossRef]
30. Brandle, J.R.; Hodges, L.; Zhou, X.H. Windbreaks in North American agricultural systems. *Agrofor. Syst.* **2004**, *61*, 65–78.
31. Mungai, N.W.; Motavalli, P.P.; Kremer, R.J.; Nelson, K.A. Spatial variation of soil enzyme activities and microbial functional diversity in temperate alley cropping systems. *Biol. Fertil. Soils* **2005**, *42*, 129–136. [CrossRef]
32. Tamang, B.; Andreu, M.G.; Rockwood, D.L. Microclimate patterns on the leeside of single-row tree windbreaks during different weather conditions in Florida farms: Implications for improved crop production. *Agrofor. Syst.* **2010**, *79*, 111–122. [CrossRef]
33. Rinnan, R.; Michelsen, A.; Bååth, E.; Jonasson, S. Mineralization and carbon turnover in subarctic heath soil as affected by warming and additional litter. *Soil Biol. Biochem.* **2007**, *39*, 3014–3023. [CrossRef]
34. Ruess, L.; Michelsen, A.; Schmidt, I.K.; Jonasson, S. Simulated climate change affecting microorganisms, nematode density and biodiversity in subarctic soils. *Plant Soil* **1999**, *212*, 63–73. [CrossRef]
35. Manimel Wadu, M.C.; Ma, F.; Chang, S.X. Phosphorus availabilities differ between cropland and forestland in shelterbelt systems. *Forests* **2019**, *10*, 1001. [CrossRef]
36. Rao, K.P.C.; Verchot, L.V.; Laarman, J. Adaptation to climate change through sustainable management and development of agroforestry systems. *J. SAT Agric. Res.* **2007**, *4*, 1–30.
37. Shukla, A.; Kumar, A.; Jha, A.; Chaturvedi, O.P.; Prasad, R.; Gupta, A. Effects of shade on arbuscular mycorrhizal colonization and growth of crops and tree seedlings in Central India. *Agrofor. Syst.* **2009**, *76*, 95–109. [CrossRef]
38. Seiter, S.; Ingham, E.R.; William, R.D. Dynamics of soil fungal and bacterial biomass in a temperate climate alley cropping system. *Appl. Soil Ecol.* **1999**, *12*, 139–147. [CrossRef]
39. Lacombe, S.; Bradley, R.L.; Hamel, C.; Beaulieu, C. Do tree-based intercropping systems increase the diversity and stability of soil microbial communities? *Agric. Ecosyst. Environ.* **2009**, *131*, 25–31. [CrossRef]
40. Banerjee, S.; Baah-Acheamfour, M.; Carlyle, C.N.; Bissett, A.; Richardson, A.E.; Siddique, T.; Bork, E.W.; Chang, S.X. Determinants of bacterial communities in Canadian agroforestry systems. *Environ. Microbiol.* **2016**, *18*, 1805–1816. [CrossRef]
41. Nelson, J.D.J.; Schoenau, J.J.; Malhi, S.S.; Gill, K.S. Burning and cultivation effects on greenhouse gas emissions and nutrients in wetland soils from Saskatchewan, Canada. *Nutr. Cycl. Agroecosyst.* **2007**, *78*, 291–303. [CrossRef]
42. Richardson, A.E.; Lynch, J.P.; Ryan, P.R.; Delhaize, E.; Smith, F.A.; Smith, S.E.; Harvey, P.R.; Ryan, M.H.; Veneklaaas, E.J.; Lambers, H.; et al. Plant and microbial strategies to improve the phosphorus efficiency of agriculture. *Plant Soil* **2011**, *349*, 121–156. [CrossRef]
43. Ehrmann, J.; Ritz, K. Plant: Soil interactions in temperate multi-cropping production systems. *Plant Soil* **2014**, *376*, 1–29. [CrossRef]
44. Adams, P.W.; Lamoureux, S. *The Use of Native Northern Plants for the Re-Vegetation of Arctic Mine Tailings and Mine Waste*; Environment and Natural Resources; Government of Northwest Territories: Yellowknife, NT, Canada, 2005; 67p.
45. Kedir, A.J.; Nyiraneza, J.; Galagedara, L.; Cheema, M.; Hawboldt, K.A.; McKenzie, D.B.; Unc, A. Phosphorus adsorption characteristics in forested and managed podzolic soils. *Soil Sci. Soc. Am. J.* **2021**, *85*, 249–262. [CrossRef]
46. Magdziak, Z.; Mleczek, M.; Kaczmarek, Z.; Golinski, P. Influence of Ca/Mg ratio and Cd^{2+} and Pb^{2+} elements on low molecular weight organic acid secretion by Salix viminalis L. roots into the rhizosphere. *Trees* **2013**, *27*, 663–673. [CrossRef]
47. Kuzovkina, Y.A.; Volk, T.A. The characterization of willow (*Salix* L.) varieties for use in ecological engineering applications: Co-ordination of structure, function and autecology. *Ecol. Eng.* **2009**, *35*, 1178–1189. [CrossRef]
48. Gordon, A.; Newman, S. *Temperate Agroforestry Systems*; CAB International: Cambridge, MA, USA, 1997; pp. 263–269.
49. Rytter, R.M. Fine-root production and turnover in a willow plantation estimated by different calculation methods. *Scand. J. For. Res.* **1999**, *14*, 526–537. [CrossRef]
50. Phillips, C.J.; Marden, M.; Suzanne, L.M. Observations of root growth of young poplar and willow planting types. *N. Z. J. For. Sci.* **2014**, *44*, 15. [CrossRef]
51. Wilton, M.J.; Karagatzides, J.D.; Tsuji, L.J.S. Fertilizing bush beans with locally made compost in a remote subarctic community. *Agrosyst. Geosci. Environ.* **2020**, *3*, e20109. [CrossRef]
52. Ghosh, U.; Chatterjee, A.; Bremer, E. Determining the moisture and plant effect on nutrient release, and plant nutrient uptake using ion exchange resin membrane. *Commun. Soil Sci. Plant Anal.* **2018**, *49*, 782–790. [CrossRef]
53. Sierra, J. Temperature and soil moisture dependence of N mineralization in intact soil cores. *Soil Biol. Biochem.* **1997**, *29*, 1557–1563. [CrossRef]
54. Evers, A.K.; Bambrick, A.; Lacombe, S.; Dougherty, M.C.; Peichl, M.; Gordon, A.M.; Thevathasan, N.V.; Whalen, J.; Bradley, R.L. Potential greenhouse gas mitigation through temperate tree-based intercropping systems. *Open Agric. J.* **2010**, *4*, 49–57. [CrossRef]
55. Aronsson, P.; Perttu, K. Willow vegetation filters for wastewater treatment and soil remediation combined with biomass production. *For. Chron.* **2001**, *77*, 293–299. [CrossRef]
56. Ryszkowski, L.; Kędziora, A. Modification of water flows and nitrogen fluxes by shelterbelts. *Ecol. Eng.* **2007**, *29*, 388–400. [CrossRef]

57. Bergeron, M.; Lacombe, S.; Bradley, R.L.; Whalen, J.; Cogliastro, A.; Jutras, M.F.; Arp, P. Reduced soil nutrient leaching following the establishment of tree-based intercropping systems in eastern Canada. *Agrofor. Syst.* **2011**, *83*, 321–330. [CrossRef]
58. Verwijst, T. Willows: An underestimated resource for environment and society. *For. Chron.* **2001**, *77*, 281–285. [CrossRef]
59. Smart, L.B.; Cameron, K.D. Genetic improvement of willow (*Salix* spp.) as a dedicated bioenergy crop. In *Genetic Improvement of Bioenergy Crops*; Vermerris, W., Ed.; Springer: New York, NY, USA, 2008; pp. 347–376. [CrossRef]
60. Myers-Smith, I.H.; Forbes, B.C.; Wilmking, M.; Hallinger, M.; Lantz, T.; Blok, D.; Hik, D.S. Shrub expansion in tundra ecosystems: Dynamics, impacts and research priorities. *Environ. Res. Lett.* **2011**, *6*, 045509. [CrossRef]
61. Sader, J.A.; Hattori, K.; Hamilton, S.; Brauneder, K. Metal binding to dissolved organic matter and adsorption to ferrihydrite in shallow peat groundwaters: Application to diamond exploration in the James Bay Lowlands, Canada. *Appl. Geochem.* **2011**, *26*, 1649–1664. [CrossRef]
62. Sundström, E.; Magnusson, T.; Hånell, B. Nutrient conditions in drained peatlands along a north-south climatic gradient in Sweden. *For. Ecol. Manag.* **2000**, *126*, 149–161. [CrossRef]
63. Jalali, M.; Arian, T.M.; Ranjbar, F. Selectivity coefficients of K, Na, Ca, and Mg in binary exchange systems in some calcareous soils. *Environ. Monit. Assess.* **2020**, *192*, 80. [CrossRef]
64. Øien, D.I.; Moen, A. Nutrient limitation in boreal plant communities and species influenced by scything. *Appl. Veg. Sci.* **2001**, *4*, 197–206. [CrossRef]
65. Tsialtas, I.T.; Shabala, S.; Baxevanos, D.; Matsi, T. Cation selectivity in cotton (*Gossypium hirsutum* L.) grown on calcareous soil as affected by potassium fertilization, cultivar and growth stage. *Plant Soil* **2017**, *415*, 331–346. [CrossRef]
66. Assunção, N.S.; Ribeiro, N.P.; da Silva, R.M.; Soratto, R.P.; Fernandes, A.M. Tuber yield and allocation of nutrients and carbohydrates in potato plants as affected by limestone type and magnesium supply. *J. Plant Nutr.* **2020**, *43*, 51–63. [CrossRef]
67. Kaya, M.; Küçükyumuk, Z.; Erdal, I. Effects of elemental sulfur and sulfur-containing waste on nutrient concentrations and growth of bean and corn plants grown on a calcareous soil. *Afr. J. Biotech.* **2009**, *8*, 4481–4489.
68. Wilhelm Scherer, H. Sulfur in soils. *J. Plant Nutr. Soil Sci.* **2009**, *172*, 326–335. [CrossRef]
69. Orlova, J.; Branfireun, B.A. Surface water and groundwater contributions to streamflow in the James Bay Lowland, Canada. *Arct. Antarct. Alp. Res.* **2014**, *46*, 236–250. [CrossRef]
70. Kayser, A.; Schröder, T.J.; Grünwald, A.; Schulin, R. Solubilization and plant uptake of zinc and cadmium from soils treated with elemental sulfur. *Int. J. Phytoremediat.* **2001**, *3*, 381–400. [CrossRef]
71. Rahman, M.M.; Soaug, A.A.; Darwish, F.H.A.; Golam, F.; Sofian-Azirun, M. Growth and nutrient uptake of maize plants as affected by elemental sulfur and nitrogen fertilizer in sandy calcareous soil. *Afr. J. Biotech.* **2011**, *10*, 12882–12889. [CrossRef]
72. Rezapour, S. Effect of sulfur and composted manure on SO_4-S, P and micronutrient availability in a calcareous saline–sodic soil. *Chem. Ecol.* **2014**, *30*, 147–155. [CrossRef]
73. Ruiz, J.M.; López-Cantarero, I.; Rivero, R.M.; Romero, L. Sulphur phytoaccumulation in plant species characteristic of gypsiferous soils. *Int. J. Phytoremediat.* **2003**, *5*, 203–210. [CrossRef] [PubMed]
74. Plaster, E.J. *Soil Science and Management*, 6th ed.; Delmar: Clifton Park, NY, USA, 2014; pp. 300–301.

 *sustainability*

Article

Effects of Domestic and Wild Ungulate Management on Young Oak Size and Architecture

Aida López-Sánchez [1,*], Sonia Roig [1], Rodolfo Dirzo [2] and Ramón Perea [1,2]

[1] Departamento de Sistemas y Recursos Naturales, Universidad Politécnica de Madrid, 28040 Madrid, Spain; sonia.roig@upm.es (S.R.); ramon.perea@upm.es (R.P.)
[2] Department of Biology, Stanford University, Stanford, CA 94305, USA; rdirzo@stanford.edu
* Correspondence: aida.lopez@upm.es

Abstract: Scattered oaks in traditional silvopastoral systems (i.e., "dehesas") provide important ecological services. However, livestock intensification applied to these systems over the last century has affected the architecture of young oak plants. This unsuitable rangeland management practice jeopardizes the long-term system sustainability. Here we examine the alterations in architecture of regenerating oak plants in Mediterranean dehesas under three representative management regimes: (1) traditional management with extensive sheep grazing; (2) commercially driven management with extensive cattle grazing, and (3) native deer grazing at moderate stocking rates (<0.11 livestock units $\times$ ha^{-1}). Plant architecture was considerably altered in cattle-grazed "dehesas", finding a 50% reduction in plant height–diameter ratios, compared to sheep-grazed dehesas where plants with higher height–diameter ratios predominated. Young oak plants, however, showed less altered architecture and less probability of damage on shoot apex (0.20-fold difference) in areas with deer grazing at moderate stocking rates. In addition, those young oak plants with multi-stemmed individual architecture were more stunted (lower values of crown height–diameter ratio) in areas with livestock grazing than wildlife areas (0.78-fold difference). Shrub presence, under all management schemes, helped to increase in plant height, except when shrubs were located under tree canopies. Conversely, plants without shrub protection showed stunted architecture with well-developed basal diameters but short stature. Appropriate sustainable practices should include cattle stocking rate reduction, traditional sheep grazing promotion, nurse shrub preservation and fencing stunted individuals along with pruning basal sprouts. Our study indicates that management may have important consequences on dehesa regeneration via alterations of plant architecture and therefore on system sustainability.

Keywords: height-diameter ratio; plant architecture; *Quercus ilex*; livestock grazing

Citation: López-Sánchez, A.; Roig, S.; Dirzo, R.; Perea, R. Effects of Domestic and Wild Ungulate Management on Young Oak Size and Architecture. *Sustainability* **2021**, *13*, 7930. https://doi.org/10.3390/su13147930

Academic Editor: Victor Rolo

Received: 25 June 2021
Accepted: 13 July 2021
Published: 15 July 2021

1. Introduction

Human-managed ecosystems featuring scattered trees occur throughout the world, providing important ecological services (e.g., seed source, nutrient enrichment, landscape connectivity functions, shelter; [1]). Many of these ecosystems are originally derived from dense forests that have been transformed into open agroforestry systems by clearing several trees and then used for grazing [1,2], supporting relatively high biodiversity [3,4]. This results in landscapes with scattered trees or shrubs (woody pastures) used for grazing. Despite their importance, these woody pastures have experienced important anthropogenic impacts such as overgrazing and land use conversion, all of which threaten their long-term persistence, particularly since the 1950s [5–7]. The lack of tree recruitment is one of the major consequences of such anthropogenic impacts [8] due to the implementation of inappropriate management practices that negatively affect natural oak (*Quercus* spp.) recruitment [9–11]. In particular, some livestock management practices such as the increment of livestock stocking rates (especially in extensive cattle grazing) or sheep replacement by cattle have been often related to oak recruitment failure [12–15].

Livestock has been regarded as an important part of ecosystem engineering that restructures ecological communities [16,17]. Generally, low levels of herbivory make woody growth and reproduction possible whereas high levels of it impede survival or affect plant development [18]. The permanent, stressing effect of high herbivory pressure, may not only affect recruitment rates but also patterns of plant recruitment (e.g., via sexual vs. asexual), as well as the architecture and growth patterns of young plants [19]. Herbivores generally destroy parts of the plants by pulling and trampling, and reduce biomass and cover of the stratum located within their grazing height [20]. The effects of large herbivores on young plant performance and architecture may in turn determine their probability of progressing from the juvenile stage to the adult (reproductive) stage, delaying or even impeding successful woodland regeneration of the woodland as a whole [21]. In addition, grazing height differs according to grazer body size. For instance, cattle can browse above 1.5 m high [22] while sheep typically browse below 1 m high [22], and red deer (*Cervus elaphus* L.) concentrates browsing at around 1 m [23]. Hence, larger herbivores are potentially able to consume greater proportions of the plant and have greater impacts damage on critical plant growth parts (e.g. apical meristems).

In Spain, ecosystems featuring scattered oak trees throughout the landscape, also known as dehesas, are internationally recognized for their ecological, socioeconomic and cultural importance, as well as for supporting high levels of biodiversity [2,24]. Yet, over the last few decades, dehesas have also suffered a dramatic increase of livestock stocking rates, which is often related to oak recruitment failure [9,13–15,25]. Research so far has primarily focused on the effect of different management schemes on oak recruitment density, both in dehesas [9,15] and other woody pasture lands [26–28]. However, little is known about the effect of different livestock management schemes (e.g., traditional vs. commercially competitive) or the use of wild vs. domestic ungulates on plant architecture and growth patterns of young oaks. Such knowledge is needed to integrate grazing into management schemes and to determine the ecological impact of herbivores on valuable human-dominated ecosystems. In similar oak wood-pastures worldwide, strong reductions in the stature of young plants (seedlings and saplings) have been observed in cattle-grazed areas compared to those without cattle for more than two decades [21,28]. The result is that oak recruits, repeatedly browsed by cattle, become bushy (multi-stemmed individuals), stunted individuals, forming coppice-like trees with multiple stems [21]. Conversely, it is known that the presence of shrubs in Mediterranean or arid ecosystems may reduce herbivory lethality [29] by reducing the probability of shoot apex damage [30], but this in turn may alter the plant architecture that facilitates the transition to plant reproductive stages. Thus, microsite location of young trees in relation to the presence of herbivore foraging patterns may play an important role not only on oak recruitment density [15,29] but also on plant architecture. Despite these relationships, oak architecture (e.g., height–diameter ratio, crown diameter, plant shape) has not been fully investigated across different microsites and under distinct management schemes.

This study brings together the effects of these two factors, impact of browsing on plant architecture, and the role of potentially protective microsites on young oak plant architecture under three representative management schemes of the Mediterranean scattered oak woodlands (dehesas). These schemes involve cattle, sheep and wildlife rearing (mostly red deer) foraging, under similar ecological conditions and for more than 30 years under the same management scheme. Specifically, we examined the architecture of *Quercus ilex* L. as this species is the dominant and most representative tree species of dehesas [31]. We predicted that (1) architectural measures (height, basal diameter and height–diameter ratio) of young plants (seedlings, saplings and oak multi-stemmed individuals) would be significantly modified under the current (>0.11 livestock units per ha, LU ha^{-1}) extensive livestock farming (especially under cattle grazing) than wildlife management with moderate stocking rates (<0.11 LU ha^{-1}); We posit, in particular, that crown height–diameter ratio of oak multi-stemmed individuals will be related to the management scheme, with gleaner oak multi-stemmed individuals under traditional sheep management; (2) the probability of finding undamaged shoot apexes would be also greater in traditional sheep management (small-sized animals) compared to other representative management schemes involving larger herbivore (wildlife or cattle) management; and (3) microsite location of young plants will affect oak architecture differently depending on the management scheme, with less impact on plant development under shrub protection.

2. Materials and Methods

2.1. Study Area

The study area (108 km^2, largely flat and open) involved different oak dehesa systems within Toledo province, Central Spain (39–40° N, 5° W). Elevation ranges from 300 to 400 m a.s.l. The climate is Mediterranean oceanic pluvioseasonal [32], with a mean annual temperature of 15.1 °C and an average annual rainfall of 571 mm. Soils are sandy and acidic with low organic matter content within topsoil (see López-Sánchez et al. [15] for details). The study area supports diverse vegetation dominated by open holm oak (*Quercus ilex* L. subsp. ballota (Desf.) Samp.) woodland with a shrub cover mostly comprised of xerophytic and evergreen species. Shrub cover is low (0.2%, 0.6% and 38.0% in cattle, sheep and wildlife-grazed areas, respectively; [33]). The herbaceous layer is dominated by sub-nitrophilous Mediterranean annual communities and therophytic oligotrophic communities (see López-Sánchez et al. [15] for details).

2.2. Study Sites

Three sites (independent estates) with distinct and representative management (grazing regime) for at least thirty years (i.e., they were not subjected to ploughing, shrub-clearing or fire) were selected within the study area. The first one (142 ha), hereafter "Cattle", consists of commercially driven management (0.33 LU ha^{-1}) of cattle (breed "Avileña negra ibérica"). The second site (140 ha), hereafter "Sheep", has been supporting traditional management (0.25 LU ha^{-1},) of extensive sheep (breed "Talaverana"). Both sites have maintained livestock rearing year-round for the last 40 years and represent the typical managements (stocking rates) for these systems [25,34]. In addition, both sites are fenced off to control livestock movements and prevent wildlife and human access (see López-Sánchez et al. [15] for details). Finally, the third site (150 ha) has not supported livestock since 1985 and its management has been devoted to recreational big game hunting, mostly red deer but also some wild boar (*Sus scrofa* L.), for the past 50 years, and represents the typical deer management (densities) of Mediterranean hunting properties in oak-dominated woodlands and dehesas [35,36] (see López-Sánchez et al. [15]).

2.3. Sample Design and Data Collection

We selected nine independent, replicated zones (three zones per site) of 5 ha, each one with similar density of holm oak trees (42.15 ± 11.44 trees ha^{-1}), with similar diameter at breast height (40.40 ± 16.23 cm, see López-Sánchez et al. [15] for details). In each zone, eighteen 4 m × 35 m belt transects (separated by 40 m) were established (total n = 162) within which we defined four different microsites: (i) under the tree canopy (hereafter tree), (ii) under the shrub canopy (hereafter shrub), (iii) under the tree and shrub canopy (hereafter tree-shrub) and (iv) in open grasslands (hereafter open). We recorded the microsite and some architectural measures of all oak young plants (height <130 cm) that were found within the transect. The architectural measures were the basal diameter (hereafter, diameter, cm) and the height (cm) of all young plants, measured on the largest diameter shoot in the event of oak multi-stemmed individuals (plants with multiple shoots). We noted whether the apical sprout was browsed, or whether there was no sign of browsing. In the event of oak multi-stemmed individuals, we also measured crossed diameters (the largest one and their perpendicular). The mean of both crossed diameters was used as a single variable (hereafter, crown diameter). Vertebrate herbivory damage was categorized according to its intensity and was assessed for all young plants. Damage categories followed a 0–4 rank of herbivory level [28]: 0 for plants with no apparent browsing evidence (hereafter, null damage), 1 for plants with low browsing (1–10% of browsable biomass damaged; hereafter low damage), 2 for plants with moderate browsing (11–40% of browsable biomass damaged; hereafter moderate damage), 3 for plants with high browsing (41–70% of browsable biomass damaged; hereafter high damage), and 4 for plants with maximum browsing (>70% of browsable biomass damaged; hereafter maximum damage, see details in López-Sánchez et al. [28] for details) which was considered as unsustainable browsing since plants at this damage level become incapable of developing onto further ontogenetic stages [28,36]. Surveys were conducted at the beginning of spring (March–April 2014), coinciding with the highest browsing damage in most Mediterranean environments [37].

2.4. Statistical Analysis

We developed generalized linear mixed models (GLMMs, [38]) for all architectural response variables (Table 1). These models are an extension of General Linear Models (e.g., regression, variance analysis) allowing different error distributions for response variables, and establishing a linear relationship through a link function. In addition, applying random effects allows grouped data to be analyzed. We established the shape of the young plants (thickest shoot in the case of multi-stemmed individual oaks) through the height–diameter ratio (hereafter, HDR) by dividing the height/basal diameter (in the same units, cm). In addition, the shape of young oak multi-stemmed individuals was measured through crown height–diameter ratio (hereafter, CHDR, in cm) by dividing the height/crown diameter (in the same units, cm). When necessary, Box–Cox transformations [39] were applied to data in order to calculate the lambda transformation (power lambda link) [40] that maximizes the likelihood. Thus, the response variables were fitted to gamma error distribution with their corresponding power lambda link function (Table 1).

Table 1. Summary of architectural (response) variables.

Response Variable	Sample Size Used (n)	Error Distribution (Power Lambda Link Function) [1]
Height	554 (all plants)	Gamma (0.70)
Diameter	174 (plants with basal diameter >1 cm)	Gamma (0.70)
Height–diameter ratio	554 (all plants)	Gamma (0.34)
Crown height–diameter ratio	253 (only oak multi-stemmed individuals)	Gamma (0.33)

[1] Power lambda link function ($g(\mu) = \mu\lambda$) is the lambda (λ, numeric value inside the brackets) used for the monotonic transformations.

All models included rangeland management, microsite and the herbivory occurrence (presence-absence) as fixed effects. The structure of the random effect was the following: transect nested within zone and, thus, nested within management regime. Moreover, we repeated the same analyses for each response variable with the same random effect and fixed effects but including herbivory intensity (using the herbivory categories described above) instead of herbivory occurrence.

In addition, the occurrence of browsed apical sprout (presence-absence, for plants with some herbivory) was analyzed as a response variable by means of GLMMs and fitted to binomial error distribution with a logit link function. The models included rangeland management and microsite as fixed effects. Transect nested within zone and within management regime was considered as random effect.

We used the model averaging approach [41] in all cases. We first fitted the maximal model, containing all the predictors. Then, we performed model comparison of all possible models by using the AIC weights. For model comparison we used the "dredge" function within the "MuMIn" package of R. Finally, we obtained the model-averaged coefficients as well as the relative importance of each predictor (from 0 to 1) by using the "model.avg" function of "MuMIn". For predicted values obtained from models, we used the "fitted" function within the "stats" basic package of R.

Data processing and statistics were performed using R 3.1.1 [42] with the modules "lme4" [43], "car" [44] and "MuMIn" [45].

3. Results

3.1. Management Effects on Young Oak Plant Architecture

Rangeland management affected the height, diameter (>1 cm) and shape (HDR) of the young oaks and multi-stemmed individuals (Table 2). In sheep-grazed areas, we found significantly greater heights of young plants (mean of 45 cm) than in cattle-grazed areas (31 cm) and wildlife areas (24 cm; Figure 1a). In contrast, the diameter of young plants was greater (3.8 cm; Figure 1a) in cattle-grazed areas than in wildlife areas with no extensive farming (2.6 cm; Figure 1a). In addition, in areas where only wildlife was present the diameter of young plants was greater than in sheep-grazed areas (2 cm; Figure 1a).

Table 2. Summary of the generalized linear mixed models to analyze the effect of management, occurrence of herbivory and microsite on plant architectural variables (height, diameter and height–diameter ratio, and crown height–diameter ratio).

Response Variable		Management		H.O. [1]	Microsite		
		Sheep	Wildlife	Presence	Shrub	Tree-Shrub	Open
Height (cm)	Estimated model coefficient	5.088	1.423	6.428	2.626	0.053	−0.539
	Standard error	1.021	0.935	0.517	1.003	0.490	0.548
	z-value	4.983	1.522	12.423	2.619	0.109	0.985
	p-value	<0.001	0.128	<0.001	0.009	0.913	0.325
Diameter (cm)	Estimated model coefficient	−0.528	−0.289	0.271	−0.662	−0.719	0.038
	Standard error	0.132	0.139	0.127	0.216	0.259	0.103
	z-value	4.000	2.076	2.124	3.065	2.779	0.363
	p-value	<0.001	0.038	0.034	0.002	0.005	0.716
Height–diameter ratio	Estimated model coefficient	0.917	0.417	0.445	0.529	0.052	−0.053
	Standard error	0.095	0.091	0.053	0.114	0.073	0.068
	z-value	9.604	4.571	8.349	4.634	0.706	0.777
	p-value	<0.001	<0.001	<0.001	<0.001	0.480	0.437
Crown Height–diameter ratio	Estimated model coefficient	0.026	0.104	0.044	0.069	−0.030	0.002
	Standard error	0.039	0.040	0.031	0.053	0.051	0.029
	z-value	0.659	2.585	1.390	1.299	0.596	0.072
	p-value	0.509	0.010	0.165	0.194	0.551	0.943

[1] H.O: Herbivory Occurrence. Reference levels from fixed effects Management, H.O. and Microsite are Cattle, Absence and Tree, respectively.

As HDR is directly proportional to height measure, we also found higher values of HDR of young plants in sheep-grazed areas (30; Figure 1a) than in areas where livestock grazing was not present (20; Figure 1a). Furthermore, in areas where only wildlife was present the HDR of young plants was higher than in cattle-grazed areas (14; Figure 1a). We found a higher proportion of young oak multi-stemmed individuals in livestock (cattle or sheep)-grazed areas (71.0% ± 40.1 and 68.3% ± 30.7, respectively) than in wildlife areas (58.3% ± 43.6). Multi-stemmed individuals were more stunted (lower CHDR) in livestock grazing areas than in wildlife areas (Table 2), but they did not differ between cattle- and sheep-grazed areas (Table 2).

Plants with greater height (50 cm) showed higher probability of being browsed (Table 2) exhibiting moderate, high and maximum damages (Figure 2). In addition, plants with greater diameter (3.5 cm) showed higher probability of being browsed (Table 2) exhibiting high and maximum damage (Figure 2).

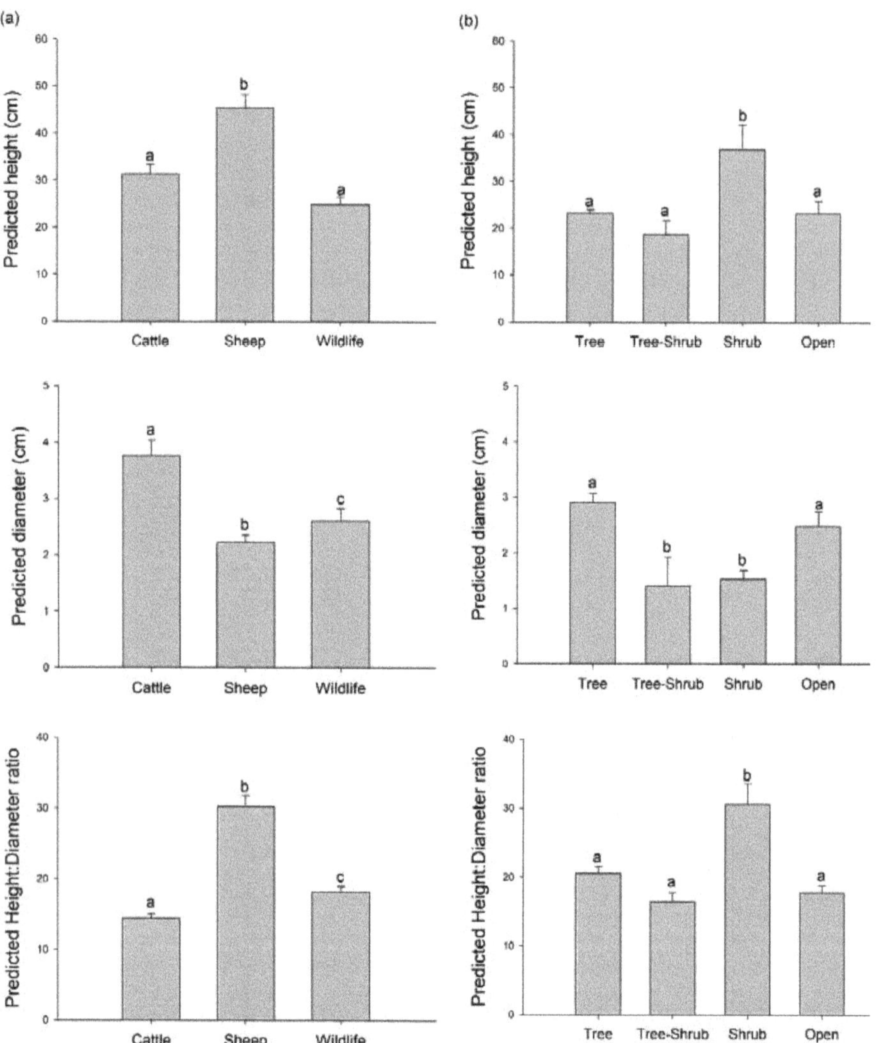

Figure 1. Architectural variables (height, diameter and height–diameter ratio) of plants depending on (**a**) the management regime (shown through the bars); and (**b**) the microsite (shown through the bars). N = 554 plants for height and height–diameter ratio, and N = 176 plants for diameter as response variables. Error lines are 95% confidence intervals. Different letters (a, b, c) indicate significant differences ($p < 0.05$) in architectural variables between (**a**) management regime and (**b**) microsite.

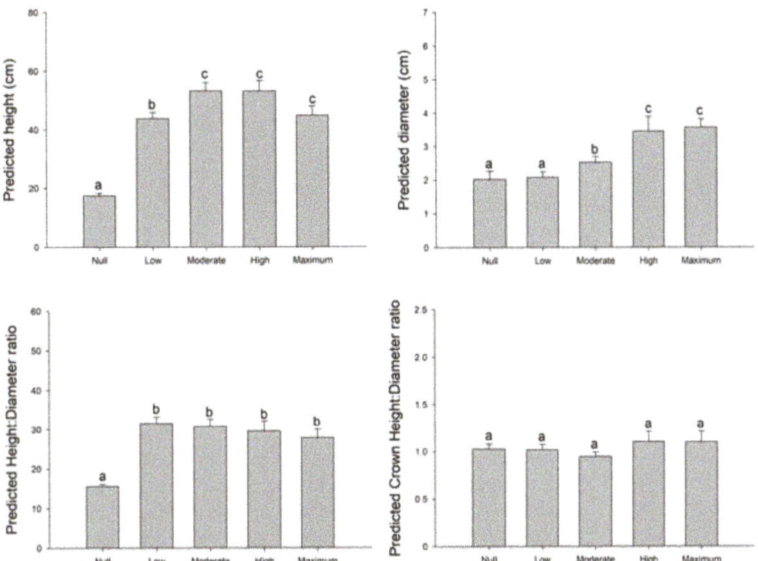

Figure 2. Probability of browsing depending on plant architectural variables (plant height, diameter and height–diameter ratio, and multi-stemmed individual crown height–diameter ratio). N = 554 plants for height and height–diameter ratio, N = 176 plants for diameter, and N = 253 multi-stemmed individuals as response variables. Bars indicate plant architectural variable value between different levels of damage intensity. Null damage: browsable biomass undamaged; Low damage: 1–10% of browsable biomass damaged; Moderate damage: 11–40% of browsable biomass damaged; High damage: 41–70% of browsable biomass damaged; Maximum damage: >70% of browsable biomass damaged. Error lines are 95% confidence intervals. Different letters (a, b, c) indicate significant differences ($p < 0.05$) in architectural variables between different levels of herbivory damage intensity.

Plants with greater HDR (30) showed higher probability of being browsed (Table 2) for any score of herbivory intensity (Figure 2). We did not find differences in oak multi-stemmed individuals CHDR depending on occurrence and intensity of herbivory (Table 1, Figure 2).

3.2. Microsite Effects on Young Holm Oak Architecture

Microsite of location affected the height, diameter (>1 cm) and shape (HDR) of the young oaks (Table 2). However, we did not find differences in shape of oak multi-stemmed individuals (CHDR) depending on the microsite (Table 2).

Young oak plants refuged under shrubs located in open areas showed greater values of heights (mean of 37 cm) than in other microsites (around 20 cm; Figure 1b). In contrast, we found significantly greater diameters in open areas and under tree canopy cover (2.8 cm) than under shrubs located both in open areas and under tree canopy cover (1.5; Figure 1b).

Additionally, we found higher HDR (30) under shrub cover than in other microsites (18; Figure 1b).

3.3. Damage on Shoot Apex

Occurrence of intact shoot apex (no apical sprout damage) in young plants varied depending on the management regime (Table 3). In livestock (cattle or sheep)-grazed areas the probability of having a damaged shoot apex was greater than in wildlife areas. We did not find differences on shoot apex damage across microsites (Table 3).

Table 3. Summary of the generalized linear mixed models to analyze the effect of management and microsite on the occurrence of apical sprout damage.

Response Variable	Predictors	Factors	Estimated Model Coefficient	Standard Error	z-Value	p-Value
Probability of occurrence of apical sprout damage	Intercept		0.706	0.739	0.954	0.340
	Management	Sheep	−0.060	0.977	0.061	0.951
		Wildlife	−3.494	0.983	3.557	<0.001
		Shrub	0.699	0.688	1.017	0.309
		Tree-shrub	−0.153	0.639	0.239	0.810
	Microsite	Open	0.695	0.393	1.769	0.077

4. Discussion

4.1. Management Effects on Young Oak Plant Architecture

The different management schemes led to a notable change in young oak recruitment architecture. We observed lower (40%) oak heights for areas with a management of commercial extensive cattle grazing in comparison to more traditional management such as sheep farming. Alternative management schemes with no extensive farming (at least 30 years without livestock) also reduced the young oak height in comparison to traditional sheep-grazed areas but the effect was less accentuated (around 30% height reduction) than with highly competitive cattle-grazing areas. Here, we found that plant architecture was modified, increasing the abundance of oak multi-stemmed individuals, with a clear shaping of the natural tree form (Figure 1). At stand level, an important consequence of abundant presence of young oak plants (height < 1.30 m) with low height–diameter ratios is the increase of the time required to reach the height threshold (1.5–2 m) that allows young trees to avoid being browsed [46,47]. In addition, stunted individuals with low height have greater probability of being killed by large browsers [48]. Importantly, we found a higher probability of shoot apex damage under permanent livestock farming (cattle or sheep), compared with representative wildlife areas, probably due to the continuous browsing activity of cattle and sheep (management under continuous grazing). Areas with no extensive farming (wildlife) usually involve lower stocking rates due to both lower animal density and greater mobility of wild ungulates. The continuous disruption of apical dominance through shoot apex removal results in shorter plants with multiple lateral meristems that facilitate the increase in width growth. This differential growth in width vs. height delays plant development and brings about a higher probability of being repeatedly damaged, either by browsing or by trampling [49]. Accordingly, we found larger basal diameters in cattle areas compared to wildlife areas, or with traditional management of sheep (Figure 1). This corroborates the idea of a disproportionally height–diameter ratio. According to Plieninger [6], dehesas with long-term silvopastoral use (>300 years) maintain adult or sub-adult trees with greater diameters than those with less historical silvopastoral management (60–100 years). Most of those adult or sub-adult trees are thick and have great diameter; but their height is not large enough to represent a well-developed adult tree. Our results showed that cattle farming management, involving continuous grazing at high stocking rates, resulted in a strong limitation for the growth of young oak plants and confirms previous studies [21,28,50].

Differences in HDR across management schemes need to be considered with caution since it may involve both management (stocking rates, presence) and animal size. Larger herbivores will reach the central shoot more easily than smaller herbivores. Most of the young oak plants in dehesas generated sprouts as a response to herbivore damage, and as soon as the horizontal extension of the stools is thick enough, the central shoot can gain height without being accessible to browsing herbivores [51]. Sheep are less able to reach the central shoot due to their smaller body size and, thus, young plants with certain height or width (e.g., wider oak multi-stemmed individuals) will escape shoot apex damage more easily, growing with higher HDR, which, in turn, results in the ability to advance to the adult stage faster. Moreover, slender young plants (higher HDR) showed a higher probability of being browsed since they were more visible and accessible than stunted plants.

Most oak regeneration found in examined livestock-grazed areas was in the form of oak multi-stemmed individuals (70%) due to high herbivore pressure, as happens in many dehesas holding livestock [9,25]. Our study reveals that livestock management exerted over more than 30 years results in more stunted plants (lower CHDR values) than in the case of wildlife grazing at moderate stocking rates (<0.11 LU ha^{-1}) due to the longer and continued consumption over oak multi-stemmed individuals. In particular, intensive commercial management of cattle sustained stunted oak multi-stemmed individuals (compared to traditional management of sheep grazing), resulting in protection of the central shoot against damage caused by cattle in this type of management. However, these extensive oak multi-stemmed individuals could not avoid constant damage in their central shoot after more than 30 years due to the current, most competitive management, involving constant presence of the cattle (year-round) at high stocking rates (>0.30 LU ha^{-1}).

4.2. Microsite Effects on Young Holm Oak Architecture

Young holm oak architecture was related to specific microsites. We found greater plant height under shrub protection in open areas. In harsh, dry environments, such as the Mediterranean, shrub cover not only provides protection against herbivore pressure [25,30], but also reduces evapotranspiration [52,53], which favors the growth of the apical shoot, leading to higher plants [52,54] with greater HDR (Figure 1). Thus, the results of our study are in line with previous research showing that the presence of some shrubs in Mediterranean dehesas may have positive effects on early recruitment of oak species, not only by facilitating seedling establishment [33,50,53,55], but also by ensuring a more natural plant architecture (reduction of apical damage) and a faster advance to the reproductive stage [15]. Plants under shrub protection located under trees were also protected against intensive browsing and desiccation. However, competition for light conditions and other resources is probably too high given the number of oak shoots under tree cover and the presence of the adult trees, which may reduce plant growth, affecting architecture. Conversely, we found higher diameter and smaller HDR under tree canopy and in open areas (without shrub protection) in comparison to those growing under shrub cover (Figure 1).

4.3. Sustainable Management Involves Well-Developed Young Oak Plants

The current age structure in dehesas (scarce adult oaks and lack of regeneration) represents a threat to the sustainability of the whole system because old-aged trees are not being replaced [9,13]. López-Sánchez et al. [15] have shown the striking differences in oak densities and herbivory role under three representative management schemes of Mediterranean dehesas. In the present study, we analyze the architecture of young oak plants subjected to these three typical management schemes in dehesas. It is not only important to evaluate the numerical oak recruitment presence but also the long-term effects of management (>30 years) on plant architecture and growth. Our results have shown that some livestock management schemes, chiefly high commercially competitive management of cattle, are generating stunted plants due to constant and heavy herbivore pressure over the same plants. However, traditional management of sheep grazing generated slender plants and promoted central shoot development compared to cattle grazing. Therefore, the management recommendation to return to a traditional livestock grazing management (e.g., using local sheep breeds) suggested by López-Sánchez et al. [15] would help plants grow faster, and with an adequate architecture that will ensure a more rapid and higher quality regeneration (i.e., incorporation of well-developed reproductive trees in less time). Furthermore, it might be helpful to reduce those Supplementary feed with high urea content that disproportionally make animals search for browse [56]. Additionally, temporary livestock exclusion with only the presence of wildlife (deer) at adequate densities (<0.11 LU ha^{-1}) will provide opportunities for holm oak regeneration and a possible encroachment of shrub cover [57] that would help to protect young plants. However, high densities of deer could produce an opposite result [36,58,59], since deer also browse and are large-bodied animals that impede oak regeneration and can modify the architecture of young oaks [16]. Therefore, under wildlife management it is also essential to monitor and control deer populations [60,61].

Young plants should not only substitute the actual old stand but also ensure the future sustainability of the whole system by producing high acorn yield. Hence, individual protection may be essential in current livestock management schemes, especially in high commercially competitive management, along with shearing epicormic and basal sprouts [31]. This management practice should promote greater HDR, reducing the disproportional horizontal extension of the sprouts. In addition, this technique will generate larger saplings in a shorter amount of time and will give rise to reproductive trees. We may remove the individual protection after the saplings reach the grazing height threshold (>2 m). Herbivores will then browse lower parts of the plant, operating as a pruning mechanism with no effect on upper parts [62]. Examples of these techniques are shown in Figure 3 and clearly deserve further investigation. The use of oak multi-stemmed individuals or stunted saplings to generate reproductive trees through protection and pruning could be more promising (both economically and environmentally) than planting new individuals. For instance, protecting oak multi-stemmed individuals and saplings instead of planting imported trees will reduce the risk of introducing important diseases (e.g., sudden oak death caused by *Phytophthora* spp.) that have been commonly brought through the introduction of plant and soil material from nurseries. Further research is needed to fully evaluate the costs and benefits of these proposed management practices.

(a)

Figure 3. *Cont.*

(b)

Figure 3. Protection of two stunted multi-stemmed individuals of oak (*Quercus ilex*) with a strong wire fence (2 m high). Only part of the oak multi-stemmed individuals was protected. Pictures show a comparison between the protected part and the original part (non-protected oak multi-stemmed individuals): (**a**) cylindrical shape of the protected part of the multi-stemmed individuals, showing an increase in height (>2 m high) and no pruning was carried out; and (**b**) small tree generated from a multi-stemmed individual after partial protection and pruning of the basal sprouts. The following Photos: Ramón Perea.

5. Conclusions

Livestock intensification (especially in high commercially competitive management with cattle grazing) applied to silvopastoral systems over the last century has affected the architecture of young oak plants, potentially reducing the probability of high acorn yield of future adult trees. This rangeland management practice jeopardizes the long-term system sustainability. Specifically, cattle-grazed "dehesas" feature considerably altered plant architecture by reducing 50% plant height–diameter ratios compared to traditional management with extensive sheep grazing. In contrast, young oak plants showed more natural architecture and less probability of damage on shoot apex in areas with deer grazing at moderate stocking rates. Shrub presence, for all management schemes, helped to increase plant height, except when shrubs were located under tree canopies.

Hence, we recommend sustainable practices such as cattle stocking rate reduction, traditional sheep grazing promotion, nurse shrub preservation and the fencing of stunted individuals along with pruning basal sprouts. Overall, our study highlights that management may have important consequences for dehesa regeneration via alterations of plant architecture and for the sustainability of the systems.

Author Contributions: Conceptualization, A.L.-S., R.P. and S.R.; formal analysis, A.L.-S.; data curation, A.L.-S.; writing—original draft preparation, A.L.-S.; writing—review and editing, R.P. and R.D.; supervision, R.P. and S.R. All authors have read and agreed to the published version of the manuscript.

Funding: This research was funded by a Study Fellowship ('FPU') of the Spanish Ministry of Education, Culture and Sport to A.L.-S., a Marie Curie fellowship from the European Union (FP7-PEOPLE-2013-IOF-627450) to R.P., and the project GLOBALFOR supported by UPM and Comunidad de Madrid (Convenio Plurianual for young researchers). RD was supported by Stanford's unrestricted funds to Dirzo Lab.

Institutional Review Board Statement: "Not applicable." Field studies did not involve endangered or protected species. Animals were only observed in the field, and were not captured or harmed in any way.

Data Availability Statement: The data presented in this study are openly available in the FigShare Repository at https://doi.org/10.6084/m9.figshare.14955483.v1 (accessed on 11 July 2021). López-Sánchez: Aida; Sonia Roig; Dirzo, Rodolfo; Perea, Ramón (2021): Effects of domestic and wild ungulate management on young oak size and architecture. figshare. Dataset. https://doi.org/10.6084/m9.figshare.14955483.v1 (accessed on 11 July 2021).

Acknowledgments: We thank the staff of "Dehesón del Encinar", especially Celia López-Carrasco and Antonio Ávila Cerrada, for their help to access the ranch and help to contact ranch owners. We thank Arturo Cobisa Pérez for access at the "Dehesa nueva" and Luis Fernando Gómez Calvo for access at the "Valdecasillas". We thank Irma Vasquez García for her assistance in the field.

Conflicts of Interest: The authors declare no conflict of interest.

References

1. Manning, A.D.; Fischer, J.; Lindenmayer, D.B. Scattered trees are keystone structures–implications for conservation. *Biol. Conserv.* **2006**, *132*, 311–321. [CrossRef]
2. Moreno, G.; Pulido, F.J. The functioning, management and persistence of dehesas. In *Agroforestry in Europe*; Springer: Berlin/Heidelberg, Germany, 2009; pp. 127–160.
3. Myers, N.; Mittermeier, R.A.; Mittermeier, C.G.; da Fonseca, G.A.; Kent, J. Biodiversity hotspots for conservation priorities. *Nature* **2000**, *403*, 853–858. [CrossRef] [PubMed]
4. Eichhorn, M.; Paris, P.; Herzog, F.; Incoll, L.; Liagre, F.; Mantzanas, K.; Mayus, M.; Moreno, G.; Papanastasis, V.; Pilbeam, D. Silvoarable systems in Europe–past, present and future prospects. *Agrofor. Syst.* **2006**, *67*, 29–50. [CrossRef]
5. Fernández-Alés, R.; Martin, A.; Ortega, F.; Ales, E.E. Recent changes in landscape structure and function in a Mediterranean region of SW Spain (1950–1984). *Landsc. Ecol.* **1992**, *7*, 3–18. [CrossRef]
6. Plieninger, T. Habitat loss, Fragmentation, and Alteration–Quantifying the Impactof Land-use Changes on a Spanish Dehesa Landscape by Use of Aerial Photography and GIS. *Landsc. Ecol.* **2006**, *21*, 91–105. [CrossRef]
7. Underwood, E.C.; Viers, J.H.; Klausmeyer, K.R.; Cox, R.L.; Shaw, M.R. Threats and biodiversity in the mediterranean biome. *Divers. Distrib.* **2009**, *15*, 188–197. [CrossRef]
8. Gibbons, P.; Lindenmayer, D.B.; Fischer, J.; Manning, A.D.; Weinberg, A.; Seddon, J.; Ryan, P.; Barrett, G. The future of scattered trees in agricultural landscapes. *Conserv. Biol.* **2008**, *22*, 1309–1319. [CrossRef]
9. Pulido, F.J.; Díaz, M.; de Trucios, S.J.H. Size structure and regeneration of Spanish holm oak Quercus ilex forests and dehesas: Effects of agroforestry use on their long-term sustainability. *For. Ecol. Manag.* **2001**, *146*, 1–13. [CrossRef]
10. Tyler, C.M.; Kuhn, B.; Davis, F.W. Demography and recruitment limitations of three oak species in California. *Q. Rev. Biol.* **2006**, *81*, 127–152. [CrossRef]
11. Plieninger, T.; Schaich, H.; Kizos, T. Land-use legacies in the forest structure of silvopastoral oak woodlands in the Eastern Mediterranean. *Reg. Environ. Chang.* **2011**, *11*, 603–615. [CrossRef]
12. Hostert, P.; Röder, A.; Hill, J.; Udelhoven, T.; Tsiourlis, G. Retrospective studies of grazing-induced land degradation: A case study in central Crete, Greece. *Int. J. Remote Sens.* **2003**, *24*, 4019–4034. [CrossRef]
13. Plieninger, T.; Pulido, F.J.; Konold, W. Effects of land-use history on size structure of holm oak stands in Spanish dehesas: Implications for conservation and restoration. *Environ. Conserv.* **2003**, *30*, 61–70. [CrossRef]
14. Cierjacks, A.; Hensen, I. Variation of stand structure and regeneration of Mediterranean holm oak along a grazing intensity gradient. *Plant Ecol.* **2004**, *173*, 215–223. [CrossRef]
15. López-Sánchez, A.; Perea, R.; Dirzo, R.; Roig, S. Livestock vs. wild ungulate management in the conservation of Mediterranean dehesas: Implications for oak regeneration. *For. Ecol. Manag.* **2016**, *362*, 99–106. [CrossRef]
16. Waller, D.M.; Alverson, W.S. The white-tailed deer: A keystone herbivore. *Wildl. Soc. Bull.* **1997**, 217–226.
17. Derner, J.D.; Lauenroth, W.K.; Stapp, P.; Augustine, D.J. Livestock as ecosystem engineers for grassland bird habitat in the western Great Plains of North America. *Rangel. Ecol. Manag.* **2009**, *62*, 111–118. [CrossRef]
18. Díaz, M.; Pulido, F.J.; Møller, A.P. Herbivore effects on developmental instability and fecundity of holm oaks. *Oecologia* **2004**, *139*, 224–234. [CrossRef]
19. Møller, A.P.; Shykoff, J.A. Morphological developmental stability in plants: Patterns and causes. *Int. J. Plant Sci.* **1999**, *160*, S135–S146. [CrossRef]
20. Milchunas, D.G.; Lauenroth, W.K. Quantitative effects of grazing on vegetation and soils over a global range of environments. *Ecol. Monogr.* **1993**, *63*, 327–366. [CrossRef]

21. Garbarino, M.; Bergmeier, E. 7 Plant and vegetation diversity in European wood-pastures. In *European Wood-Pastures in Transition: A Social-Ecological Approach*; Hartel, T., Plieninger, T., Eds.; Routledge: Abingon, UK, 2014; pp. 113–131.
22. Sanon, H.; Kaboré-Zoungrana, C.; Ledin, I. Behaviour of goats, sheep and cattle and their selection of browse species on natural pasture in a Sahelian area. *Small Rumin. Res.* **2007**, *67*, 64–74. [CrossRef]
23. Renaud, P.-C.; Verheyden-Tixier, H.; Dumont, B. Damage to saplings by red deer (*Cervus elaphus*): Effect of foliage height and structure. *For. Ecol. Manag.* **2003**, *181*, 31–37. [CrossRef]
24. Plieninger, T.; Wilbrand, C. Land use, biodiversity conservation, and rural development in the dehesas of *Cuatro lugares*, Spain. *Agrofor. Syst.* **2001**, *51*, 23–34. [CrossRef]
25. Plieninger, T.; Pulido, F.J.; Schaich, H. Effects of land-use and landscape structure on holm oak recruitment and regeneration at farm level in *Quercus ilex* L. dehesas. *J. Arid Environ.* **2004**, *57*, 345–364. [CrossRef]
26. McClaran, M.P.; Bartolome, J.W. Effect of Quercus douglasii (Fagaceae) on herbaceous understory along a rainfall gradient. *Madroño* **1989**, *36*, 141–153.
27. Dufour-Dror, J.-M. Influence of cattle grazing on the density of oak seedlings and saplings in a Tabor oak forest in Israel. *Acta Oecol.* **2007**, *31*, 223–228. [CrossRef]
28. López-Sánchez, A.; Schroeder, J.; Roig, S.; Sobral, M.; Dirzo, R. Effects of cattle management on oak regeneration in northern Californian Mediterranean oak woodlands. *PLoS ONE* **2014**, *9*, e105472. [CrossRef] [PubMed]
29. Rolo, V.; Plieninger, T.; Moreno, G. Facilitation of holm oak recruitment through two contrasted shrubs species in Mediterranean grazed woodlands. *J. Veg. Sci.* **2013**, *24*, 344–355. [CrossRef]
30. Perea, R.; Gil, L. Shrubs facilitating seedling performance in ungulate-dominated systems: Biotic versus abiotic mechanisms of plant facilitation. *Eur. J. For. Res.* **2014**, *133*, 525–534. [CrossRef]
31. Serrada, R.; San Miguel, A. *Selvicultura en Dehesas*; Serrada, R., Montero, G., Reque, J.A., Eds.; INIA-Fundación Conde del Valle de Salazar: Madrid, Spain, 2008; Volume 1178, pp. 861–876.
32. Rivas-Martínez, S.; Rivas-Saenz, S. Worldwide Bioclimatic Classification System 1996–2017. Available online: http://www.globalbioclimatics.org (accessed on 22 November 2017).
33. Perea, R.; López-Sánchez, A.; Roig, S. The use of shrub cover to preserve Mediterranean oak dehesas: A comparison between sheep, cattle and wild ungulate management. *Appl. Veg. Sci.* **2016**, *19*, 244–253. [CrossRef]
34. Escribano, M.; Rodríguez, A.; Mesías, F.J.; Pulido, F. Niveles de cargas ganaderas en la dehesa extremeña. *Arch. Zootec.* **2002**, *51*, 315–326.
35. Acevedo, P.; Ruiz-Fons, F.; Vicente, J.; Reyes-García, A.; Alzaga, V.; Gortázar, C. Estimating red deer abundance in a wide range of management situations in Mediterranean habitats. *J. Zool.* **2008**, *276*, 37–47. [CrossRef]
36. Perea, R.; Girardello, M.; San Miguel, A. Big game or big loss? High deer densities are threatening woody plant diversity and vegetation dynamics. *Biodivers. Conserv.* **2014**, *23*, 1303–1318. [CrossRef]
37. Perea, R.; Perea-García-Calvo, R.; Díaz-Ambrona, C.G.; San Miguel, A. The reintroduction of a flagship ungulate Capra pyrenaica: Assessing sustainability by surveying woody vegetation. *Biol. Conserv.* **2015**, *181*, 9–17. [CrossRef]
38. Zuur, A.F.; Ieno, E.N.; Walker, N.J.; Saveliev, A.A.; Smith, G.M. Mixed effects modelling for nested data. In *Mixed Effects Models and Extensions in Ecology with R*; Springer: Berlin/Heidelberg, Germany, 2009; pp. 101–142.
39. Box, G.E.; Cox, D.R. An analysis of transformations. *J. R. Stat. Soc. Ser. B Methodol.* **1964**, *26*, 211–252. [CrossRef]
40. Crawley, M.J. *The R Book*; John Wiley & Sons: Hoboken, NJ, USA, 2012.
41. Burnham, K.P.; Anderson, D.R. *Model Selection and Multi-Model Inference: A Practical Information-Theoretic Approach*, 2nd ed.; Springer: New York, NY, USA, 2002.
42. R Development Core Team. *R: A Language and Environment for Statistical Computing*; Foundation for Statistical Computing: Vienna, Austria, 2015.
43. Bates, D.; Maechler, M.; Bolker, B.; Walker, S. Fitting Linear Mixed-Effects Models Using lme4. *J. Stat. Softw.* **2015**, *67*, 1–48. [CrossRef]
44. Fox, J.; Weisberg, S. *An {R} Companion to Applied Regression*; Sage: Thousand Oaks, CA, USA, 2011.
45. Barton, K. *MuMIn: Multi-Model Inference. R Package Version 1.15.1*, 2015.
46. Pérez-Férnandez, M.A.; Gomez-Gutiérrez, J.M. Evolution of the tree cover (Quercus pyrenaica Willd and Quercus ilex subspecies ballota (Desf. Samp) in a dehesa over the last 100 years. *Cah. Options Méditerranéennes* **1995**, *12*, 259–266.
47. Zamora, R.; Gómez, J.M.; Hódar, J.A.; Castro, J.; García, D. Effect of browsing by ungulates on sapling growth of Scots pine in a Mediterranean environment: Consequences for forest regeneration. *For. Ecol. Manag.* **2001**, *144*, 33–42. [CrossRef]
48. Bartolome, J.W.; McClaran, M.P.; Allen-Diaz, B.H.; Ford, L.D.; Standiford, R.B.; McDougald, N.K.; Forero, L.C. Effects of fire and browsing on regeneration of blue oak. In *Oaks in California's Changing Landscapes, Proceedings of the 5th Symposium on Oak Woodlands*; Standiford, R., McCreary, D., Purcell, K., Eds.; General Technical Report PSW-GTR-184; PSRS, Forest Service: Albany, CA, USA; USDA: San Diego, CA, USA, 2001; pp. 281–286.
49. Aarssen, L.W. Hypotheses for the evolution of apical dominance in plants: Implications for the interpretation of overcompensation. *Oikos* **1995**, *74*, 149–156. [CrossRef]
50. Carmona, C.P.; Azcárate, F.M.; Oteros-Rozas, E.; González, J.A.; Peco, B. Assessing the effects of seasonal grazing on holm oak regeneration: Implications for the conservation of Mediterranean dehesas. *Biol. Conserv.* **2013**, *159*, 240–247. [CrossRef]

51. San Miguel, A. *La Dehesa Española: Origen, Tipología, Características y Gestión*; Fundación Conde del Valle Salazar: Madrid, Spain, 1994; p. 96.
52. Castro, J.; Zamora, R.; Hódar, J.A.; Gómez, J.M. Use of shrubs as nurse plants: A new technique for reforestation in Mediterranean mountains. *Restor. Ecol.* **2002**, *10*, 297–305. [CrossRef]
53. Gómez-Aparicio, L.; Zamora, R.; Gómez, J.M.; Hódar, J.A.; Castro, J.; Baraza, E. Applying plant facilitation to forest restoration: A meta-analysis of the use of shrubs as nurse plants. *Ecol. Appl.* **2004**, *14*, 1128–1138. [CrossRef]
54. Pulido, F.J.; Díaz, M. Regeneration of a Mediterranean oak: A whole-cycle approach. *Ecoscience* **2005**, *12*, 92–102. [CrossRef]
55. Pulido, F.; García, E.; Obrador, J.J.; Moreno, G. Multiple pathways for tree regeneration in anthropogenic savannas: Incorporating biotic and abiotic drivers into management schemes. *J. Appl. Ecol.* **2010**, *47*, 1272–1281. [CrossRef]
56. San Miguel, A. La vegetación arbórea. In *Manual de Buenas Prácticas de Gestión en Fincas de Montes Mediterráneo de la Red Natura 2000*; González, L.M., San Miguel, A., Eds.; Dirección General para la Biodiversidad; Ministerio de Medio Ambiente: Madrid, Spain, 2005; pp. 157–176.
57. Ramírez, J.A.; Díaz, M. The role of temporal shrub encroachment for the maintenance of Spanish holm oak *Quercus ilex* dehesas. *For. Ecol. Manag.* **2008**, *255*, 1976–1983. [CrossRef]
58. Mysterud, A. The concept of overgrazing and its role in management of large herbivores. *Wildl. Biol.* **2006**, *12*, 129–141. [CrossRef]
59. Gill, R.M.A.; Morgan, G. The effects of varying deer density on natural regeneration in woodlands in lowland Britain. *Forestry* **2010**, *83*, 53–63. [CrossRef]
60. Olea, L.; San Miguel, A. The Spanish Dehesa. A Traditional Mediterranean Silvopastoral System Linking Production and Nature Conservation. In *the 21st General Meeting of the European Grassland Federation*; European Grassland Federation: Kassel, Germany, 2006; pp. 3–11.
61. Martínez-Jaúregui, M.; Arenas, C.; Herruzo, A.C. Understanding long-term hunting statistics: The case of Spain (1972–2007). *For. Syst.* **2011**, *20*, 139–150. [CrossRef]
62. Perea, R. Bellotas: El esperado maná. *Trofeo. Caza Conserv.* **2012**, *511*, 60–65.

Article

Assessment of a Cocoa-Based Agroforestry System in the Southwest of Colombia

William Ballesteros-Possú [1,*], Juan Carlos Valencia [2] and Jorge Fernando Navia-Estrada [1]

1 Department of Natural Resources and Agroforestry Systems, Universidad de Nariño, Ciudadela Universitaria Torobajo, San Juan de Pasto 52001, Colombia; jornavia@gmail.com
2 Agricultural Sciences School, Universidad Abierta y a Distancia—UNAD, Tumaco 528509, Colombia; 77valencia@hotmail.es
* Correspondence: wballesterosp@gmail.com; Tel.: +57-315-4019-339

Abstract: Cocoa-based agroforests play an important role in farmer livelihood and the global environment; however, despite these facts, their low yields and tree aging put at risk their fate. This project investigated the carbon storage potential, productivity, and economics of different agroforestry arrangements of cocoa (*Theobroma cacao*) with Melina (*Gmelina arborea*) trees, in the southwest of Colombia. We established the experiment under a Randomized Complete Blocks design with seven treatments and three repetitions. Different allometric models were tested. Allometric models were made for *G. arborea* trees with dbh, ranging between 30.24 and 50.11 cm. The total carbon accumulation fluctuated between 49.2 (Treatment 4) and 88.5 t ha^{-1} (Treatment 2), soil organic matter (SOM) ranged between 9 and 17%, bulk density decreased from 0.83 to 0.77 g cm^{-3}. Cocoa yield ranged between 311 kg ha^{-1} year^{-1} (Treatment 7, traditional farm) and 922 kg ha^{-1} year^{-1} (Treatment 6). Treatment 6 showed the best performance with a net present value (NPV) of COP 1,446,467 (US $337.6), an internal rate of return (IRR) of 42%, and a cost-benefit ratio (B/C) of 1.67%. The benefits of AFS were also evidenced in some of the physical and chemical soil properties. Despite local marginality, these cocoa agroforest arrangements are a viable alternative to improve the traditional (local) cocoa systems because cacao agroforest arrangements increased cacao yield and carbon storage becoming a suitable alternative to improve traditional systems.

Keywords: climate change; tropical rain forest; traditional cocoa systems; allometric models; tree biomass

Citation: Ballesteros-Possú, W.; Valencia, J.C.; Navia-Estrada, J.F. Assessment of a Cocoa-Based Agroforestry System in the Southwest of Colombia. *Sustainability* **2022**, *14*, 9447. https://doi.org/10.3390/su14159447

Academic Editor: Antonio Boggia

Received: 8 June 2022
Accepted: 25 July 2022
Published: 2 August 2022

1. Introduction

Food security and climate change represent one of the most worrisome threats to the global environment and humanity due to the impacts caused on human health, biodiversity, food security, economy, natural resources, and infrastructure [1]. These troubles will significantly threaten human fate because of their social, environmental, and economic impacts and the high cost of implementing adaptation and mitigation measures [2]. Therefore, the 2030 agenda for 'Sustainable Development members' calls attention to protecting the planet from degradation and taking urgent action to tackle climate change concerns [3].

Fortunately, many of these hazards can be reduced by increasing resilience and resource-efficient use in agricultural production systems [4], which will improve the local farmers' adaptive capacity. Similarly, the FAO proposed a 'climate-smart agriculture' (CSA) approach, a tool to improve agricultural productivity and reduce yield variability over time under adverse climatic scenarios [4,5].

GHGs could be mitigated in two ways: by reducing anthropogenic CO_2 emissions or by creating and improving carbon sequestration sinks [6]. Agroforestry systems (AFS) and the mixture of crops and trees are a C sink and an alternative to mitigate the effect of the agricultural production systems [7] because they integrate agricultural landscapes with rural communities [8]. AFS, the art and science of farming with trees [9,10], is a multifunctional land-use system that offers environmental services in agricultural landscapes [9,11]. For instance, cacao agroforests preserve the natural soil biota, control soil erosion, and provide sustained land use land while maintaining suitable yields [12–14].

These systems can sequester and store atmospheric carbon, in agricultural lands at a low cost [7,8]. Growing windbreaks, shelterbelts, silvopastoral systems, forest farms, riparian buffers, woodlots, promoting natural regeneration of vegetation, and the conservation of forests make an enormous contribution to the local carbon budgets [11–18].

Trees in the AFS also provide numerous goods and environmental services that benefit farmers, landscapes, and infrastructure and reduce the susceptibility to extreme climatic events, making it a tool for Smart agriculture approaches [18–23]. Likewise, Agroforestry is considered a sustainable and efficient land management system, the AFS can contribute to the implementation of 9 out of the 17 SDGs, and special attention is paid to food security, climate change adaptation and mitigation, and biodiversity [24].

Environmentally sustainable practices are one of the highest priorities for agricultural systems in the tropics [25,26]. An overview of the main technologies currently used by cocoa growers, in Colombia, shows that cocoa growth under traditional production systems [27,28], with low technology, old cacao trees, and high shading. These conditions trigger fungal diseases such as frosty pod rot (*Moniliophthora roreri*) and Witch's broom (*Moniliophthora perniciosa*) [29], which reduce substantially cocoa yields. In this scenario, traditional farms average a yield of 300 kg ha/year [27,30]. The outcome of various researchers conclude that these systems are not viable for cocoa farmers [16]; however, other researchers identify important social and economic co-benefits of environmentally-friendly cocoa production regarding reduced pesticide use, resource conservation, long-term productivity, soil health, and human welfare [31].

On the other hand, cocoa farming, however, is associated with numerous environmental, social, and economic concerns, such as deforestation, child labor, and farmers' poverty [31], that lead to public and consumer pressure [32] and a growing claim for sustainably-produced cocoa [33]. Thus, cocoa traders and chocolate companies have started addressing farm-level sustainability through supply chain mechanisms [34].

Past research on cocoa sustainability has mostly focused only on child labor in the value cocoa chain [35], environmental impacts, and the relationship between environmentally-friendly farming practices and profitability [36].

Full-sun cocoa farming is currently the most widespread cocoa cultivation system in the world [37]; however, its negative effects including biodiversity loss, soil fertility depletion, and soil quality degradation, have incredibly received low attention. Some ecofriendly purposes must be studied to revert the ecological effects of full-sun cocoa systems.

T. cacao-G. arborea association has been little studied on the pacific coast of Colombia despite its good performance; thus, a better understanding of its economic and environmental effects is required [30], to offer different strategies to cocoa farmers' decision makers about the new cocoa production systems in the country.

The potential of AFS to store aboveground carbon is estimated to be 1.9×10^9 Mg C year^{-1} intemperate, and 2.1×10^9 Mg C year^{-1} in tropical biomes [38]. Carbon accumulation in monoculture of trees and crops shows 40% and 84% less than AFS, demonstrating that AFS is an important carbon sink in agricultural lands [39–41].

This research is focused on answering what are the carbon storage potential and profitability of different cocoa-based agroforestry arrangements compared to traditional cocoa systems? Therefore, the purpose of this research is to assess cocoa-agroforestry arrangements' potential to increase the yield and storage in atmospheric carbon compared to traditional cocoa systems. For that, understanding the agroforestry arrangements' performance was important for adoption purposes taking into account the 'view of the farmer' which facilitates the implementation of improved agroforest arrangements [42].

2. Materials and Methods

2.1. Study Area

The research was carried out in Tumaco, Nariño, southwest of Colombia, located at 01°39′12″ north latitude and 78°41′49″ west longitude (Figure 1). The experimental zone, located in a Tropical Rain Forest (bh-T) [43] had an average temperature of 27 °C, relative humidity of 80%, precipitation of 2800 mm/year, and a solar brightness of 1008 h/year [44].

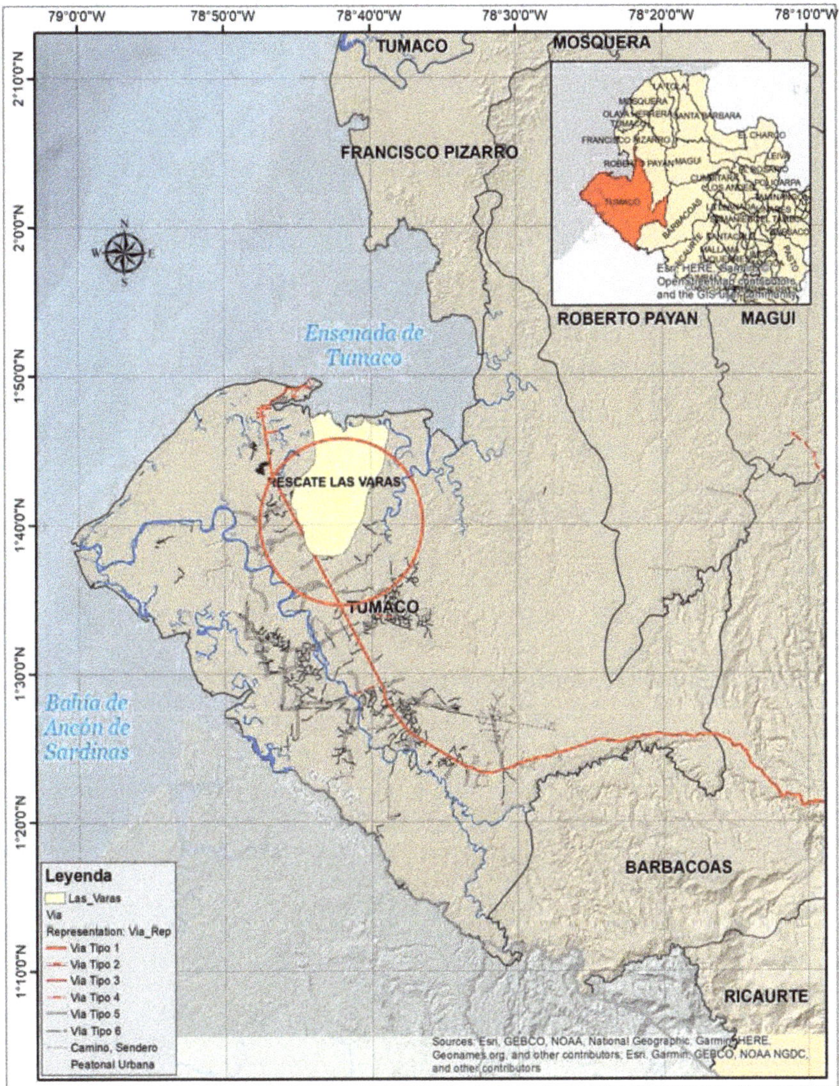

Figure 1. Study area: Consejo Comunitario Rescate Las Varas.

The agroforestry system was 10 years old, established under an experimental plot of four ha, with a Complete Randomized Blocks Design (CRBD), and encompassing three blocks and seven treatments. Each trial unit had 1600 m^2. Table 1 gives more details about treatments. The *G. arborea* trees have a dbh range between 30.24 and 54.11 cm, and a height range between 57.03 cm and 27.61 m, respectively.

Table 1. Treatments, spacing, and density of cacao and Melina trees in the experimental design.

Treatments	Cacao		Melina	
	Spacing (m^2)	Trees/ha	Spacing (m^2)	Trees/ha
1	3 × 3	1111	3 × 3	1111
2	4 × 4	625	4 × 4	625
3	3 × 4	833	8 × 6	208
4	3 × 4	833	12 × 6	139
5	3 × 4	833	16 × 6	104
6	3 × 4	833	3 × 3 × 7.5	440
7 *	7 × 7	200	12 × 12	69

* Traditional cocoa production system (control).

2.2. Tree Sampling

The tree sample size was 70 trees (35 of *G. Melina* and 35 of *T. cocoa*), five for treatment chosen randomly to replicate the whole range of diameters and height in the treatments. Tree biomass was attained according to McDicken, and Picard et al. approach [45,46].

We cut Melina trees (five for treatment) at ground level and immediately: (i) separated four fractions of aboveground biomass (i.e., stem, branches, twigs, and leaves); (ii) weighted each fraction (total fresh weight) using a hanging field-scale (Golden Lark model, precision 200 g), and (iii) collected a sample of 500 g from each fraction of aboveground biomass and stored each sample in a paper bag, which was subsequently weighed on a precision balance (Kern 440-55N, precision 0.2 g, Ebingen, Germany).

In the lab, the samples were dried at a temperature of 64 °C until they reached a constant weight. the moisture content of the sample was determined with Formula (1). These values were extrapolated to the biomass of different components of the cut trees; then, using the 0.47 conversion factor [47] biomass-carbon carbon stock was attained.

$$Biomass = \frac{Total\ fresh\ weight \times Dry\ weight}{Fresh\ weight} \times 100 \quad (1)$$

2.3. Data Processing

A total of 17 models were evaluated; eight equations were proposed by different authors (Table 2) and nine by the study authors. Some of these regression models included transformed data to estimate biomass based on trunk dbh and height. Although dbh is currently used for most local or regional biomass estimations, some researchers have suggested that both dbh and height should be included for larger-scale applications [48,49]. As such, we included height in some models for estimating biomass in these open-grown trees.

Before and after evaluating the best-fit model to the *G. arborea* biomass data, the homoscedasticity (constant variability of errors), normality errors, linear independence, and linearity of the residuals were tested. Homoscedasticity was accomplished by eliminating three observations (outliers) and applying the Breusch–Pagan test for constant variance. Normality was attained by using the QQ-plots and the normal curve, in this, some deviated points were observed in the tails, but they did not affect normality. Linear independence was found with Pearson's correlation coefficient, which was not significant between explanatory variables. Finally, linearity was in some cases met using the exact valor of lambda [51] or log transforming response variables. A correction factor prevented us from bias in biomass prediction when log-transformed variables are used [52,53] (Equation (2)).

$$CF = \exp\left(\frac{SEE^2}{2}\right) \quad or \quad \exp\left(\frac{variance}{2}\right) \quad (2)$$

where *CF* is the correction factor, *SEE* the standard error of regression estimate, and variance is the square of the root-mean-square error (RMSE2) in logarithmic form. Finally, the validation process consisted of two steps. First, the database was split into two parts, one for training, which corresponded to 80% of the data, and the other for testing, which retained the remaining 20%. Second, the final competing models were evaluated with the variance inflation factor (VIF), the R2, and the mean squared prediction error (MSPE) [54–56]. After selecting the best-fit equation, graphic analysis of the student residuals was performed to test for normality and heteroscedasticity in the errors.

Table 2. Allometric equations used to estimate the biomass potential of AFS.

Number	Author	Equation
1	Berkhout	bm = a + b × dbh
2	Kopezky	bm = a + b × dbh^2
3	Hohenadl-Krenn	bm = a + b × dbh + c × dbh^2
4	Husch	Ln bm = a + b × ln dbh
5	Spurr	bm = a + b × dbh^2 × ht
6	Stoate	bm = a + b × dbh^2 + c × dbh^2 × ht + d × ht
7	Meyer	bm = a + b × dbh^2 + c × dbh^2 × ht + d × dbh^2 × h
8	Schumacher-Hall	Log bm = a + b × ln dbh + c × ln × ht
9	Brenack	Log bm = a + b × dbh + c × 1/dbh
10	This study	Sqrt (bm) = a + b × dbh
11	This study	bm = a + b ln(dbh) + ln(ht)
12	This study	Sqrt (bm) = a + b × dbh + c × ht
13	This study	bm = a × exp(b × dbh)
14	This study	bm = a × dbh^b
15	This study	bm = a + b ln(dbh))

Source: Loetsch et al. [50]. bm = biomass (kg/tree) or C (kg/tree); dbh = diameter at the breast height (1.30 m), ht = total height (m); a, b, c, d = model parameters; ln = logarithm base e = 2.718282.

On the other hand, cocoa aboveground biomass was calculated by using a regression equation [57] which is based on the tree collar diameter at 30 cm as given in Equation (3).

$$LogAGB = -1.625 + 2.626 * log\ (D30) \quad (3)$$

where *AGB* = Aboveground biomass; *D*30 = tree diameter at 30 cm aboveground.

Cocoa belowground biomass was attained with Equation (4) [58].

$$BGB = \exp[-1.0587 + 0.8836 * Ln(AB) \quad (4)$$

where *BGB* = root biomass (t ha^{-1}); *Ln* = natural logarithm; exp: power base e.

$$AB = \text{Total aboveground biomass (t ha}^{-1}\text{) of cocoa}$$

Concerning cocoa yield, in all treatments, it was measured monthly.

2.4. Data Analysis

Tree carbon storage and cocoa yield were compared with the One-way ANOVA procedure and the adjusted Tukey test to detect differences among treatments. Data processing was carried out with SAS 9.3 and R 3.5.3. (R Core Team, Vienna, Austria) software [59].

2.5. Economic Analysis

Production cost and cocoa yield from each treatment was monthly registered, while *G. arborea* trees were valued in situ at USD 15.79. The economic analysis was performed on a hectare basis over the 10-year period using the Net Present Value (*NPV*) (5), the Internal Rate of Return (IRR) (6), and the Benefit-Cost Ratio (*BCR*) (7) indexes.

$$NPV = \sum_{t=1}^{n} \frac{CF_t}{(1+r)^t} - Co \quad (5)$$

$$\text{IRR} = 0 = NPV = \sum_{t=1}^{n} \frac{CF_t}{(1+r)^t} - Co \quad (6)$$

$$BCR = \frac{\sum PV\ benefits}{\sum PV\ costs} \quad (7)$$

where:

CF_t = net cash inflow-outflows during period t
r = internal rate of return that could be earned in alternative investments
t = time period cash flow is received
n = number of individual cash flows
Co = Total initial investment cost
PV = Present value

3. Results

3.1. Carbon Storage Potential of Agroforestry Arrangements

The agroforestry system with cocoa showed statistical differences (Pr < 0.0001) for aboveground C stored by the agroforestry system (*T. cocoa* and *G. arborea*). The highest C for the stock arrangements was treatment 7 (14.02 t Cha^{-1} $year^{-1}$), while for the improved arrangements, it was treatment 2 (11.32 t Cha^{-1} $year^{-1}$) (Figure 2C). The outcomes of the other treatments ranged between 8.73 and 6.29 tCha^{-1} $year^{-1}$; accordingly, the total carbon accounted (below- and aboveground) by the agroforestry systems in 10 years was between 140.226 and 62.85 tCha^{-1} (Figure 2D), which is equivalent to a 14.02 to 6.29 tCha^{-1} $year^{-1}$.

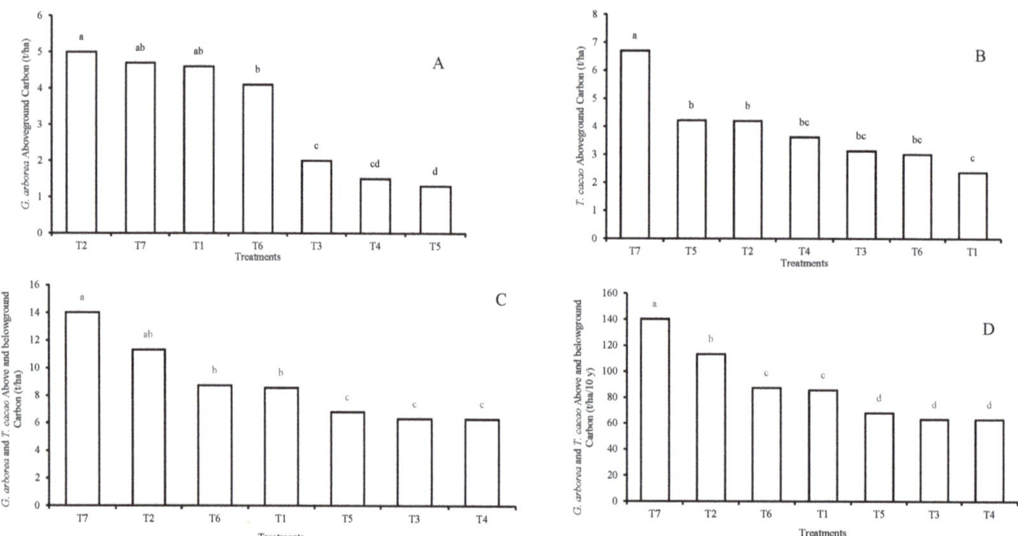

Figure 2. Aboveground carbon storage potential of *G. arborea* (**A**), and *T. cacao* (**B**) agroforestry arrangements. Aboveground carbon storage potential of agroforestry arrangements (**C**), and total above and belowground carbon storage potential of agroforestry system (**D**), a–d: statistical significance indices: "Means with the same letter are not significantly different".

G. arborea aboveground carbon had statistical differences ($p < 0.0001$) among treatments, treatment 2 displayed the highest amount of biomass stored with 5 t Cha^{-1} (Figure 2A). *T. cacao* aboveground biomass had statistical differences ($p < 0.0001$) among treatments too. Treatment 7 showed the highest potential (6.7 t Cha^{-1}) (Figure 2B). The difference in carbon stored by traditional cocoa agroforest compared to improved arrangements was around 20%.

3.2. Biomass Allometric Equations

Comparing different allometric models using data from destructively sampled *G. arborea* trees, 15 models fitted the data rationally well, fulfilling almost all information criteria (Table 3).

Table 3. Performance of the different competing allometric models in the information criteria indexes.

Model	R^2	AIC	BIC	PRESS	Cp	VIF	FI
1	0.93	399.7	404	560.3	14.4	1.0	3.9
2	0.93	401.6	406.4	55.8	16.5	1.0	59.1
3	0.93	400	405	48.2	5.8	14.5	33.8
4	0.91	399.8	404.5	153.6	2.00	1.0	65.8
5	0.93	70.8	66.00	134. 8	16.4	139.6	59.8
6	0.94	405.5	410.2	0.30	2.80	144.8	19.4
7	0.94	401.6	407.9	0.30	15.2	66.1	3.8
8	0.91	70.8	66.00	155.5	18.7	13.7	67.5
9	0.91	68.8	62.5	150.1	1.80	144.8	65.5
10	0.92	118.0	125.9	143.1	2.80	1.0	62.6
11	0.92	403.5	411.4	134.8	17.4	124.5	60.6
12	0.92	114.6	119.3	147.7	4.80	13.7	63.3
13	0.93	402.9	412.4	155.5	2.60	1.0	7.4
14	0.93	402.8	413.9	134.9	5.60	1.0	33.24
15	0.93	399.8	404.5	160	16.4	1.0	68.7

R^2 = Determination coefficient; AIC = Akaike information criteria; BIC = Bayesian information criteria; PRESS = predicted residual sum of squares; VIF = Variance inflation factor; FI = Furnival index.

According to the results in Table 3, models 1, 2, 4, 10, 13, 14, and 15 were selected. These had the highest determination coefficient, VIF less than 10, lowest Akaike information criteria (AIC); Bayesian information criteria (BIC); predicted residual sum of squares (PRESS); Mallows' Cp (Cp); Variance inflation factor (VIF) and Furnival index (FI). These competing models were tested to determine their accuracy using mean square prediction error (MSPE) (Table 4). The best predictor model was Berkouth with MSPE = 5.74 which was very close to the 'This study 3' model.

Table 4. Competing allometric models selected to make biomass estimates for *G. arborea*.

Model Number	Model [1]	R^2	α	β	MSPE	Rank
1	Berkouth	0.93	−534.339	33.282	5.74	1
14	This study 3	0.93	1.8048	1.6469	5.78	2
4	Husch	0.91	0.3982	1.6982	5.82	3
10	This study 1	0.92	4.64771	0.58219	5.84	4
2	Kopezky	0.93	141.54096	0.39794	5.89	5
15	This study 4	0.93	−4153.61	1346.99	6.11	6
13	This study 2	0.93	164	0.0386	6.21	7

[1] 35 trees for model training and 1 tree drawn at random (testing) without replacement to evaluate the precision of each model repeated in 8 different trees. α, and β are regression coefficients, R^2 = Adjusted determination coefficient, MSPE = mean square prediction error.

3.3. Cocoa Yield

The variance analysis showed significant statistical differences ($p > 0.0001$) in cocoa yield by the treatments. Treatments 2, 3, 4, 5, 6 and 1 showed the highest yields (922, 845, 838, 837, 831, 710 kg ha^{-1}) (Figure 3A); treatment 1 presented the lowest. When the improved cocoa arrangements were compared to traditional cocoa farms (treatment 7), the improved arrangements showed significant differences. The traditional cocoa farm had the lowest yield, with an average of 310.67 kg ha^{-1} $year^{-1}$.

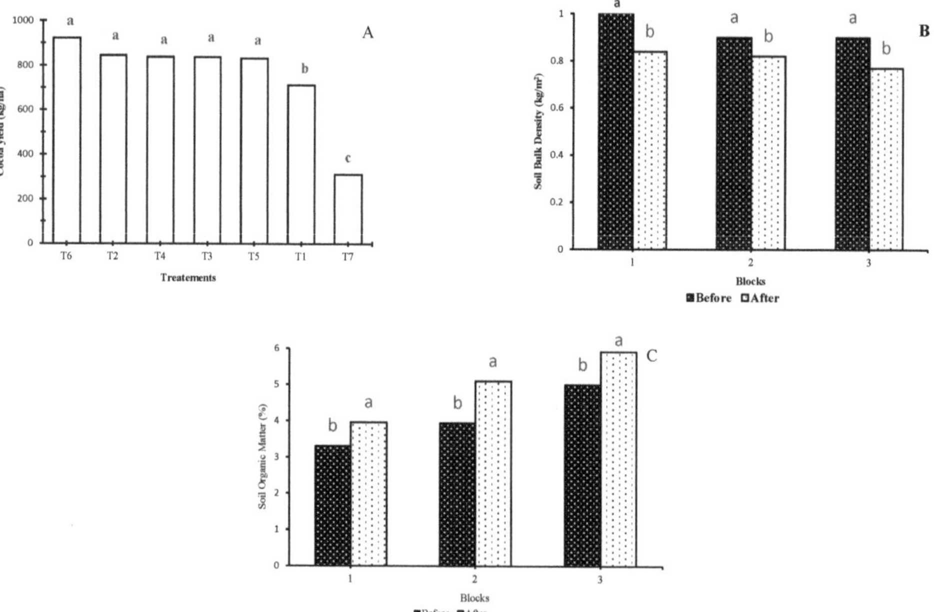

Figure 3. Average cocoa yield (**A**), soil bulk density (**B**), and soil organic matter (**C**) in a different AFS. a,b,c: Statistical significance indices: "Means with the same letter are not significantly different".

There was also a positive effect of all AFS in reducing soil bulk density (Figure 3B) where there were no significant differences in the treatments within the blocks, but between the three blocks studied, there was a reduction. This reduction was equivalent to an average of 23%. These results show that the AFS is an alternative for improving soil health.

In the soil organic matter (SOM), there were no significant differences in the treatments within the blocks but between the three blocks studied there was an increase (approx. 17%) in the amount of organic matter transferred by the AFS to the soil (Figure 3C).

3.4. Economic Analysis

The economic analysis shows that all treatments had positive values in the indicators analyzed (Table 5). Compared to all improved cocoa arrangements, treatment 6 offers the best profitability with an NPV of 1446.45 (Colombian pesos—$ COP), an IRR of 42%, and a BCR of 1.67. The lowest performance treatment was treatment 2 with an NPV of 1262, an IRR of 25%, and a BCR of 1.16. Relating these values with those obtained by the cocoa traditional farm (treatment 7), the traditional farm displayed even lower values in these indices (NPV 651.65, IRR 13%, and a BCR of 1.02).

Table 5. Metrics of the financial analysis to estimate the profitability of the different agroforestry arrangements in San Luis Robles—Tumaco, Nariño.

Treatments	NPV[1]	IRR (%)	BCR
1	1,737,024	26	1.14
2	1,262,004	25	1.16
3	1,308,798	30	1.22
4	1,302,763	29	1.20
5	1,287,711	29	1.18
6	1,446,457	42	1.67
7	651,650	13	1.02

NPV[1] = Net Present Value (Colombian pesos -$ COP); IRR = Internal Rate of Return; BCR = Benefit-Cost Ratio.

4. Discussions

4.1. Carbon Stored by the AFS

The carbon capture and storage potential depend on the agroforestry framework's structure and functions dictated by various socio-economic and environmental factors [60]. In this study, total biomass (aboveground and belowground) in all treatments was lower than the control (traditional coca farm). The conceivable explanation behind this could be the effect of microclimatic interactions in cocoa-based agroforestry systems, the spacing [61], the age/diameter, and the miscellaneous structure of traditional farms. It is reported that radiant energy and rainfall interception by foliage and the temperature and humidity of the air surrounding the foliage [62] impact plant growth. This study detected that treatments with fewer woody trees and cocoa (traditional farm) but with multiple understory species growing with them stored more biomass. In the same way, the traditional cocoa agroforest was older than all modern cocoa arrangements, which affected the carbon storage budget. Despite cocoa's high-density planting (HDP) being reported to reach higher cocoa yield than traditional ones (less than 700 trees/ha) [63], its effects on carbon storage are limited [64].

Many authors report the potential of C storage by *G. arborea*. For example, total biomass C stocks (above- and belowground) increased five-fold from the monoculture to the multi-shade tree system (from 11 to 57 Mg ha^{-1}), this increase was associated with a six-fold increase in aboveground biomass, but only a 3.5-fold increase in root biomass, indicating a clear shift in C allocation to aboveground tree organs with increasing shade for both cacao and shade trees [65].

G arborea plantation stored 41.6 t C ha^{-1} at one year and three months old, in Costa Rica [66]; 22.9 t C ha^{-1} at six years, 145.4 t C ha^{-1} at nine years old in Madhya Pradesh, India [67], 8.31 t ha^{-1} at 3 years old in Mexico [61], 15.54 t ha^{-1} at four-year-old, in Mexico [68], 24.39 t ha^{-1} at six years old in Colombia and [69], 41.6 t ha^{-1} all five years old [70]. The C storage potential from these outcomes ranges from 2.77 to 40 t C ha^{-1}, reflecting the effect of environment, site index, and spacing on the C stored by *G. arborea*. Therefore, the range from 16.2 to 20.9 t ha^{-1} yr^{-1} [71] does not encompass the total variability of the *G. melina* in different scenarios.

T. cacao carbon storage potential in AFS seems to be in the range of agroforestry systems. For instance, a 13-year-old gmelina-cacao agroforestry system stored 185 Mg C/ha in the Philippines [72], which is lower than a pure gmelina plantation (294 Mg/ha). This could be due to the spacing because it favored the development of cacao trees. However, we found that in the opposite case, the trees of treatment 1, despite having the highest planting density (1111 trees ha^{-1}), reported the lowest C storage with a potential of 2.36 t tCha^{-1} year^{-1}. One reason is that possibly the high humidity and the low luminosity in the tropical forest increased by the planting density caused more longitudinal growth than that in diameter, generating more foliage than wood. The average C storage potential ranging from 1.7 to 3.3 t tCha^{-1} year^{-1} [73,74] is lower than in this study.

Cocoa planted in very shady environments leads to lower rates of CO_2 assimilation and lower production of fruits and biomass. Regarding water availability, there are already reports that cocoa plants adjust their physiology [74]. Therefore, light and water are one of the factors that can favor or stress cocoa plants, becoming an important factor in plant growth and photosynthetic rate [75].

The high carbon stored in this study could partly be explained by the fast-growth tree rate and the high stand density in some of the AFS. Furthermore, site conditions in the study area support optimum growth for *G. arborea* and *T. cacao*. Therefore, it appears that the high nutrient requirement of *G. arborea* [76] is adequately met by the soils of the study area. This is to be expected since they were planted on humid tropical rainforest soils where biomass production is higher, and it is associated with soils with pH > 6 and C:N ratio > 20 [71]. Thus, soil organic matter is an important indicator of agricultural productivity and environmental resilience and is vital for soil structure stabilization, improvement of water-holding capacity, and retention of crucial plant nutrients [1].

The introduction of trees improves the potential of agricultural lands to store soil organic carbon [9,77–80]. Numerous authors report greater stocks of carbon in agroforestry than in field crops [9,81].

The values for bulk density found in this study corroborate the positive effects of AFS on soils. These values represent an improvement in the soil quality over the first 20 cm, which is very important to understory plants [22].

4.2. Allometric Models

We tested and proposed linear regression models of the square root, logarithm, and power of the biomass using coefficients extracted from tropical AFS biomass data. Our optimal models have only dbh as the independent variable, which met all statistical parameters [54]. Tree height was left out because of its effect on multicollinearity [54]. The models explain more than 90% of the sample's variance with MSPE lower than 7%. These results are within the range of those reported by similar studies in AFS [11,81]. The response variable without transformation got the best accuracy in biomass prediction. However, the power regression function (formula as: $y = a \times \exp(b \times dbh)$) had a prediction accuracy very close to untransformed data.

Logarithmic allometric models are used in tree biomass studies [79]; in this research, log-transformed model fitted the data well but with lower R2. The advantage of these models is that they can be derived by linear regression and be generalized more easily than linear or weighted regression models [79]. Their disadvantage is that the sum of component estimation usually does not equal estimates from real data and has to be adjusted when the response variable is log-transformed. In this case, the correction factor was 1.032. In some cases, logarithmic transformation of the dependent variables stabilizes the variance and fits well with the model meeting the information indexes' requirements [80,81].

The initial observation was that all the models generated in this study met accuracy criteria, having low AIC, BIC, PRESS, IF, IV, and very high R2 (Table 3). It indicates the independent variable (dbh or dbh-height) explained a very well high proportion of the variation in C.

Generic models met most of the information criteria. Unfortunately, when they were tested for variance inflation (VIF) and R2 indexes, some failed. High VIF values greater than 10 suggest that the predictor variables considered in a regression model are highly correlated with themselves and are synonymous with multicollinearity [80].

The allometric models showed a similar prediction of the total aboveground biomass with some models [81] but disagreed with others [41,82], which over- or underestimated the total aboveground biomass for trees with higher dbh. These outcomes call attention to reviewing the background of each model before using it. The two winner models performed the best in the validation process (Table 4), but some warning must be taken when using them. In this regard, before using any allometric model, it must be ensured that the data are within the validity range of the selected models [41].

From this call, it can be inferred that the derived allometric models can be correctly and efficiently used to estimate the aboveground biomass and carbon stocks in AFS. To sum up, there is an attractive opportunity for farmers to benefit financially from traditional and AFS with *G. arborea* if the cocoa growers can access the carbon markets because these arrangements had a high potential to produce environmental services.

4.3. Cocoa Yield in the Agroforestry Arrangements

The agroforestry arrangements evaluated showed variability in cocoa yield. The high planting densities (lower spacing) increased humidity triggering cocoa diseases and demanding high labor to control them [23,27]. The low yields of the traditional farm (treatment 7) indicate that traditional cocoa farms are not profitable if their only objective is cocoa yield [23,27]. However, if evaluated as a system, including all good and environmental services such as biodiversity, the results may differ. In this sense, some cacao growers state that the institutional trend to increase cacao planting densities goes against the traditional farmers' view because the low cocoa trees' density (400 trees ha^{-1}) allows them growth miscellaneous crops for their livelihood [23]. However, with some new practices, they can increase their yield to 1000 kg ha^{-1} keeping these traditional cocoa arrangements [83,84].

In the traditional system, the energy investment is less than commercial ones; for every 1 megajoule invested, 23 are produced (1:23 ratio) [27]. Therefore, the energy balance is highly efficient. On average, cacao yields were, 41% higher in monocultures, but the revenues derived from traditional cocoa agroforest economically overcompensated for this difference. Indeed, the return on labor across the years was roughly twice higher in the agroforestry systems than in the monocultures [84].

When analyzing yield in each of the treatments, the same trend can be observed with the traditional farm reporting the lowest values. However, in the dry season, which occurred in the municipality between July to October, there was a slight separation of the treatments, suggesting that the shading produced a differential effect.

For good performance and yield of cocoa-based systems, the companion trees should provide a shade below 50%; the most recommended is 30% [85] because shade greater than 60% causes yield limitations [86,87]. However, cacao has been traditionally grown under shade regardless of yield [86]. In an experiment in Ghana, strong reductions in yield were observed due to shading. Cacao trees growing in full sunlight had a higher yield than the shading ones [87]. Conversely, scientists and environmentalists alike are beginning to realize this switch's error for the ecology of cocoa plantations [88]. Canopy cover-up to 39% raised yields at the field level compared to cacao grown in full sun, but increasing shade above this level can limit yields [89,90]. Whereas in Cameroon, increasing canopy cover-up to 47% enhanced yield but greater than 60% reduced cocoa yield [91].

Therefore, growing in sunlight will require more intensive management, including increased fertilizer inputs [85]. Shading may not be required in all cacao-farm regions, its applicability is dependent on local climate conditions. That is why the control of both the growth and density of cacao trees, the increases in aeration, light perception, and fertilization [88,92] are necessary to raise the yield.

4.4. Economic Assessment of the Agroforestry Arrangements

The low profitability of the traditional cocoa agroforests agrees with most studies [93]. They indicate that traditional cocoa-based AFS with few forest species, and few investments generate low incomes. However, lower yields, in some cases, can be offset by cacao's premium prices [94]. This is an important finding because traditional/indigenous production systems make a high contribution to the environment and rural welfare, more economic research is needed. In other scenarios, the revenues derived from agroforestry by-crops, economically overcompensate due to self-consumption of companion crops, such as bananas, fruits, herbs, and peach palm, among others [94,95]. There is a bias, when assessing cacao production systems because most studies focus on cacao yields without quantitatively evaluating the economic returns from companion crops [86]. Other drawbacks of the cacao farms' competitiveness are a lack of access to the market and low density of trees (400 per ha). Therefore, despite their potential to increase farmers' revenues, crop products have been mainly used for self-consumption [96].

Results show that cacao yields were, on average, 41% higher in improved cacao agroforestry, but the revenues from agroforestry companion crops have great potential. The explanation about high IRR in treatment 6 could be the result of the optimum performance of *G. arborea* trees under these spacing and alley arrangements and that *G. arborea* is a sunlight-demanding, drought-resistant tree, which has fairly good tolerance to moist soils [97]. Under these conditions on the Colombian pacific coast, *G. arborea* trees reached dbh and heights higher than 50 cm and 11 m, respectively, which were sold at $70.000 COP, (USD 17.5). Indeed, the best economic performance was attained with intermediate diversity (three to seven species) and a density of 100 to 400 companion plants ha^{-1}) [98]. The same trend was reported in another study where *G. arborea* agroforest was 2–3 times more profitable in terms of net present value compared to sole *G. arborea* or mango or pigeon pea [99], these tendencies are very close to the results in this study. When comparing the cost of all improved cocoa arrangements they had no significant differences; however, when compared to traditional systems the differences were outstanding. Traditional agroforestry budgets are very low (1 million COP, USD 250, approximately), especially for labor, which allows farmers to develop miscellaneous activities and the use of on-farm inputs. Sustainable production systems such as agroforestry and organically managed systems are expected to yield less cacao but by-crops and premium prices, respectively, might economically compensate for the lower yields [84].

The IRR of 41% was high compared to the banks' discount rate but there is no information about the IRR for other agricultural systems in the region.

These findings provide a roadmap for optimizing crop diversity and density to adjust the economic balance between sales and self-consumption in cocoa AFS.

Finally, if agroforestry is to be used in carbon capture and storage schemes such as the clean development mechanisms, better information is required to determine above- and belowground carbon stocks and soil carbon in agroforestry systems landscapes [100]. Thus, there is a need for biomass estimation, which is sensitive to a specific arrangement, objectives, spacing, companion plants, and management practices. The AFS assessed in this study could be a good alternative for the Colombian government to full fill its commitments to COP 21 and the Sustainable Development Goals– SDG [100], or the 2030 agenda in SDG 13 (climate action) [101].

Further research is needed to investigate the impact of different ways of integrating miscellaneous trees with cocoa on wider environmental, social and economic sustainability aspects, that are driving increasing interest in the SDGs, climate change, and food security [102–104]. Likewise, it is needed to design the strategies for adopting [105] these results in the region.

5. Conclusions

The carbon stock in the different agroforestry arrangements was statistically different. The lowest accumulation happened in treatment 4 and the highest in treatment 2. To estimate the biomass produced in the different agroforestry arrangements, two allometric models were developed for *G. arborea* based on dbh. Cocoa yield showed significant statistical differences in the different arrangements. The proposed agroforestry arrangements had positive effects on soil bulk density (Da) and organic matter, the first was reduced and the second had a significant increase. All agroforestry arrangements had positive financial benefits. Overall, treatment 6 presented the best performance in all financial indexes.

Author Contributions: Conceptualization, W.B.-P. and J.C.V.; methodology, W.B.-P., J.C.V. and J.F.N.-E.; software, W.B.-P.; validation, W.B.-P., J.C.V. and J.F.N.-E.; formal analysis, W.B.-P.; investigation, W.B.-P., J.C.V. and J.F.N.-E.; resources, J.F.N.-E.; data curation, W.B.-P.; writing—original draft preparation, J.C.V.; writing—review and editing, W.B.-P. and J.F.N.-E.; visualization, J.F.N.-E. and J.C.V.; supervision, J.F.N.-E.; project administration, W.B.-P.; funding acquisition, J.F.N.-E. All authors have read and agreed to the published version of the manuscript.

Funding: This research was funded by the Project 'Fondo Colombiano de Ciencia Tecnología e Innovación-FCTI del Sistema General de Regalías, grant number BEPIN 2015000100001' and 'The APC was funded by MDPI discount vouchers'.

Institutional Review Board Statement: Not applicable.

Informed Consent Statement: Not applicable.

Acknowledgments: We would like to thank all the people who collaborated in this study: mainly the farmers of the Rescate las Varas Village, and the countless people who cooperated selflessly in the fieldwork, including the Cortepaz cocoa growers' association.

Conflicts of Interest: The authors declare no conflict of interest.

References

1. Wheeler, T.; von Braun, J. Climate change impacts on global food security. *Science* **2013**, *341*, 508. [CrossRef] [PubMed]
2. IPCC. *Climate Change: Synthesis Report. Contribution of Working Groups I.; II and III to the Fifth Assessment Report of the Intergovernmental Panel on Climate, Change*; Core Writing Team, Pachauri, R.K., Meyer, L.A., Eds.; IPCC: Geneva, Switzerland, 2014; 151p.
3. United Nations Organization—UNO. Sustainable Development Goals. Available online: https://www.un.org/sustainabledevelopment/sustainable-development-goals/ (accessed on 23 January 2021).
4. Lipper, L.; Thornton, P.; Campbell, B.; Baedeker, T.; Braimoh, A.; Bwalya, M.; Caron, P.; Cattaneo, A.; Garrity, D.; Henry, K.; et al. Climate-smart agriculture for food security. *Nat. Clim. Chang.* **2014**, *4*, 1068–1072. [CrossRef]
5. FAO. *Climate-Smart Agriculture Case Studies. Successful Approaches from Different Regions*; FAO: Rome, Italy, 2018; 44p.
6. Anguiano, J.M.; Aguirre, J.; y Palma, J.M. Cunnigham y Pennisetum purpureum. Cuba CT—115. *Av. Investig. Agropecu.-Aia* **2013**, *17*, 149–160.
7. Udawatta, R.P.; Jose, S. Carbon sequestration potential of agroforestry practices in temperate North America. In *Carbon Sequestration Potential of Agroforestry Systems: Opportunities and Challenges, Advances in Agroforestry*, 1st ed.; Kumar, B., Mohan Nair, P.K.R., Eds.; Springer: Dordrecht, The Netherlands, 2011; Volume 17, pp. 17–42.
8. Pardos, J.A. *Los Ecosistemas Forestales y el Secuestro de Carbono Ante el Cambio Climático*; Instituto Nacional de Investigación y Tecnología Agraria y Alimentaria (INIA): Madrid, Spain, 2010; 253p.
9. Van Noordwijk, M.; Coe, R.; Sinclair, F.L. *Central Hypotheses for the Third Agroforestry Paradigm within Common Definition*; Working Paper, No 233; World Agroforestry Center (ICRAF) Southeast Asia Regional Program: Bogor, Indonesia, 2016. Available online: https://www.cabdirect.org/cabdirect/abstract/20163398164 (accessed on 8 February 2021).
10. Torquebiau, E. A renewed perspective on agroforestry concepts and classification. *Life Sci.* **2000**, *323*, 1009–1017. [CrossRef]
11. Nair, P.K.R. Carbon sequestration studies in agroforestry systems: A reality-check. *Agrofor. Syst.* **2012**, *86*, 243–253. [CrossRef]
12. Alvarado, J.; Andrade, H.; Segura, M. Almacenamiento de carbono orgánico en suelos en sistemas de producción de café (*Coffea arabica* L.) en el municipio del Líbano, Tolima, Colombia. *Colomb. For.* **2013**, *16*, 21–31.
13. Nair, P.K.R.; Nair, V.D.; Kumar, B.M.; Showalter, J. Carbon sequestration in agroforestry systems. *Adv. Agron.* **2010**, *108*, 237–307. [CrossRef]
14. Brandle, J.R.; Hintz, D.L.; Sturrock, J.W. *Windbreak Technology*; Elsevier: Amsterdam, The Netherlands, 1988; 598p.
15. Koohafkan, P.; Altieri, M.A. *Globally Important Agricultural Heritage Systems: A Legacy for the Future*; UN-FAO: Rome, Italy, 2010; 41p.
16. Arévalo-Gardini, E.; Canto, M.; Alegre, J.; Loli, O.; Julca, A.; Baligar, V. Changes in soil physical and chemical properties in long term improved natural and traditional agroforestry management systems of cacao genotypes in Peruvian Amazon. *PLoS ONE* **2015**, *10*, e0132147. [CrossRef]
17. Arias, R.M.; Heredia, A.G. Fungal diversity in coffee plantation systems and in a tropical montane cloud forest in Veracruz, Mexico. *Agrofor. Syst.* **2014**, *88*, 921–933. [CrossRef]
18. Solis, R.; Vallejos-Torres, G.; Arévalo, L.; Marín-Díaz, J.; Ñique-Alvarez, M.; Engedal, T.; Bruun, T. Carbon stocks and the use of shade trees in different coffee growing systems in the Peruvian Amazon. *J. Agric. Sci.* **2020**, *158*, 450–460. [CrossRef]
19. Mortimer, R.; Saj, S.; David, C. Supporting and regulating ecosystem services in cacao agroforestry systems. *Agrofor. Syst.* **2018**, *92*, 1639–1657. [CrossRef]
20. Saj, S.; Durot, C.; Mvondo Sajouma, K.; Tayo Gamo, T.; Avana-Tientcheu, M.L. Contribution of associated trees to long-term species conservation, carbon storage and sustainability: A functional analysis of tree communities in cacao plantations of Central Cameroon. *Int. J. Agr. Sustain.* **2017**, *15*, 282–302. [CrossRef]
21. Toledo-Hernández, M.; Wanger, T.C.; Tscharntke, T. Neglected pollinators: Can enhanced pollination services improve cocoa yields? A review. *Agr. Ecosyst. Environ.* **2017**, *247*, 137–148. [CrossRef]
22. Murray, N.R.M.; Bojórquez, S.J.; Hernández, J.A.; Orozco, M.G.; García., J.D.; Gómez, A.R.; Ontiveros, G.H.; Aguirre, O.J. Efecto de la materia orgánica sobre las propiedades físicas del suelo en un sistema agroforestal de la llanura costera norte de Nayarit, México. *Rev. Bio Cienc.* **2011**, *1*, 27–35.

23. Schroth, G.; Jeusset, A.; da Silva Gomez, A.; Florence, C.T.; Pinto Coelho, A.; Faria, D.; Laderach, P. Climate friendliness of cocoa agroforests is compatible with productivity increase. *Mitig. Adapt. Glob. Chang.* **2016**, *21*, 67–80. [CrossRef]
24. Andersson, L. Achieving the Global Goals through Agroforestry. Stocolm. Available online: http://agroforestrynetwork.org.hemsida.eu/wp-content/uploads/2018/09/Achieving-the-Global-Goals-through-agroforestry.pdf (accessed on 18 July 2022).
25. Harvey, C.A.; Komar, O.; Chazdon, R.; Ferguson, B.G.; Finegan, B.; Griffith, D.M.; Martinez-Ramos, M.; Morales, H.; Nigh, R.; Soto-Pinto, L.; et al. Integrating agricultural landscapes with biodiversity conservation in the Mesoamerican hotspot. *Conserv. Biol.* **2008**, *22*, 8–15. [CrossRef]
26. Nair, P.K.R. Agroforestry systems and environmental quality: Introduction. *J. Environ. Qual.* **2011**, *40*, 784–790. [CrossRef]
27. Preciado, O.; Ocampo, C.; Possú, W.B. Caracterización del sistema de producción tradicional de cacao de cacao (*Theobroma cacao* L.) en seis núcleos productivos del municipio de Tumaco, Nariño. *Rev. Cienc. Agrícolas* **2011**, *20*, 58–69.
28. Espinosa-Alzate, J.A.; Ríos-Osorio, L.A. Caracterización de sistemas agroecológicos para el establecimiento de cacao (*Theobroma cacao* L.); en comunidades afrodescendientes del Pacífico Colombiano (Tumaco-Nariño, Colombia). *Acta Agronómica* **2016**, *65*, 211–217. [CrossRef]
29. Cardenas-Pardo, N.J.; Darghan, A.; Sosa-Rico, M.D.; Rodriguez, A. Spatial analysis of diseases incidence in different cocoa genotypes (*Theobroma cacao* L.) in Yopal (Casanare).; Colombia. *Acta Biol. Colomb.* **2017**, *22*, 209–220. [CrossRef]
30. Agronet. Agricultural Statistics, Cocoa, Production, Yield Share in the Department. Available online: https://www.agronet.gov.co/Documents/6-CACAO_2017.pdf (accessed on 8 February 2021).
31. Tennhardt, L.; Lazzarini, G.; Weisshaidinger, R.; Schader, C. Do environmentally-friendly cocoa farms yield social and economic co-benefits? *Ecol. Econ.* **2022**, *197*, 107428. [CrossRef]
32. Fountain, A.; Huetz-Adams, F. Cocoa Barometer. 2018. Available online: https://voicenetwork.cc/wp-content/uploads/2019/07/2018-Cocoa-Barometer.pdf (accessed on 18 July 2022).
33. Mithöfer, D.; Roshetko, J.M.; Donovan, J.A.; Nathalie, E.; Robiglio, V.; Wau, D.; Blare, T. Unpacking 'sustainable' cocoa: Do sustainability standards, development projects and policies address producer concerns in Indonesia, Cameroon and Peru? *Int. J. Biodivers. Sci. Ecosyst. Serv. Manag.* **2017**, *13*, 444–469. [CrossRef]
34. Meier, C.; Sampson, G.; Larrea, C.; Schlatter, B.; Voora, V.; Dang, D.; Bermudez, S.; Wozniak, J.; Willer, H. *The State of Sustainable Markets 2020: Statistics and Emerging Trends*; ITC: Geneva, Switzerland, 2020.
35. Grabs, J.; Carodenuto, S.L. Traders as sustainability governance actors in global food supply chains: A research agenda. *Strateg. Environ.* **2021**, *30*, 1314–1332. [CrossRef]
36. Busquet, M.; Bosma, N.; Hummels, H. A multidimensional perspective on child labor in the value chain: The case of the cocoa value chain in West Africa. *World Dev.* **2021**, *146*, 105601. [CrossRef]
37. Ebagnerin Tondoh, J.; N'guessan Kouamé, F.; Martinez Guéi, A.; Sey, B.; Wowo Koné, A.; Gnessougou., N. Ecological changes induced by full-sun cocoa farming in Côte d'Ivoire. *Glob. Ecol. Conserv.* **2015**, *3*, 575–595. [CrossRef]
38. Goswami, S.; Verma, K.S.; Kaushal, R. Biomass and carbon sequestration in different agroforestry systems of a Western Himalayan watershed. *Biol. Agric. Hortic.* **2013**, *30*, 88–96. [CrossRef]
39. Oelbermann, M.; Voroney, R.P.; Gordon, A.M. Carbon sequestration in tropical and temperate agroforestry systems: A review with examples from Costa Rica and southern Canada Agriculture. *Ecosyst. Environ.* **2004**, *104*, 359–377. [CrossRef]
40. Dhyani, S.K.; Ram, A.; Dev, I. Potential of agroforestry systems in carbon sequestration in India. *Indian J. Agric. Sci.* **2016**, *86*, 1103–1112.
41. Ballesteros Possu, W.; Brandle, J.R.; Domke, G.M.; Schoeneberger, M.; Blankenship, E. Estimating carbon storage in windbreak trees on U.S. agricultural lands. *Agrofor. Syst.* **2016**, *90*, 889–904. [CrossRef]
42. Chirwa, P.W.; Mala, W. Trees in the landscape: Towards the promotion and development of traditional and farm forest management in tropical and subtropical regions. *Agrofor. Syst.* **2016**, *90*, 555–561. [CrossRef]
43. Holdridge, L. *Ecología Basada en Zonas de Vida*; IICA: San José, Costa Rica, 1982; p. 216.
44. Instituto de Hidrología, Meteorología y Estudios Ambientales-IDEAM. Cartas Climatológicas del Aeropuerto la Florida, Tumaco. Nariño. Available online: www.ideam.gov.co (accessed on 10 March 2020).
45. MacDicken, K.G. *A guide to monitoring carbon storage in forestry and agroforestry projects. Report of the forest carbon monitoring program*; Winrock Internationl Institute for Agricultural Development: Arlington, VA, USA, 1997; p. 92.
46. Picard, N.; Saint-André, L.; Henry, M. *Manual for Building Tree Volumen and Biomass Allometric Equations: From Field Measurement to Prediction*; Food and Agricultural Organization of the United Nations: Rome, Italy; Centre de Coopération Internationale en Recherche Agronomique pour le Développement: Montpellier, France, 2012; p. 215.
47. Skog, K.E.; Nicholson, G.A. Carbon cycling through wood products: The role of wood and paper products in carbon sequestration. *For. Prod. J.* **1998**, *48*, 75.
48. Jenkins, J.C.; Birdsey, R.A.; Pan, Y. Biomass and NPP estimation for the mid-Atlantic region (USA) using plot-level forest inventory data. *Ecol. Appl.* **2001**, *11*, 1174–1193. [CrossRef]
49. Crow, T.R. Biomass and production in three contiguous forests in northern Wisconsin. *Ecology* **1978**, *59*, 265–273. [CrossRef]
50. Loetsch, F.; Zohrer, F.; Haller, K.E. *Forest Inventory*; BLV Verlagsgesellschaft: Munchen, Germany, 1973; p. 469.
51. Box, G.E.P.; Cox, D.R. An Analysis of Transformations. *J. R. Stat. Soc. Ser. B* **1964**, *26*, 211–252. [CrossRef]
52. Furnival, G.M. An index for comparing equations used in constructing volume tables. *For. Sci.* **1961**, *7*, 337–341.

53. Nakamura, T.; Judd, K.; Mess, M.I.; Small, M. A comparative Study of Information Criteria for Model Selection. Available online: http://staffhome.ecm.uwa.edu.au/~{}00027830/pdf/IJBC16-2.pdf (accessed on 9 February 2021).
54. Kutner, M.H.; Nachtsheim, C.J.; Neter, J. *Applied Linear Regression Models*, 4th ed.; Irwin Inc.: New York, NY, USA, 2004; p. 701.
55. Whitesell, C.D.; Miyasaka, S.C.; Strand, R.F.; Schubert, T.H.; McDuSe, K.E. *Equations for Predicting Biomass in 2- to 6-year-old Eucalyptus Saligna in Hawaii;* USAD Forest Service Research Notes, PSW-402; Pacific Southwest Forest and Range Experiment Station: Albany, CA, USA, 1988; p. 5.
56. Busing, R.T.; Clebsch, E.E.C.; White, P.S. Biomass production of southern Appalachian cove forests reexamined. *Can. J. For. Res.* **1993**, *23*, 60–65. [CrossRef]
57. Andrade, H.; Segura, M.; Somarriba, E.; Villalobos., M. Valoración biofísica y financiera de la fijación de carbono por uso del suelo en fincas cacaoteras indígenas de Talamanca, Costa Rica. *Agroforestería En Las Américas* **2008**, *46*, 45–50.
58. Cairns, M.A.; Brown, S.; Helmer, E.H.; Baumgardner, G.A. Root biomass allocation in the world's upland forests. *Oecologia* **1997**, *111*, 1–11. [CrossRef]
59. R Core Team. *R: A Language and Environment for Statistical Computing;* R Foundation for Statistical Computing: Vienna, Austria, 2013. Available online: https://www.R-project.org/ (accessed on 9 February 2020).
60. Albrecht, A.; Kandji, S.T. Carbon sequestration in tropical agroforestry systems. *Agric. Ecosyst. Environ.* **2003**, *99*, 15–27. [CrossRef]
61. Telles Antonio, R.; Alanís Rodríguez, E.; Jiménez Pérez, J.; Aguirre Calderón, O.A.; Treviño Garza, E.J. Accumulated carbon estimation in *Gmelina arborea* Roxb. from Tlatlaya.; Estado de México with allometric equations. *Rev. Mex. Cienc. For.* **2019**, *10*, 135–153. [CrossRef]
62. Monteith, J.; Chin, L.; Ong, K.; Corlett, J.E. Microclimatic interactions in agroforestry system. *For. Ecol. Manag.* **1991**, *45*, 31–44. [CrossRef]
63. Ayegboyin Kayode, O.; Famaye Amos, O.; Adeosun Seun, A.; Idrisu, M.; Ugioro, O.; Asowata, F.E.; Okunade Favour, A. Effect of high density planting on the vigour and yield of Theobroma cacao L. in the Southwest of Nigeria. *World J. Adv. Res. Rev.* **2020**, *8*, 217–223. [CrossRef]
64. Spaggiari Souza, C.A.; dos Santos Dias, L.A.; Galeas Aguila, M.A.; Songheti, S.; Oliveira, J.; Andrade Costa, J.L. Cacao yield in different planting densities. *Braz. Arch. Biol. Technol.* **2009**, *52*, 1313–1320. [CrossRef]
65. Abou Rajab, Y.; Leuschner, C.; Barus, H.; Tjoa, A.; Hertel, D. Cacao cultivation under diverse shade tree cover allows high carbon storage and sequestration without yield losses. *PLoS ONE* **2016**, *11*, e0149949. [CrossRef] [PubMed]
66. Vega, Y. Análisis mensual de acumulación de biomasa y fijación de carbono en una plantación de Gmelina arborea Roxb. Los Chiles, Alajuela, Costa Rica. Ph.D. Thesis, Ingeniero forestal-Instituto tecnológico de Costa Rica, Escuela de Ingeniería Forestal Costa Rica, Alajuela, Costa Rica, 2016; p. 39.
67. Bohre, P.; Chaubey, O.P.; Singhal, P.K. Biomass Accumulation and Carbon Sequestration in *Tectona grandis* Linn. f. and *Gmelina arborea* Roxb. *Int. J. Bio-Sci. Bio-Technol.* **2013**, *5*, 153–174.
68. Cámara, C.D.C.; Arias, M.; Martínez, S.J.L.; Castillo, A.O. Carbono almacenado en selva mediana de *Quercus oleoides* y plantaciones de *Eucalyptus urophylla* y *Gmelina arborea* en Huamanguillo, Tabasco. In *Estado Actual del Conocimiento del Ciclo del Carbono y Sus Interacciones en México: Síntesis a 2013*; Pellat, F.P., Julio, W.G., Maira y, B., Saynes, V., Eds.; Universidad Autónoma de Chapingo: Texcoco, Mexico, 2013; p. 702.
69. Melo Cruz, O.A. Modelación del Crecimiento.; Acumulación de Biomasa y Captura de Carbono en Árboles de *Gmelina arborea* Roxb., Asociados a Sistemas Agroforestales y Plantaciones Homogéneas en Colombia. Ph.D. Thesis, Facultad de Ciencias Agrarias, Universidad Nacional de Colombia, Sede Medellín, Medellín, Colombia, 2015; 166p.
70. Patiño Forero, S.; Suárez Santos, L.N.; Andrade Castañeda, H.J.; Segura Madrigal, M.A. Captura de carbono en biomasa en plantaciones forestales y sistemas agroforestales en Armero-Guayabal, Tolima, Colombia. *Rev. De Investig. Agrar. Y Ambient.* **2018**, *9*, 121–134. [CrossRef]
71. Onyekwelu, C.J. Above-ground biomass production and biomass equations for even-aged *Gmelina arborea* (ROXB) Plantations in South-Western Nigeria. *Biomass Bioenergy* **2004**, *26*, 39–46. [CrossRef]
72. Lasco, R.; Come, R.; Estrella, R.; Saplaco, S.R.; Castillo, A.S.A.; Cruz, R.; Pulhin, F. Carbon stock assessment of two agroforestry systems in a tropical forest reserve in the Philippines. *Philipp. Agric. Sci.* **2001**, *84*, 401–407.
73. Owusu, S.; Anglaaere, L.C.N.; Abugre, S. Aboveground biomass and carbon content of a cocoa—*Gliricida sepium* agroforestry system in Ghana. *Ghana Jnl Agric. Sci.* **2018**, *53*, 45–60. [CrossRef]
74. Aristizábal, J.; Guerra, A. Estimación de la Tasa de Fijación de Carbono en el Sistema Agroforestal Nogal Cafetero *Cordia alliodora*, cacao *Theobroma cacao* y Plátano *Musa paradisíaca*. Ph.D. Thesis, Ingeniero Forestal, Universidad Distrital, Bogotá, Colombia, 2002; 108p. Available online: http://www.sidalc.net/repdoc/a4836e/a4836e.pdf (accessed on 8 February 2021).
75. Vespa, A. Relaciones Hídricas e Intercambio de Gases en *Theobroma cacao* var. Guasare Bajo Períodos de Déficit Hídrico. *Rev. Fac. Agron. Univ. Zulia* **2008**, *2*, 112–120.
76. Chijioke, E.O. *Impart on Soils of Fast Growing Species in Lowland Humid Tropics;* FAO Forestry Paper, No.21; Food and Agricultural Organization: Rome, Italy, 1980; 111p.
77. Schoeneberger, M.M. Agroforestry: Working trees for sequestering carbon on agricultural lands. *Agrofor. Syst.* **2009**, *75*, 27–37. [CrossRef]

 *sustainability*

Article

Does Shade Impact Coffee Yield, Tree Trunk, and Soil Moisture on *Coffea canephora* Plantations in Mondulkiri, Cambodia?

Lenka Ehrenbergerová [1,*], Marie Klimková [1], Yessika Garcia Cano [1], Hana Habrová [1], Samuel Lvončík [1], Daniel Volařík [1], Warbota Khum [1,2], Petr Němec [1], Soben Kim [2], Petr Jelínek [1] and Petr Maděra [1]

[1] Department of Forest Botany, Dendrology and Geobiocoenology, Mendel University in Brno, 613 00 Brno, Czech Republic; klimkova.marie@gmail.com (M.K.); yessika.agarcia@gmail.com (Y.G.C.); hana.habrova@mendelu.cz (H.H.); samuel.lvoncik@mendelu.cz (S.L.); daniel.volarik@seznam.cz (D.V.); khumwarbota@gmail.com (W.K.); petr.nemec@mendelu.cz (P.N.); jelen@mendelu.cz (P.J.); petr.madera@mendelu.cz (P.M.)

[2] Faculty of Forestry Science, Royal University of Agriculture, Phnom Penh P.O. Box 2696, Cambodia; kimsoben@gmail.com

* Correspondence: lenka.ehrenbergerova@mendelu.cz; Tel.: +420-545-134-064; Fax: +420-545-212-293

Abstract: Shade is a natural condition for coffee plants; however, unshaded plantations currently predominate in Asia. The benefits of shading increase as the environment becomes less favorable for coffee cultivation, e.g., because of climate change. It is necessary to determine the effects of shade on the yield of *Coffea canephora* and on the soil water availability. Therefore, three coffee plantations (of 3, 6, and 9 ha) in the province of Mondulkiri, Cambodia, were selected to evaluate the effect of shade on *Coffea canephora* yields, coffee bush trunk changes, and soil moisture. Our study shows that shade-grown coffee delivers the same yields as coffee that is grown without shading in terms of coffee bean weight or size (comparing average values and bean variability), the total weight of coffee fruits per coffee shrub and the total weight of 100 fruits (fresh and dry). Additionally, fruit ripeness was not influenced by shade in terms of variability nor in terms of a possible delay in ripening. There was no difference in the coffee stem diameter changes between shaded and sunny sites, although the soil moisture was shown to be higher throughout the shaded sites.

Keywords: agroforestry systems; bean size; bean homogeneity; ripening

Citation: Ehrenbergerová, L.; Klimková, M.; Cano, Y.G.; Habrová, H.; Lvončík, S.; Volařík, D.; Khum, W.; Němec, P.; Kim, S.; Jelínek, P.; et al. Does Shade Impact Coffee Yield, Tree Trunk, and Soil Moisture on *Coffea canephora* Plantations in Mondulkiri, Cambodia?. *Sustainability* **2021**, *13*, 13823. https://doi.org/10.3390/su132413823

Academic Editor: Victor Rolo

Received: 27 October 2021
Accepted: 9 December 2021
Published: 14 December 2021

1. Introduction

Coffee is one of the most traded crops, and it has a total yearly consumption of 9 billion kilograms around the world. Coffee available on international markets originates from the species *Coffea arabica* (arabica) and *Coffea canephora* (robusta) [1]. Coffee crops are highly sensitive to climate change [2]. Areas suitable for growing coffee are expected to decrease by 21% in Ethiopia [3] and by more than 90% in Nicaragua under global warming [4]. Arabica crops have already declined in the Tanzanian highlands and India under increasing temperatures [5,6]. In the face of climate change, robusta is more adaptable than arabica and able to thrive in warmer and lower land conditions [7]; however, it is more susceptible to low temperatures than arabica [8]. The optimum mean annual temperature range for arabica (18–23 °C) is lower than that for robusta (22–26 °C) [7]. Robusta is popular in blends; it is used to produce instant coffee [7] and it can be grown successfully from sea level to an elevation of 1400 m [7]. Arabica trees are generally less vigorous and productive and have higher production costs than robusta trees; however, arabica is a higher-quality beverage and dominates the high-quality specialty coffee market [9,10]. For both species, a short dry period lasting 2–4 months, corresponding to the quiescent growth phase, is important for triggering flowering [11]. The roots of robusta coffee are shallower [8]; thus, it is less tolerant of dry conditions than arabica but grows well in soils that are wetter and shallower. In each case, changes in the prevailing climatic conditions will have a significant effect on the condition of the arabica and robusta coffee plants and the prevalence of pests and diseases [7].

Similar to arabica coffee trees, robusta trees have a natural adaptation to shade because the species originates from the understory of African forests [8]; therefore, growing coffee in agroforestry systems naturally brings many benefits. Coffee agroforestry systems promote the conservation of natural resources, landscape diversity [12] and biodiversity [13,14]. Moreover, trees in agroforestry systems mitigate excess carbon [15], are sources of economic diversification by providing fuel wood, timber or fruits, and protect the soil [8,16]. Although tree species diversity in agroforestry coffee plantations was lower than that in forest reserves [17], agroforestry coffee plantations are still areas of greater diversity than monocultural plantations or other agricultural land. Moreover, shaded plants can avoid heat stress and high light intensity [18]. Shading reduces the mean air and soil temperatures and increases the air humidity; furthermore, the soil temperatures fluctuate more markedly in non-shaded areas [19]. Shade trees may compete with coffee for resources such as light, water, and soil nutrients [20]. Our prior research [19] reported that the soil conditions were drier at the beginning and end of the dry season in the shaded areas. Although water supply is important for coffee production, it is difficult to know how future changes in rainfall patterns will impact coffee production because consolidated models of rainfall changes have a high degree of uncertainty [21].

One of the reasons why farmers changed agroforestry coffee plantations to monoculture ones was the incidence of diseases. The occurrence of the coffee berry borer and coffee leaf rust in coffee crops in agroforestry plantations highlights the benefits of monoculture coffee planting, which enables higher chemical inputs and pruning and high coffee plant density [13,22]. However, a consensus has not been reached on the higher incidence of these diseases in agroforestry coffee plantations, with some authors indicating that the coffee leaf rust incidence decreased with increasing shade and others recommending intermediate levels of shading [23–25], whereas other studies did not detect any interdependence between coffee leaf rust and shade [26–28].

An increase in production is another reason for reducing agroforestry coffee areas. The effect of shade on coffee quality has been widely documented for arabica coffee and appears to be highly site-dependent [29]. Some studies [30,31] reported that arabica coffee shrubs grown in agroforestry produce lower yields than those grown under full sun. Intense light has been shown to reduce fruit loads and cause lower yields because of longer internodes, fewer fruiting nodes, lower flower induction, and larger bean size [32–34]. On the other hand, arabica coffee grown under shade (60%) gave a higher yield (419 kg ha^{-1}) than sun-grown coffee (259 kg ha^{-1}) [35]; shade cover between 23% and 38% had a positive effect on arabica coffee yield and increasing yield could be maintained with cover up to 48%, while production may decrease under shade cover >50% [12]. The optimal cover percentage for high yields from arabica was between 30 and 45% [36], and arabica coffee yield was higher in plants grown under shade than in those grown under sun, with the main benefits of shading including higher coffee bean weight, larger beans, higher antioxidant activity, higher total phenolic content, and higher chlorogenic acid content [18]. According to the authors of [37], shaded arabica beans are heavier and have a better liquor taste than those developed in direct sunlight. It is assumed that shade delays the ripening of berries and makes it more uniform, thus yielding beans of better quality than sun-grown arabica coffee [35,38,39]. Previous studies have focused on the influence of shade on arabica yield; however, empirical studies on the impact of shade on robusta coffee plants are notoriously few [40].

Agroforestry is seen as an opportunity to maintain the fertility of productive areas in the face of a changing climate; therefore, knowledge of the impact of shading on the yield of robusta coffee is extremely important as decreasing yield is one of the reasons why farmers do not grow coffee within the agroforestry system. On the other hand, changes in coffee stem diameter can indicate if the coffee plant is suffering from a lack of water, which can be caused by water competition with shade trees. For this reason, three coffee plantations were chosen for the investigation, in the province of Mondulkiri, Cambodia. They all grow robusta on both shaded and sunny sites. The main questions of this study were as follows. (i) Is the coffee yield and coffee bean size or weight higher or lower in the shaded parts of the plantations? (ii) Is the size and weight of coffee beans more homogenous in the shade? (iii) Is there any difference in coffee fruit ripeness between shaded and sunny sites? (iv) How does shade influence coffee trunk diameter changes and soil moisture?

2. Materials and Methods

2.1. Study Area Description

This study was carried out on three coffee plantations in Senmonorom Municipality in the province of Mondulkiri, Cambodia (Figure 1). The dominant soil type in all research areas is latosol. The soil texture is clayey throughout the profile [41]. Senmonorom is situated in a tropical climate where the annual mean temperature is 22.9 °C, and the mean annual rainfall is 2203 mm. The highest amount of rainfall occurs in August; the highest temperatures are observed in April, and the lowest temperatures occur in January. According to information collected from coffee growers, the coffee flowering period in the study area generally occurs in March at the end of the dry season and the harvest is in November at the beginning of the dry season.

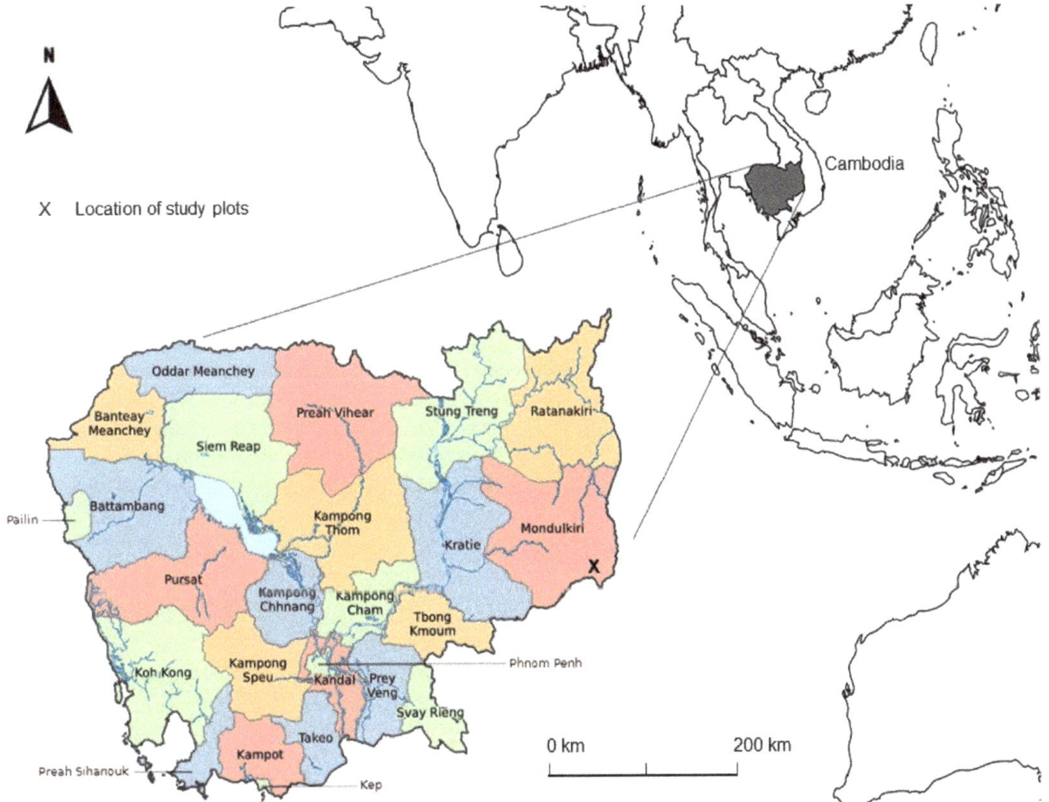

Figure 1. Map of the study area and plantations where the research was conducted.

All three plantations selected for this research grow robusta coffee in monoculture and agroforestry systems. The average altitude of the studied plantations is 700 m above sea level, and the slope on which the coffee plants are grown is 5%. Slope exposure to sunshine is from the southwest. The distance between coffee shrubs was 3 × 3 m in all studied plantations. Shade trees are irregularly distributed on the agroforestry sites, and they are not pruned.

Study Plantation 1 is located 7.2 km northeast of the municipality of Senmonorom (12°29′ N, 107°13′ E). Coffee in this study plot is grown in an area of 3 hectares (ha). The main shade trees in the plantation are lychee (*Litchi chinensis* Sonn), durian (*Durio zibethinus* Merr.), and avocado (*Persea americana* Mill.). The average quantity of shade was 44 ± 21% (mean ± standard deviation). Study Plantation 2 is located 3.5 km northeast of the municipality of Senmonorom (12°28′ N, 107°12′ E). In total, there are 9 ha of coffee plantations. The main shade species in the plantation are banana (*Musa* spp.), durian, and avocado. The average quantity of shade was 44 ± 13%. Study Plantation 3 is located 3.9 km northeast of the municipality of Senmonorom (12°27′ N, 107°12′ E). Coffee in this study plot is grown in an area of 6.3 ha. The main shade trees in the plantation are cassia tree (*Senna siamea* (Lam.) Irwin et Barneby) and durian. The average quantity of shade range was 43 ± 16%.

Shaded and sunny sites were selected according to a visual evaluation in the field. Selected study plots in the shaded sites ranged from 20–70% shade. The levels of shade were determined by hemispherical photographs from fisheye photos taken from the center of each coffee measurement plot, on a shaded site at a height of 2 m (using a Canon EOS 550D camera with a Sigma 4.5mm f/2.8 EX DC lens) at the same times that data were collected. The photographs were intentionally underexposed by two stops (−2EVs), as shown by the authors of [42]. Photographs were classified using the IsoData method in ImageJ version 1.48v. Finally, canopy openness values were calculated from each photograph, employing CIMES v9 [43].

2.2. Coffee Measurement

The coffee yield measurements were performed in November and December 2017. A total of four research plots were selected in each plantation, with two in the shaded sites and two in the sunny sites. Five coffee shrubs were randomly selected in each study plot, a total of 10 shrubs in the shaded and 10 shrubs in the sunny sites. All fruits of the selected coffee shrubs were picked up to determine the total yield from each coffee shrub. The total weight of the fresh fruit was determined immediately after harvest. A sample of one hundred randomly selected fruits was taken from each harvested coffee shrub. The weight of the coffee samples was measured as the fresh weight (immediately after harvest) and dry weight (dried to a constant weight). Of the total sample of 100 dry fruits, 10 fruits were randomly selected and peeled to measure the weight, length, and width of each coffee bean. To assess the maturity of the fruits, 5 coffee plants in shaded sites and 5 coffee plants in sunny sites were selected for each plantation. A sample of 100 randomly selected fruits was picked up from each of 5 coffee shrubs; subsequently, the number of ripe fruits was determined.

2.3. Microclimatic Data

Meteorological data collection took place for one year (27 November 2017 to 23 November 2018). In each study plot, 2 soil moisture sensor TMS-3 dataloggers (TOMST s.r.o., Prague, Czech Republic) were installed at a depth of 0.5 m. Rainfall was continuously measured at an open plot in Study Plantation 3 with a rain gauge (Rain-O-Matic Professional (mm), (Pronamic, Ringkøbing, Denmark)). Based on the measurements and the studied literature, the dry and wet seasons were determined as follows: the dry season was from 27 November 2017 to 16 March 2018 and the rainy season was from 17 March 2018 to 23 November 2018. Air temperature (°C) and relative air humidity (%) were continuously monitored (Minikin THI (EMS Brno, Czech Republic)) at a sunny site in Study Plantation 3, and the interval between measurements was one hour.

The total rainfall in the study plot during the measurement period was 2427 mm. The highest rainfall amount recorded over a calendar month at the site was 553 mm in May 2018, with the peak of highest daily rainfall of 125 mm on 11 May. Additionally, the total number of days in the year with ≥1 mm of precipitation was 143 days. The measured rainfall is shown in Figure 2. The annual average temperature was 22.9 °C, the minimum temperature was observed on 27 December 2017 (15 °C), and the maximum temperature was observed on 14 April 2018 (36 °C). Regarding air humidity, the annual average was 87.2% (Figure 3).

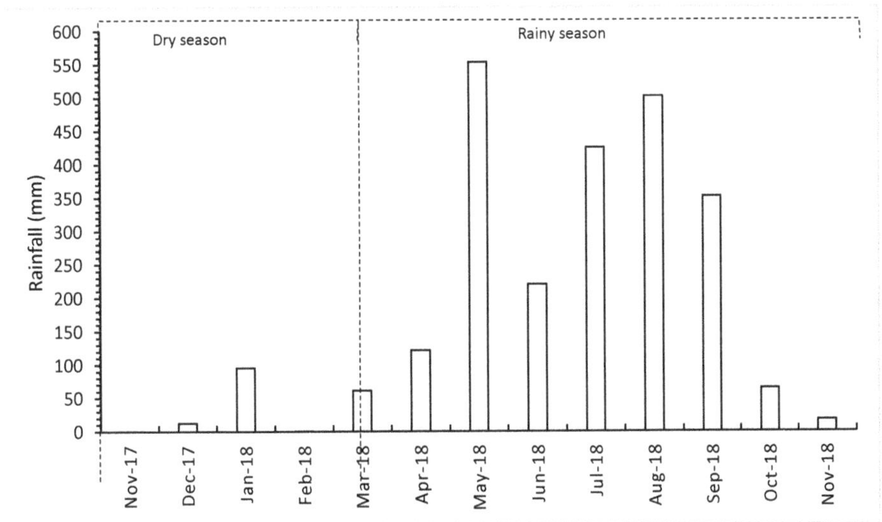

Figure 2. Monthly precipitation from November 2017 until November 2018 in Study Plantation 3.

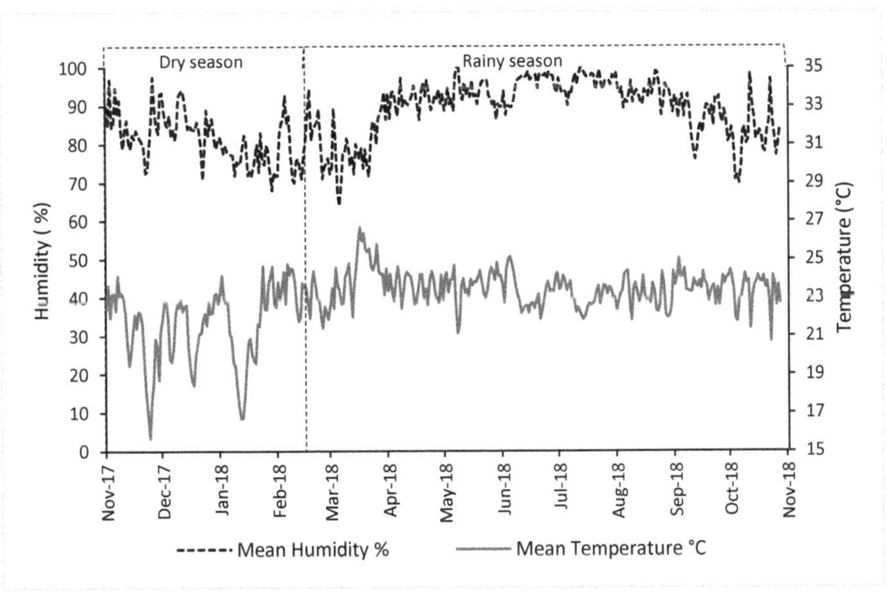

Figure 3. Daily average air temperature and humidity in Study Plantation 3 at the sunny site.

2.4. *Coffee Stem Diameter Increment and Soil Moisture Measurement*

Data on the coffee diameter increments were measured for a period of one year (from 27 November 2017 to 23 November 2018). The dendrometers were installed in each study plantation on 2 coffee plants in the shaded sites and on 2 coffee plants in the sunny sites. Measurements were performed for coffee plants with synchronized phenology stages. The increments in the coffee stem diameters were monitored using stem-diameter sensors (PDS40 SDI, (EMS Brno, Czech Republic)), which were noninvasively fixed to the coffee shrubs at a height of approximately 1.5 m. The measurements were taken at intervals of one hour.

To measure the soil moisture in the study area, one TMS 4 datalogger for soil moisture measurements was installed in each study plantation near one of the coffee plants in the shaded and sunny sites. The depth necessary for installation was 0.5 m. All data could not be downloaded from the microdendrometer and Tomst datalogger in Study Plantation 2 and from one microdendrometer in Study Plantation 3 because of the loss of measuring instruments.

2.5. *Statistic Evaluation*

Statistical analyses focused on the effect of shade on measured parameters (bean weight, projected area, length, and width) of coffee beans. As our data were collected in multiple levels (plantations–plot–shrub–fruit–bean), we used mixed-effect models that allowed us to specify an appropriate multilevel structure. Moreover, we found during our exploratory analysis that variance varied within plantations or shade/sun plots. We used a Bayesian approach to analyze these complex models and assumed a normal distribution of errors. Shade effect, plantation, and their interaction were used as fixed effect variables. Their effect was studied on both mean value and standard deviation (sigma). We tried two variants of random effect—the complex variant encompasses all three levels (plot/tree/fruit) and a simpler variant with just tree/fruit levels. Finally, we decided to use a simpler variant in all reported analyses as we found that the effect of plot was small relative to tree/fruit.

All models were fitted using statistical modeling and the high-performance statistical computation platform Stan [44] through the brms package, version 2.14.4 [45,46], within an R software environment [47]. We used default brms priors, 4 chains and 6000 iterations for each model. For final models, we reported the effect estimate, its 89% highest-density intervals (HDI), and the probability of direction. Additional analyses were provided for bean ripeness. These are proportional data; therefore, binomial (Bernoulli) distribution was used. Moreover, the structure of the random effect was simpler as the effect of fruit was no more applicable. For bean ripeness, we were interested also in whether there is a difference in variability between sun/shade shrubs. We analyzed this by specifying three variants of random effect and we compared those three models using the leave-one-out information criterion (LOOIC). To deal with high K Pareto values, we used the moment_match argument or reloo option when necessary. Again, Bayesian models were used within the R software environment [47]. The final model formula in brms syntax was specified (response ~ shade + plantation + (1 | tree/fruit), sigma ~ shade + plantation), where the response was one of weight, area, length or width.

3. Results

3.1. *Coffee Yield*

Shade did not have an influence on the weight or size of the coffee beans (comparing average values and bean variability). However, the beans differed significantly based on the plantation (in mean values and mostly in variability). Plantation 2 yielded significantly larger and heavier beans than Plantations 1 and 3 (Tables 1–4; Figure 4b).

Table 1. The influence of shade on the mean value and variability of coffee bean weight and the differences between plantations. Sigma = standard deviation.

Parameters Impacting Coffee Bean Weight (mg)	Estimate	HDI 89%	Probability of Direction (%)
Intercept—mean value for sun and Plantation 3	161.63	(149.20, 174.06)	100.0
Difference for shade	3.76	(−8.65, 15.91)	68.9
Difference for Plantation 1	**−27.93**	**(−43.16, −12.46)**	**99.8**
Difference for Plantation 2	**43.97**	**(28.77, 58.66)**	**100.0**
Sigma (log link): estimate for sun and Plantation 3	2.79	(2.70, 2.89)	100.0
Sigma (log link): difference for shade	−0.09	(−0.18, 0.00)	93.1
Sigma (log link): Difference for Plantation 1	**−0.35**	**(−0.47, −0.24)**	**100.0**
Sigma (log link): Difference for Plantation 2	**0.17**	**(0.05, 0.29)**	**99.0**
Model with interaction shade: plantation			
Difference for shade: Plantation 1 (Interaction)	5.44	(−26.4, 34.74)	61.8
Difference for shade: Plantation 2 (Interaction)	−26.87	(−58.45, 1.83)	92.4

Random effect: tree SD 28.19 (23.59, 33.31); tree/fruit SD 26.24 (24.65, 27.78). Significant differences are indicated in bold.

Table 2. Influence of shade on the mean value and variability of coffee bean area and the differences between plantations. Sigma = standard deviation.

Parameters Impacting Coffee Bean Area (mm^2)	Estimate	HDI 89%	Probability of Direction (%)
Intercept—mean value for sun and Plantation 3	45.11	(42.66, 47.66)	100.0
Difference for shade	0.62	(−1.83, 3.18)	65.3
Difference for Plantation 1	**−5.68**	**(−8.74, −2.60)**	**99.7**
Difference for Plantation 2	**10.64**	**(7.46, 13.73)**	**100.0**
Sigma (log link): estimate for sun and Plantation 3	1.40	(1.29, 1.50)	100.0
Sigma (log link): difference for shade	0.04	(−0.06, 0.13)	72.78
Sigma (log link): Difference for Plantation 1	**−0.69**	**(−0.81, −0.58)**	**100.0**
Sigma (log link): Difference for Plantation 2	**−0.22**	**(−0.33, −0.10)**	**99.9**
Model with interaction shade: plantation			
Difference for shade: Plantation 1 (Interaction)	3.34	(−2.89, 9.37)	81.0
Difference for shade: Plantation 2 (Interaction)	−1.81	(−8.38, 4.26)	68.8

Random effect: tree SD 5.79 (4.84, 6.78); tree/fruit SD 5.24 (4.93, 5.55). Significant differences are indicated in bold.

Table 3. Influence of shade on the mean value and variability of coffee bean length and the differences between plantations. Sigma = standard deviation.

Parameters Impacting Coffee Bean Length (mm)	Estimate	HDI 89%	Probability of Direction (%)
Intercept—mean value for sun and Plantation 3	8.48	(8.23, 8.73)	100.0
Difference for shade	0.17	(−0.07, 0.44)	86.6
Difference for Plantation 1	**−0.82**	**(−1.11, −0.49)**	**100.0**
Difference for Plantation 2	**1.17**	**(0.85, 1.47)**	**100.0**
Sigma (log link): estimate for sun and Plantation 3	−0.81	(−0.92, −0.71)	100.0
Sigma (log link): difference for shade	0.09	(0.00, 0.19)	93.2
Sigma (log link): Difference for Plantation 1	**−0.56**	**(−0.68, −0.45)**	**100.0**
Sigma (log link): Difference for Plantation 2	**−0.28**	**(−0.40, −0.17)**	**100.0**
Model with interaction shade: plantation			
Difference for shade: Plantation 1 (Interaction)	0.59	(0.00, 1.24)	93.7
Difference for shade: Plantation 2 (Interaction)	0.28	(−0.32, 0.93)	76.8

Random effect: tree SD 0.58 (0.48, 0.68); tree/fruit SD 0.55 (0.52, 0.59). Significant differences are indicated in bold.

Table 4. Influence of shade on the mean value and variability of coffee bean width and the differences between plantations. Sigma = standard deviation.

Parameters Impacting Coffee Bean Width (mm)	Estimate	HDI 89%	Probability of Direction (%)
Intercept—mean value for sun and Plantation 3	6.70	(6.48, 6.91)	100.0
Difference for shade	−0.00	(−0.21, 0.23)	50.1
Difference for Plantation 1	**−0.20**	**(−0.47, 0.06)**	**89.2**
Difference for Plantation 2	**0.61**	**(0.33, 0.86)**	**99.9**
Sigma (log link): estimate for sun and Plantation 3	−1.02	(−1.12, −0.92)	100.0
Sigma (log link): difference for shade	0.02	(−0.07, 0.12)	66.4
Sigma (log link): Difference for Plantation 1	**−0.50**	**(−0.61, −0.39)**	**100.0**
Sigma (log link): Difference for Plantation 2	**−0.18**	**(−0.30, −0.07)**	**99.2**
Model with interaction shade: plantation			
Difference for shade: Plantation 1 (Interaction)	0.01	(−0.48, 0.58)	51.5
Difference for shade: Plantation 2 (Interaction)	−0.54	(−1.07, −0.01)	94.5

Random effect: tree SD 0.50 (0.42, 0.58); tree/fruit SD 0.40 (0.37, 0.42). Significant differences are indicated in bold.

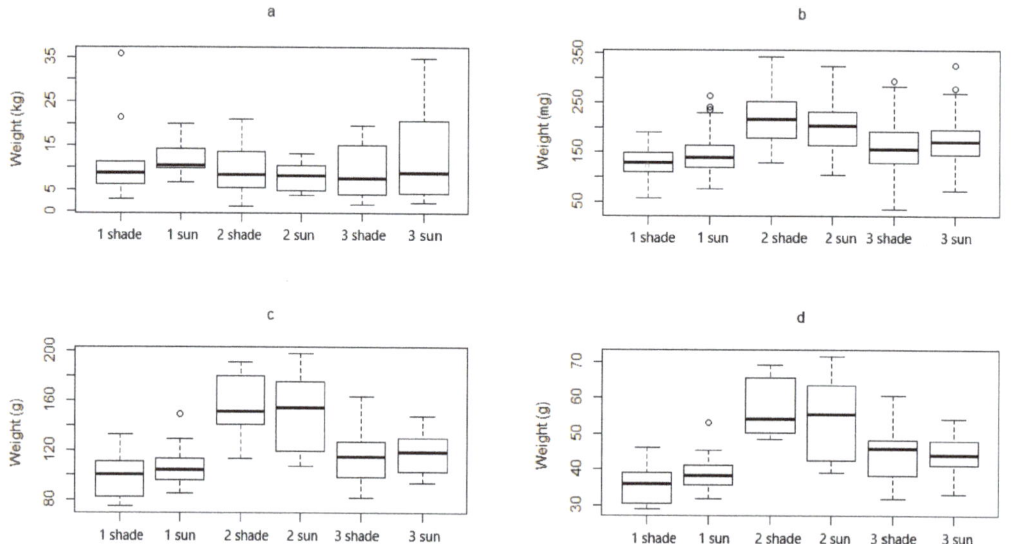

Figure 4. (**a**) The total weight of coffee fruits per coffee shrub (kg), (**b**) weight of individual coffee beans (mg), (**c**) weight of 100 fresh coffee fruits (g), (**d**) weight of 100 dry coffee fruits (g) for shaded and sunny sites for all studied plantations.

There was no significant difference in the total weight of fresh coffee fruits per coffee shrub between the shaded and sunny sites or between coffee plantations (Figure 4a, Table 5). There was also no difference in weight of the 100 fresh or dry fruits between the shaded and sunny sites. We found that on Plantation 2, the coffee fruits were heavier both in the dry and wet states (see Figure 4c,d and Table 5).

Table 5. Effect of shade and plantation on the total weight of coffee fruits per coffee shrub and on the wet and dry weights of 100 fruits. Analyses are based on a two-way frequentists ANOVA, including the interaction between shade and plantation.

Parameter	Estimate	95% CI	*p* Value
Total weight of coffee fruits per coffee shrub (kg)			
Intercept—mean value for sun and Plantation 3	9.08	(4.30, 13.90)	<0.0001
Difference for shade	3.88	(−2.88, 10.60)	0.2548
Difference for Plantation 1	2.66	(−4.10, 9.42)	0.4334
Difference for Plantation 2	0.54	(−6.22, 7.30)	0.8733
Difference for shade: Plantation 1 interaction	−3.51	(−13.2, 6.18)	0.4709
Difference for shade: Plantation 2 interaction	−5.51	(−15.1, 4.05)	0.2528
Weight of 100 fruits—wet (g)			
Intercept—mean value for sun and Plantation 3	116	(102, 131)	<0.0001
Difference for shade	2.1	(−18.6, 22.8)	0.8390
Difference for Plantation 1	−18.7	(−39.4, 2.1)	0.0764
Difference for Plantation 2	**38.9**	**(18.2, 59.6)**	**0.0004**
Difference for shade: Plantation 1 interaction	8.2	(−21.1, 37.5)	0.5750
Difference for shade: Plantation 2 interaction	−4.4	(−33.7, 24.9)	0.7630
Weight of 100 fruits—dry (g)			
Intercept—mean value for sun and Plantation 3	44.7	(39.6, 49.9)	<0.0001
Difference for shade	−0.4	(−7.7, 6.9)	0.9130
Difference for Plantation 1	**−8.9**	**(−16.2, −1.6)**	**0.0178**
Difference for Plantation 2	**11.8**	**(4.5, 19.1)**	**0.0020**
Difference for shade: Plantation 1 interaction	3.8	(−6.5, 14.2)	0.4600
Difference for shade: Plantation 2 interaction	−2.2	(−12.5, 8.1)	0.6700

Significant differences are indicated in bold.

3.2. *Coffee Fruit Ripeness*

A general trend in fruit ripeness differences was not demonstrated between shaded and unshaded coffee bushes. The exception is Plantation 1, which had fewer ripe grains on the shaded shrubs (Figure 5, Table 6). A strong effect of plantations was observed on the occurrence of mature fruits. Plantation 2 had a lower proportion of overall mature grains than the other plantations. There is no influence of shade on the ripeness variability (Table 7).

Table 6. Influence of shade, plantation, and their interaction on the proportion of ripe fruits among the total fruit number (model #m.Ripeness.shade.shrub). Mean values and their 95% credible intervals are shown. Estimates of differences and their HDIs are on the logit scale.

Parameter	Estimate	HDI 95%	Probability of Direction (%)
Shade—sun	0.56	(−1.20, 2.26)	75.4
Plantation 1–3	1.03	(−0.77, 2.68)	87.7
Plantation 2–3	−1.35	(−3.14, 0.25)	94.9
Shade 1: Plantation 1	−1.61	(−4.16, 0.75)	90.0
Shade 1: Plantation 2	−0.19	(−2.57, 2.11)	56.3

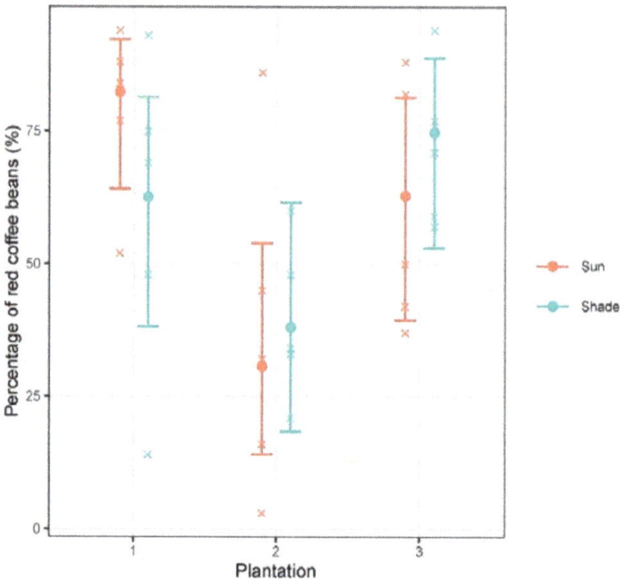

Figure 5. Proportion of mature fruits from the total number of fresh fruits on the coffee shrub, by shade and plantation. Mean values, their 89% credible intervals, and measured values are shown.

Table 7. Influence of shade on the ripeness variability.

Model ID	Model Specification	ELPD Difference	LOOIC	LOOIC SE
m.Ripeness.shrub	shade × plantation + (1 | Shrub_ID)	0.0	212.2	3.7
m.Ripeness.shade.shrub	shade × plantation + (shade | Shrub_ID)	−14.3	240.8	9.7
m.Ripeness.simple	shade × plantation	−345.6	903.4	182.3

3.3. *Coffee Stem Diameter Increment and Soil Moisture*

There was no observed difference in coffee-stem diameter increment between sunny and shaded sites, nor between plantations ($p > 0.05$, using ANOVA). However, a significant difference was observed between the diameter increment of the coffee shrubs among seasons ($p = 0.021$, using Student's t-test), which is also evident from Figure 6. Some coffee bushes in both shaded and sunny sites showed a negative increment. The monthly average coffee-stem diameter changes (during the day) in December 2017 (dry season) (Figure 7a–c) were higher than those in May 2018 (rainy season) (Figure 8a–c). The changes were slightly more pronounced in the sunny sites than in the shade. The changes in coffee stem sizes reached the maximum diameter early in the morning and the minimum in the evenings, which findings correspond with the daily temperature and humidity trends (Figure 7d).

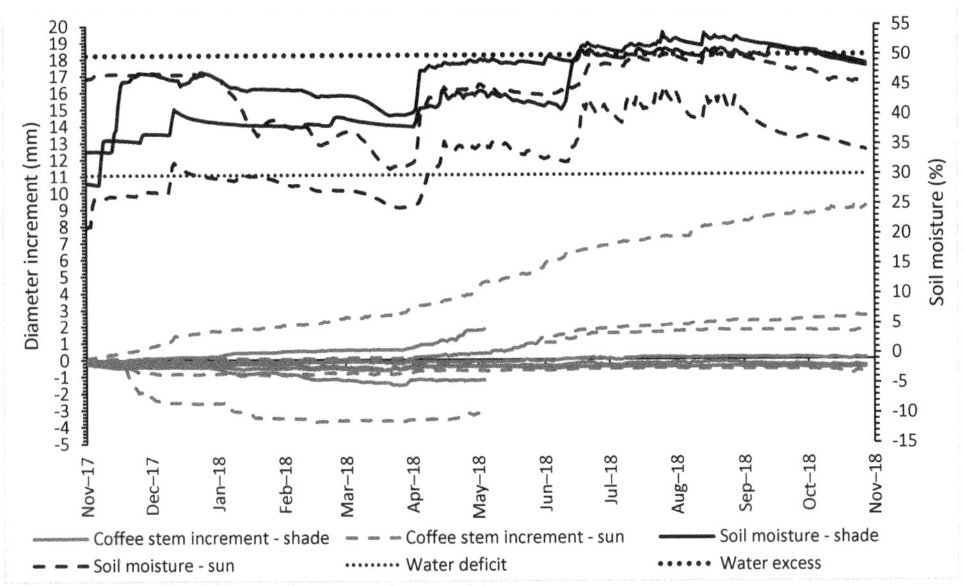

Figure 6. Coffee stem diameter increment and soil moisture in studied plantations divided by shaded and sunny sites from 27 November 2017 to 3 May 2018. Water deficit <30% and water excess as >50% of soil moisture.

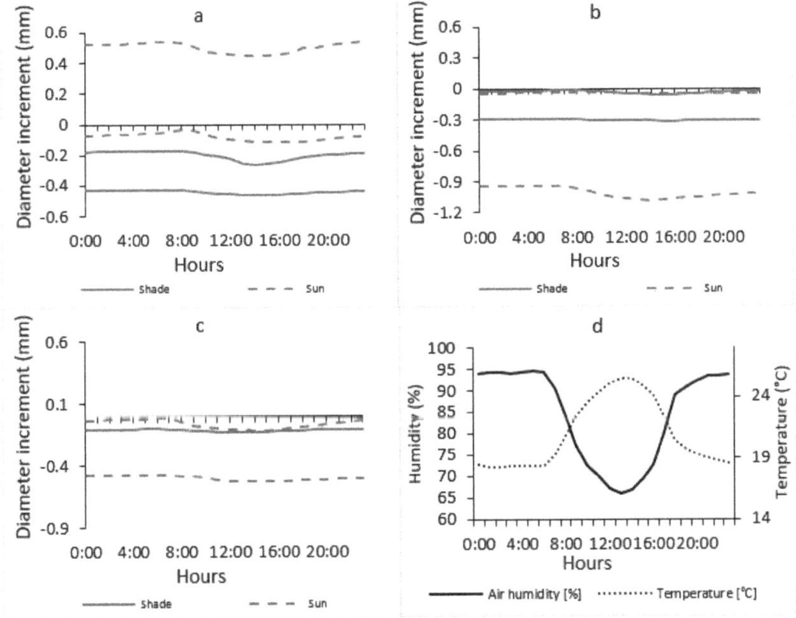

Figure 7. The monthly average coffee stem diameter changes during the day for each plantation (**a**–**c**) and the monthly average air humidity and temperature (**d**) during the day in December 2017 (dry season).

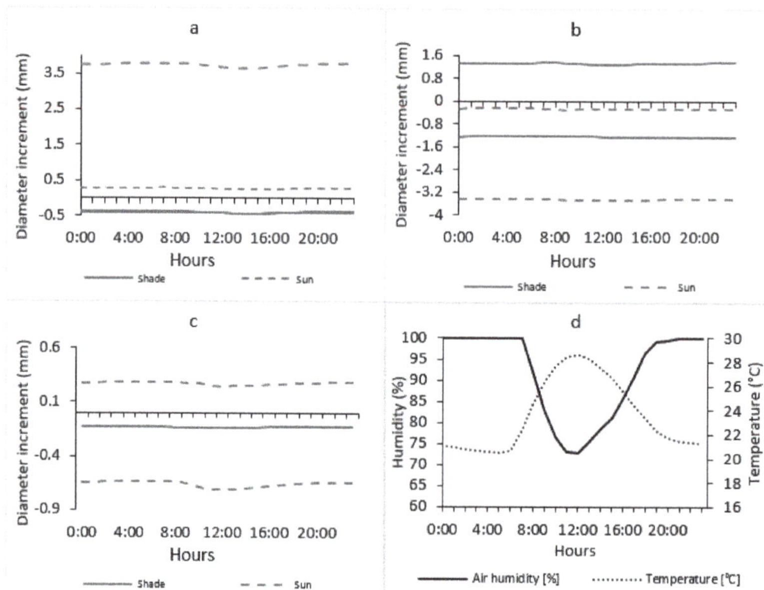

Figure 8. The monthly average coffee-stem diameter changes during the day in shaded and sunny sites on the studied plantation (**a**–**c**) and the monthly average air humidity and temperatures (**d**) during the day in May 2018 (wet season).

The soil moisture was higher in the shaded sites. The soil moisture was below the limit for water deficit (30%) in the sunny site in Plantation 3 almost constantly throughout the measurement period from 27 November 2017 to 3 May 2018. On the other hand, the soil moisture in shaded sites oscillated around 50%, which is the limit for water excess, from July to November (Figure 6). Water deficit was defined as <30% of volumetric soil moisture, and water excess at >50%, as in the study by [48].

4. Discussion

The main questions of this study were whether the size and weight of coffee beans and fruits are influenced by shade and if the yield per coffee shrub is higher for shaded or sunny sites. The results indicate that there is no difference in these parameters between shaded and sunny sites. We found that some yield parameters, such as the weight and size of coffee beans, the weight of 100 fresh and dry coffee fruits, and the proportion of mature fruits in the samples were influenced by the plantation conditions. However, within the same plantation, differences were not observed between shaded and sunny sites. These findings indicate that planting robusta coffee in agroforestry systems will not cause losses in yield for individual plants. These findings are very important and should be considered when deciding whether to grow coffee in agroforestry or monocultural systems because the loss of yield is one of the main arguments for growing coffee in monoculture plantations.

The number of studies examining the effect of shade on robusta yield and bean size is limited. For example, a meta-analysis [40] found that shade trees positively affected the growth and yield of *Coffea canephora* plants. According to this synthesis, providing shade had positive impacts on older robusta shrubs (mean of 16 years) but had either insignificant or negative effects on younger shrubs. On the other hand, several studies have examined the effect of shade on arabica yield although their conclusions are not the same. Some authors point to a higher yield in coffee monoculture. For example, in Brazil, arabica yield in an agroforestry system (514.8 kg ha^{-1} of dry berries) was lower than that in a monoculture system (2442.8 kg ha^{-1}) [30]. A Venezuelan study [49] found that by increasing the number of shade trees from 259 to 353 per ha, the yield was reduced by 26%, and by increasing the number of shade trees from 419 to 561 per ha, the yield was reduced by 100%. Researchers in Costa Rica concluded that a density of 100 trees of *Cordia alliodora* per ha maintained the same yields of arabica coffee found in plots without shade control; with higher densities, a decrease in yield was observed [31]. Other authors found that a certain degree of shading increased the arabica coffee yield. A study conducted in Mexico concluded that shade tree cover from 23% to 38% had a positive effect on coffee yield, shade ranging from 38 to 48% maintained the yield, and shade cover greater than 50% decreased the yield [12]. Similarly, a study in Costa Rica found coffee yields were optimized at 40% shade cover.

Looking at coffee-bean size, our findings correspond to a southern Colombia study on arabica that found that shade did not affect coffee-bean size [50]. However, other authors found that arabica plants grown under shade developed larger and heavier beans than those grown in full sun [18,37,39].

Coffee cultivation in agroforestry systems represents the potential of diversifying production and, thereby, farming incomes [51]. According to the authors of [52], intercropping coffee with shade trees shows no negative relationship with economic performance for smallholder coffee systems. Moreover, income from other products, including income from timber, can provide an extra source of income, which is an opportunity to increase economic resilience [16,52]. For example, intercropping coffee and banana appears to be more profitable than coffee mono-cropping [53]. On the other hand, labor costs can be a major barrier to the adoption of complex agroforestry systems [54].

Another question in our study was whether there was any difference in the coffee fruit ripeness in shaded and sunny sites. During the dry processing of coffee, beans are classified, among other variables, according to size and weight [55]. For this reason, variability in seed size and weight is very important for growers. Multiple studies [35,38,39] found that growing under shade delays the ripening process of arabica. According to the authors of [39], arabica berries ripen faster in full sun than in shade due to exposure to higher temperatures. This trend was not confirmed for Plantation 1 but was noticeable for the remaining two studied plantations. According to Muschler [35], at low-elevation coffee sites with high temperatures, the reduction of temperature extremes through shading may have played a dominant role in the uniform growth and ripening of arabica berries. However, our work does not show that shade has an effect on the variability of robusta fruit ripening.

Another question we sought to answer was how shade influences coffee stem diameter increment and soil moisture in shaded and sunny sites. We did not observe differences in the coffee stem diameter between the two conditions. However, soil moisture was higher in the shaded sites. These results show that the plants were not affected by the lower soil moisture in sunny sites or higher soil moisture in shaded sites, which may indicate that competition for water did not occur between coffee tree and shade tree roots in the studied plantations. Our prior research [19] indicated that competition for water can occur between shade trees and coffee tree roots at a depth of 40 cm, which was based on measurements performed in agroforestry plantations in Peru, where the soil water availability was lower on shaded sites than on sunny sites during the dry season. This hypothesis was not confirmed during our measurement period in Cambodia, where higher soil moisture was present in the shaded site at a depth of 0.5 m, even during the dry season. Our results, showing that soil moisture was higher in shaded sites, are in accordance with other authors' results for arabica coffee [56,57]; this fact could be explained by the litter layer reducing the evaporative loss of soil water and shading decreasing the solar energy available for evaporation [58]. Another possible explanation is that shade increases moisture inputs through the horizontal interception of mist or clouds [59].

Some coffee plants from both a shaded and a sunny site presented a decrease in stem diameter. Water stress could be an indicator of stem diameter fluctuation in perennial trees [60]. Microclimate temperatures can impact the stem diameter of coffee plants [61].

Even though the annual average temperature in the area of study was within the optimal range suitable for coffee cultivation, during the dry season, temperatures dropped as low as 15 °C. Past research found that the stem diameter of *Coffea canephora* both narrowed and widened when temperatures dropped to 19 °C [62]. Some studies presented a higher increment of coffee-stem diameter during the rainy season [61,63] and restrained growth during the dry period [61]. In addition, the authors of [64] related the early contraction of nectarine trunk diameter as a result of water stress. According to the authors of [61], in active phase of vegetative growth of coffee, during the morning (8 AM to 9 AM) stomatal conductance was relatively high and decreased gradually. During the inactive growth phase, the stomata were closed for most of the diurnal period. In our study, in the morning the trunk of the coffee tree shrinks and, after 12 PM, it grows again. The changes in coffee trunk diameter are visibly connected (especially in the dry season) with the curves of air humidity and temperature on studied localities. These daily changes were more pronounced on sunny sites.

Durian, litchi, avocado, and cassia trees are multi-purpose tree species found in studied agroforestry plantations. All provide for income diversification by giving fruits and leaves for human consumption or fodder. Litchi is a deep-rooted, evergreen fruit crop that conserves degraded lands [65] and increases soil-available zinc (Zn), manganese (Mn), copper (Cu), and iron (Fe) [66]. Cassia is a Fabaceae species and has the ability to fix nitrogen (N) and improve soil fertility [67], which are properties comparable with *Inga* species (also Fabaceae) and are widely used as shading trees in coffee plantations in the Americas.

Some researchers [12] have advised arabica producers to grow arabica coffee under shade trees with approximately 50% shade cover; although this practice does not lead to a significant decrease in yields, it has additional economic benefits derived from other products extracted from the plantations. Our results also suggest planting robusta on shaded plantations because in our study, coffee agroforestry systems neither decreased the coffee bean size nor the yield per plant and they did not influence the level of coffee fruit ripeness. In any case, it is necessary to find out which level of shade is most suitable for robusta coffee. Additionally, our study shows that there is no difference in the coffee stem diameter changes between shaded and sunny sites, although the soil moisture was lower at sunny sites.

Agroforestry coffee plantations have many non-commercial functions that help to increase biodiversity, decrease erosion, and contribute to higher landscape values. Moreover, coffee has proven to be highly sensitive to climate change [2] and the mitigation of its effects is another reason for adopting agroforestry. The importance of robusta coffee is increasing, as evidenced by its growing share of world coffee production: it rose from 37% in 2016 to 47% in 2021 [68]. Based on our findings, it seems unreasonable to change robusta agroforestry plantations to monocultural ones in order to increase yield. Agroforestry can increase labor costs [54] but the adoption of appropriate planting schemes and suitable shade species can minimize these costs. Growing coffee in the shade has many benefits, and our study is another in a series confirming that agroforestry systems are a good choice for today's world.

5. Conclusions

Our results imply that shading does not affect coffee yield (in terms of coffee bean weight or size, total weight of coffee fruits per shrub, and weight per 100 fruits), the variability of coffee bean sizes, and the fruit weight or ripeness of *Coffea canephora* in the studied plantations (of 3, 6, and 9 ha). Soil moisture plays a crucial role under conditions of higher temperatures and low precipitation, and it was higher throughout shaded sites. Although the soil moisture in sunny sites was for some periods lower than 30% (water deficit) and in shaded sites was higher than 50% (water access), the coffee-stem diameter changes were the same in the shaded and sunny sites. Taking into account the ecological attributes and ecosystem service benefits of agroforestry, such as increased biodiversity [13,14] and landscape diversity [12], providing soil protection [8], increasing the carbon stock [15], and providing sources of fuel wood and construction material [8], we suggest maintaining and planting *Coffea canephora* in agroforestry shade-based systems, as coffee producers transition away from monoculture plantations.

Author Contributions: Conceptualization, L.E., H.H., P.M., M.K. and Y.G.C.; methodology, L.E., H.H., P.M. and D.V.; software, D.V.; validation, P.M.; formal analysis, D.V. and L.E.; investigation, M.K., Y.G.C., H.H., S.L., W.K., P.N., S.K., P.J. and P.M.; resources, S.K.; data curation, L.E. and D.V.; writing—original draft preparation L.E., M.K. and Y.G.C.; writing—review and editing, H.H., P.M., W.K. and S.K.; visualization, L.E.; supervision, H.H. and P.M.; project administration, H.H. and L.E.; funding acquisition, H.H. and L.E. All authors have read and agreed to the published version of the manuscript.

Funding: This research was funded by Mendel University in Brno and contributions from the Czech Development Agency, grant numbers DP-2019-047-DO-11420 (Support for Practical Education and Final Thesis Elaboration at the Faculty of Forestry, Royal University of Agriculture, Cambodia) and 22/2017/08 (Strengthening of Practical Education, Field Research and Scientific Publication on Faculty of Forestry, Royal University of Agriculture, Cambodia), and from the Czech Ministry of Foreign Affairs, grant number 2020-PKVV-014 (Innovation of Study Program and Support for Practical Final Theses at the Faculty of Forestry at the Royal University of Agriculture, Cambodia).

Institutional Review Board Statement: Not applicable.

Informed Consent Statement: Not applicable.

Acknowledgments: The authors are grateful to the students and employees of the Royal University of Agriculture, Cambodia for their support during the fieldwork at the time of data collection. We thank the farmers Chay Mao, Om Pheap (Coffee Plantation), and Nget Chan Raksmey for their willingness to participate and for providing their plantations for research. Finally, we thank Kay Van Damme for help with language corrections.

Conflicts of Interest: The authors declare no conflict of interest.

References

1. International Coffee Organization. The Current State of the Global Coffee Trade | #CoffeeTradeStats. 2016. Available online: http://www.ico.org/monthly_coffee_trade_stats.asp (accessed on 3 January 2020).
2. Bunn, C.; Läderach, P.; Rivera, O.O.; Kirschke, D. A bitter cup: Climate change profile of global production of Arabica and Robusta coffee. *Clim. Chang.* **2015**, *129*, 89–101. [CrossRef]
3. Moat, J.; Williams, J.; Baena, S. Resilience potential of the Ethiopian coffee sector under climate change. *Nat. Plants* **2017**, *3*, 1–14. [CrossRef]
4. Läderach, P.; Ramirez–Villegas, J.; Navarro-Racines, C.; Zelaya, C.; Martinez–Valle, A.; Jarvis, A. Climate change adaptation of coffee production in space and time. *Clim. Chang.* **2017**, *141*, 47–62. [CrossRef]
5. Craparo, A.C.W.; van Asten, P.J.A.; Läderach, P.; Jassogne, L.T.P.; Grab, S.W. Coffea arabica yields decline in Tanzania due to climate change: Global implications. *Agric. For. Meteorol.* **2015**, *207*, 1–10. [CrossRef]
6. Jayakumar, M.; Rajavel, M.; Surendran, U.; Gopinath, G.; Ramamoorthy, K. Impact of climate variability on coffee yield in India—With a micro-level case study using long-term coffee yield data of humid tropical Kerala. *Clim. Chang.* **2017**, *145*, 335–349. [CrossRef]
7. Waller, J.M.; Bigger, M.; Hillocks, R.J. *Coffee Pests, Diseases and Their Management*; CABI: Wallingford, UK, 2007.
8. Wintgens, J.N. *Coffee: Growing, Processing, Sustainable Production. A Guidebook for Growers, Processors, Traders, and Researchers*; WILEY-VCH Verlag GmbH & Co. KGaA: Corseaux, Switzerland, 2004.
9. DaMatta, F.M.; Ronchi, C.P.; Maestri, M.; Barros, R.S. Coffee: Environment and Crop Physiology. In *Ecophysiology of Tropical Tree Crops*; Nova Science Publishers: New York, NY, USA, 2010; pp. 181–216.
10. DaMatta, F.M.; Avila, R.T.; Cardoso, A.A.; Martins, S.C.; Ramalho, J.C. Physiological and agronomic performance of the coffee crop in the context of climate change and global warming: A review. *J. Agric. Food Chem.* **2018**, *66*, 5264–5274. [CrossRef]
11. Haarer, A.E. *Modern Coffee Production*; Leonard Hill: London, UK, 1958.
12. Soto-Pinto, L.; Perfecto, I.; Castillo-Hernández, J.; Caballero-Nieto, J. Shade effect on coffee production at the northern Tzeltal zone of the state of Chiapas, Mexico. *Agric. Ecosyst. Environ.* **2000**, *80*, 61–69. [CrossRef]
13. Perfecto, I.; Rice, R.A.; Greenbergr, R.; Van der Voort, M.E. Shade Coffee: Disappearing Refuge for Biodiversity. *Bioscience* **1996**, *46*, 596–608. [CrossRef]
14. Moguel, P.; Toledo, V.M. Biodiversity Conservation in Traditional Coffee Systems of Mexico. *Conserv. Biol.* **1999**, *13*, 11–21. [CrossRef]

15. Ehrenbergerová, L.; Cienciala, E.; Kučera, A.; Guy, L.; Habrová, H. Carbon stock in agroforestry coffee plantations with different shade trees in Villa Rica, Peru. *Agrofor. Syst.* **2016**, *90*, 433–445. [CrossRef]
16. Ehrenbergerová, L.; Šeptunová, Z.; Habrová, H.; Tuesta, R.H.P.; Matula, R. Shade tree timber as a source of income diversification in agroforestry coffee plantations, Peru. *Bois Forets Trop.* **2019**, *342*, 93–103. [CrossRef]
17. Negawo, W.J.; Beyene, D.N. The role of coffee based agroforestry system in tree diversity conservation in eastern Uganda. *Landsc. Ecol.* **2017**, *10*, 1–18. [CrossRef]
18. Somporn, C.; Kamtuo, A.; Theerakulpisut, P.; Siriamornpun, S. Effect of shading on yield, sugar content, phenolic acids and antioxidant property of coffee beans (*Coffea arabica* L. cv. Catimor) harvested from north-eastern Thailand. *J. Sci. Food Agric.* **2012**, *92*, 1956–1963. [CrossRef] [PubMed]
19. Ehrenbergerová, L.; Šenfeldr, M.; Habrová, H. Impact of tree shading on the microclimate of a coffee plantation: A case study from the Peruvian Amazon. *Bois Forets Trop.* **2017**, *4*, 13–22. [CrossRef]
20. Siles, P.; Harman, J.M.; Vaast, P. Effects of *Inga densiflora* on the microclimate of coffee (*Coffea arabica* L.) and overall biomass under optimal growing conditions in Costa Rica. *Agrofor. Syst.* **2010**, *78*, 186–269. [CrossRef]
21. Intergovernmental Panel on Climate Change. An IPCC Special Report on Climate Change, Desertification, Land Degradation, Sustainable Land Management, Food Security, and Greenhouse Gas Fluxes in Terrestrial Ecosystems. Available online: https://www.ipcc.ch/srccl/ (accessed on 10 November 2020).
22. Coyner, M.S. *Agriculture and Trade in Nicaragua*; Foreign Agriculture Service: Washington, DC, USA, 1960.
23. Eskes, A.B. The effect of light intensity on incomplete resistance of coffee to *Hemileia vastatrix*. *Neth. J. Plant Pathol.* **1982**, *88*, 191–202. [CrossRef]
24. Kushalappa, A.C. Advences in coffee rust researche. *Annu. Rev. Phytopathol.* **1989**, *27*, 503–531. [CrossRef]
25. Staver, C.; Guharay, F.; Monterroso, D.; Muschler, R.G. Designing pest-suppressive multistrata perennial crop systems: Shade-grown coffee in Central America. *Agrofor. Syst.* **2001**, *53*, 151–170. [CrossRef]
26. Soto-Pinto, L.; Perfecto, I.; Caballero-Nieto, J. Shade over coffee: Its effects on berry borer, leaf rust and spontaneous herbs in Chiapas, Mexico. *Agrofor. Syst.* **2002**, *55*, 37–45. [CrossRef]
27. Salgado, B.G.; Macedo, R.L.G.; Carvalho, V.L.D.; Salgado, M.; Venturin, N. Progress of rust and coffee plant cercosporiose mixed with grevilea, with ingazeiro and in the full sunshine in Lavras MG. *Cienc. Agrotecnologia* **2007**, *31*, 1067–1074. [CrossRef]
28. Ehrenbergerová, L.; Kučera, A.; Cienciala, E.; Trochta, J.; Volařík, D. Identifying key factors affecting coffee leaf rust incidence in agroforestry plantations in Peru. *Agrofor. Syst.* **2018**, *92*, 1551–1565. [CrossRef]
29. Tolessa, K.; D'heer, J.; Duchateau, L.; Boeckx, P. Influence of growing altitude, shade and harvest period on quality and biochemical com- position of Ethiopian specialty coffee. *J. Sci. Food Agric.* **2017**, *97*, 2849–2857. [CrossRef] [PubMed]
30. Campanha, M.M.; Santos, R.H.S.; De Freitas, G.B.; Martinez, H.E.P.; Garcia, S.L.R.; Finger, F.L. Growth and yield of coffee plants in agroforestry and monoculture systems in Minas Gerais, Brazil. *Agrofor. Syst.* **2004**, *63*, 75–82. [CrossRef]
31. Hernández, G.O.; Beer, J.; von Platen, H. Rendimiento de café (*Coffea arabica* cv Caturra), producción de madera (*Cordia alliodora*) y análisis financiero de plantaciones con diferentes densidades de sombra en Costa Rica. *Agrofor. Am.* **1997**, *4*, 8–13.
32. Avelino, J.; Barboza, B.; Araya, J.C.; Fonseca, C.; Davrieux, F.; Guyot, B.; Cilas, C. Effects of slope exposure, altitude and yield on coffeequality in two altitude terroirs of Costa Rica, Orosi and Santa María de Dota. *J. Sci. Food Agric.* **2005**, *85*, 1869–1876. [CrossRef]
33. Guyot, B.; Gueule, D.; Manez, J.C.; Perriot, J.J.; Giron, J.; Villian, L. Influence de l'altitude et de l'ombrage sur la qualitedes cafes Arabica. *Plant. Rech. Dev.* **1996**, *3*, 272–283.
34. Salazar, E.I. Calidad de *Coffea arabica* Bajo Sombra de *Erythrina poeppigiana* a Diferentes Elevaciones en Costa Rica. Master's Thesis, Centro Agronómico Tropical de Investigación y Enseñanza, Turrialba, Costa Rica, 1999.
35. Muschler, R.G. Shade improves coffee quality in a sub-optimal coffee zone of Costa Rica. *Agrofor. Syst.* **2001**, *85*, 131–139. [CrossRef]
36. Muschler, R.G. Sombra o sol para un cafetal sostenible: Un nuevo enfoque de una vieja discusión. In *Memorias del 18vo Simposio Latinoamericano de Caficultura*; Editorama S.A.: San José, Costa Rica, 1997; pp. 471–476.
37. Bote, A.D.; Struik, P.C. Effects of shade on growth, production, and quality of coffee (*Coffea arabica*) in Ethiopia. *J. Hortic. For.* **2011**, *3*, 336–341.
38. Alemu, M.M. Effect of Tree Shade on Coffee Crop Production. *J. Sustain. Dev.* **2015**, *8*, 66–70. [CrossRef]
39. Vaast, P.; Van Kanten, R.; Siles, P.; Dzib, B.; Franck, N.; Harmand, J.M. *Shade: A Key Factor for Coffee Sustainability and Quality*; Association Scientifique Internationale du Café (ASIC): Allenwinden, Switzerland, 2006; pp. 887–896.
40. Piato, K.; Lefort, F.; Subía, C.; Caicedo, C.; Calderón, D.; Pico, J.; Norgrove, L. Effects of shade trees on robusta coffee growth, yield and quality. A meta-analysis. *Agron. Sustain. Dev.* **2020**, *40*, 1–13. [CrossRef]
41. White, P.F.; Oberthur, T.; Pheav, S. *The Soils Used for Rice Production in Cambodia*; International Rice Research Institute: Manila, Philippines, 1997; ISBN 978-971-22-0094-6.
42. Beckschäfer, P.; Seidel, D.; Kleinn, C.; Xu, J. On the exposure of hemispherical photographs in forests. *iForest* **2013**, *6*, 228–237. [CrossRef]
43. Gonsamo, A.; Walter, J.M.; Pellikka, P. CIMES: A package of programs for determining canopy geometry and solar radiation regimes through hemispherical photographs. *Comput. Electron. Agric.* **2011**, *79*, 207–215. [CrossRef]

44. Stan Development Team. Stan Modeling Language Users Guide and Reference Manual 2020. Available online: https://mc-stan.org (accessed on 11 December 2020).
45. Bürkner, P.-C. brms: An R Package for Bayesian Multilevel Models Using Stan. *J. Stat. Softw.* **2017**, *80*, 1–28. [CrossRef]
46. Bürkner, P.-C. Advanced Bayesian Multilevel Modeling with the R Package brms. *R J.* **2018**, *10*, 395–411. [CrossRef]
47. R Core Team. *R: A Language and Environment for Statistical Computing*; R Foundation for Statistical Computing: Vienna, Austria, 2020; p. 12. Available online: https://www.R-project.org/ (accessed on 20 January 2021).
48. Bermúdez-Florez, L.N.; Cartagena-Valenzuela, J.R.; Ramírez-Builes, V.H. Soil humidity and evapotranspiration under three coffee (*Coffea arabica* L.) planting densities at Naranjal experimental station (Chinchiná, Caldas, Colombia). *Acta Agron.* **2018**, *67*, 402–413. [CrossRef]
49. Escalante, E.E. Coffee and agroforestry in Venezuela. *Agrofor. Today (ICRAF)* **1995**, *7*, 5–7.
50. Bosselmann, A.S.; Dons, K.; Oberthur, T.; Olsen, C.S.; Ræbild, A.; Usma, H. The influence of shade trees on coffee quality in small holder coffee agroforestry systems in Southern Colombia. *Agric. Ecosyst. Environ.* **2009**, *129*, 253–260. [CrossRef]
51. Rice, R.A. Agricultural intensification within agroforestry: The case of coffee and wood products. *Agric. Ecosyst. Environ.* **2008**, *128*, 212–218. [CrossRef]
52. Jezeer, R.E.; Santos, M.J.; Boot, R.G.; Junginger, M.; Verweij, P.A. Effects of shade and input management on economic performance of small-scale Peruvian coffee systems. *Agric. Syst.* **2018**, *162*, 179–190. [CrossRef]
53. Van Asten, P.J.; Wairegi, L.W.; Mukasa, D.; Uringi, N.O. Agronomic and economic benefits of coffee–banana intercropping in Uganda's smallholder farming systems. *Agric. Syst.* **2011**, *104*, 326–334. [CrossRef]
54. Craswell, E.T.; Sajjapongse, A.; Howlett, D.J.B.; Dowling, A.J. Agroforestry in the management of sloping lands in Asia and the Pacific. In *Directions in Tropical Agroforestry Research. Forestry Sciences*; Nair, P.K.R., Latt, C.R., Eds.; Springer: Dordrecht, The Netherlands, 1998; Volume 53. [CrossRef]
55. Luna González, A.; Macías Lopez, A.; Taboada Gaytán, O.R.; Morales Ramos, V. Cup quality attributes of Catimors as affected by size and shape of coffee bean (*Coffea arabica* L.). *Int. J. Food Prop.* **2019**, *22*, 758–767. [CrossRef]
56. Suárez de Castro, F.; Montenegro, L.; Aviles, C.; Moreno, M.; Bolaños, M. Efecto del sombrío en los primeros años de vida de un cafetal. *Café El Salvador* **1961**, *31*, 317–350.
57. Morais, H.; Caramori, P.H.; Maria, A.; Ribeiro, D.A.; Gomes, J.C. Microclimatic characterization and productivity of coffee plants grown under shade of pigeon pea in Southern Brazil. *Pesqui. Agropecuária Bras.* **2006**, *41*, 763–770. [CrossRef]
58. Beer, J.W. Production and Competitive Effects of the Shade Trees *Cordia alliodora* and *Erythrina poeppigiana* in an Agroforestry System with *Coffea arabica*. Ph.D. Thesis, University of Oxford, Oxford, UK, 1992.
59. Willey, R.W. The use of shade in coffee, cocoa and tea. *Hortic. Abstr.* **1975**, *45*, 791–798.
60. Diaz, Y.; Torrecillas, A.; Rodriguez, P. Stem diameter fluctuations as stress indicator in fruit trees and its use in irrigation management. *Cultiv. Trop.* **2015**, *36*, 59–66.
61. Barros, R.S.; da Se Mota, J.W.; Da Matta, F.M.; Maestri, M. Decline of vegetative growth in *Coffea arabica* L. in relation to leaf temperature, water potential and stomatal conductance. *Field Crops Res.* **1997**, *54*, 65–72. [CrossRef]
62. Partelli, F.L.; Marré, W.B.; Falqueto, A.R.; Vieira, H.D.; Cavatti, P.C. Seasonal Vegetative Growth in Genotypes of *Coffea canephora*, as Related to Climatic Factors. *J. Agric. Sci.* **2013**, *5*, 108–116. [CrossRef]
63. Solorzano, N.; Querales, D. Crecimiento y desarrollo del café (*Coffea arabica*) bajo la sombra de cinco especies arbóreas. *Rev. For. Latinoam.* **2010**, *25*, 61–80.
64. Rosa Sánchez, J.M.; Conesa Saura, M.D.; Domingo Miguel, R.; Pérez Pastor, A. Contracción temprana de diámetro del tronco. Un nuevo indicador altamente sensible al estrés hídrico en nectarinos extratempranos. In *III Workshop en Investigación Agroalimentaria—WiA3.14, Cartagena, Colombia, 12–13 May 2014*; Universidad Politécnica de Cartagena, Servicio de Documentación: Cartagena, Colombia, 2014; pp. 143–146. ISBN 978-84-697-1358-7.
65. Rathore, A.C.; Mehta, H.; Sharma, N.K.; Gupta, A.K.; Alam, N.M.; Islam, S.; Dogra, P. Performance of litchi (*Litchi chinensis* Sonn.) based agri-horticultural land uses in rainfed condition on degraded lands in North Western Himalayas, India. *Agrofor. Syst.* **2020**, *94*, 2225–2236. [CrossRef]
66. Sarkar, S.; Das, D.K.; Singh, A. Soil micronutrients status of different agroforestry systems in north Bihar. *J. Pharmacogn. Phytochem.* **2020**, *9*, 355–358.
67. Logah, V.; Tetteh, E.N.; Adegah, E.Y.; Mawunyefia, J.; Ofosu, E.A.; Asante, D. Soil carbon stock and nutrient characteristics of *Senna siamea* grove in the semi-deciduous forest zone of Ghana. *Open Geosci.* **2020**, *12*, 443–451. [CrossRef]
68. Foreign Agricultural Service. *Coffee: World Markets and Trade*; United States Department of Agriculture, 2021. Available online: https://apps.fas.usda.gov/psdonline/circulars/coffee.pdf (accessed on 15 November 2020).

Review

Traditional Subsistence Farming of Smallholder Agroforestry Systems in Indonesia: A Review

Budiman Achmad [1], Sanudin [2], Mohamad Siarudin [3,4,*], Ary Widiyanto [2,*], Dian Diniyati [1], Aris Sudomo [5], Aditya Hani [3,5], Eva Fauziyah [2,3], Endah Suhaendah [5], Tri Sulistyati Widyaningsih [6,7], Wuri Handayani [4], Dewi Maharani [5], Suhartono [3,4], Maria Palmolina [8], Dila Swestiani [5], Harry Budi Santoso Sulistiadi [5], Aji Winara [9], Yudha Hadian Nur [9], Muthya Diana [9], Dewi Gartika [9] and Agus Ruswandi [9]

1 Research Center for Social Welfare, Village and Connectivity, BRIN, Jakarta 10330, Indonesia; budi071@brin.go.id (B.A.); dian075@brin.go.id (D.D.)
2 Research Center for Population, BRIN, Jakarta 12710, Indonesia; sanu003@brin.go.id (S.); evaf001@brin.go.id (E.F.)
3 Forest Science Graduate Program, Faculty of Forestry, Universitas Gadjah Mada, Yogyakarta 55281, Indonesia; adit029@brin.go.id (A.H.); har436@gmail.com (S.)
4 Research Center for Ecology and Ethnobiology, BRIN, Bogor 16911, Indonesia; wuri003@brin.go.id
5 Research Center for Plant Conservation, Botanic Gardens, and Forestry, Bogor 16122, Indonesia; aris.sudomo@brin.go.id (A.S.); enda051@brin.go.id (E.S.); maharanid858@gmail.com (D.M.); dila001@brin.go.id (D.S.); harr014@brin.go.id (H.B.S.S.)
6 Leadership and Policy Innovation Program, The Graduate School, Universitas Gadjah Mada, Yogyakarta 55281, Indonesia; tri.sulistyati.widyaningsih@brin.go.id
7 Bureau for Organization and Human Resources, BRIN, Gedung B.J. Habibie, Jl. M.H. Thamrin No. 8, Jakarta Pusat 10340, Indonesia
8 Research Center for Society and Culture, BRIN, Jakarta 12710, Indonesia; mari039@brin.go.id
9 Research and Development Agency of West Java Province, Jl. Kawaluyaan Indah Raya No. 6, Kelurahan Jatisari, Kecamatan Buah Batu, Kota Bandung 40286, Indonesia; ajiwinara@jabarprov.go.id (A.W.); yudhahadian@jabarprov.go.id (Y.H.N.); muthyadiana@jabarprov.go.id (M.D.); dewigartika@jabarprov.go.id (D.G.); agusruswandi@jabarprov.go.id (A.R.)
* Correspondence: mohamad.siarudin@brin.go.id (M.S.); aryw002@brin.go.id (A.W.)

Citation: Achmad, B.; Sanudin; Siarudin, M.; Widiyanto, A.; Diniyati, D.; Sudomo, A.; Hani, A.; Fauziyah, E.; Suhaendah, E.; Widyaningsih, T.S.; et al. Traditional Subsistence Farming of Smallholder Agroforestry Systems in Indonesia: A Review. *Sustainability* **2022**, *14*, 8631. https://doi.org/10.3390/su14148631

Academic Editor: Victor Rolo

Received: 14 June 2022
Accepted: 12 July 2022
Published: 14 July 2022

Abstract: Agroforestry has been practiced for decades and is undoubtedly an important source of income for Indonesian households living near forests. However, there are still many cases of poverty among farmers due to a lack of ability to adopt advanced technology. This literature review aims to identify the characteristics and factors causing the occurrence of agricultural subsistence and analyze its implications for the level of farmer welfare and the regional forestry industry. The literature analysis conducted reveals that small land tenure, low literacy rates, and lack of forest maintenance are the main causes of the subsistence of small agroforestry farmers. Another reason is that subsistence-oriented agroforestry practices are considered a strong form of smallholder resilience. All of these limitations have implications for low land productivity and high-sawn timber waste from community forests. To reduce the subsistence level of farmers, government intervention is needed, especially in providing managerial assistance packages, capital assistance, and the marketing of forest products. Various agroforestry technologies are available but have not been implemented consistently by farmers. Therefore, it is necessary to develop an integrated collaboration between researchers, farmers, and regionally owned enterprises (BUMD) to increase access to technology and markets. Although it is still difficult to realize, forest services, such as upstream–downstream compensation and carbon capture, have the potential to increase farmer income.

Keywords: agroforestry; collaboration; farmers; government intervention; subsistence

1. Introduction

Agroforestry has long been practiced in Indonesia. As a sustainable land-use practice [1], it increases overall agricultural productivity by combining woody perennial with food crops, including livestock on the same land [2]. When the concept is practiced appropriately, it can provide economic, ecological, social, and cultural benefits. The crop diversity in intercropping can provide a good income on a daily, monthly, and yearly basis. However, it depends on the agroforestry pattern adopted, such as agrisilviculture (forestry–agriculture), silvopasture (forestry–animal husbandry), agrisilvopasture (forestry–agriculture–livestock), silvofishery (forestry–fisheries), bee-forestry, sericulture (silkworm forestry), or multipurpose forest tree production systems (fusion complex). Ecologically, agroforestry supports soil and water conservation. With multi-strata plants, of course, it can minimize the occurrence of erosion, reduce runoff, and increase the effectiveness of water absorption. The spatial pattern of agroforestry can also function as a windbreak [3,4].

Sustainable agroforestry is thought to be a future agricultural practice as an alternative to unsustainable conventional agriculture [5]. In conventional agriculture, maximum tillage accelerates the decomposition of organic matter, thereby reducing its presence in the soil, while in the agroforestry pattern, the rate of decomposition of organic matter is slower due to a minimum tillage balance with the input of organic matter from trees as a complementary strategy [6]. Maximum tillage has the potential to reduce mycorrhizal fungi and increase runoff so that soil organic matter is reduced, while conservation tillage, by minimizing soil damage, can increase the presence of mycorrhizal fungi, as well as the absorption of phosphorus and soil aggregates [6,7]. The conservation agriculture principle is to have minimum soil disturbance and crop rotation while maximizing cover crops to obtain increased yields (30–200%) and labor efficiency (60%) [6]. Conservation agriculture with trees (CAWT) avoids maximum tillage to prevent the negative effects of intensive tillage, such as from ploughing (barren soil, erosion, heating, decomposition of organic matter, and damage to structures and nature) [6]. The top layer of soil, which is responsible for supporting crop life but is also the most susceptible to erosion and degradation, must be protected with particular care [6]. Crop rotations should include legumes, deep-rooted crops, and high-residue crops that have fixed nitrogen in the soil, and their biomass should add nitrogen through decomposition [6]. The litter and roots of tree components continuously add plant nutrients to the soil [8].

The function of trees in CAWT is potentially positive for agricultural crop production [6] and contributes to soil nutrient enrichment and crop production [8]. The function of trees in agroforestry, such as in fertilizer trees, can optimize the supply of native soil N and increase land productivity. The advantage of using fertilizer trees (*Gliricidia* sp., *Calliandra* sp., *Leucaena* sp., etc.) in agroforestry is that they ensure a multifunctional farm that provides wood, fodder, shade, soil improvement, and watershed breeding [9]. As an integrated, tree-based farming system, agroforestry is a reliable system due to its potential to address land degradation with additional environmental and social benefits [10,11]. In addition, agroforestry supports biodiversity conservation [12] and has higher financial returns than conventional agriculture [13]. Agroforestry also contributes significantly to climate change mitigation by increasing carbon sequestration and storage in the biosphere [14]. Furthermore, in contrast to conventional systems, agroforestry systems can better maintain biodiversity and provide food security, land security, and financial security [15].

Fertilization technology increases food crop production and provides additional income for households through sources such as the selling of tree seeds and firewood [16]. The choice of technology is driven by the size of the landownership, and more benefits are associated with larger landholdings. The adoption of agroforestry in subsistence agriculture is often limited by local social conditions and natural endowments [17].

Economically, agroforestry practices are a part of the livelihood strategies of farmers. In some cases, smallholder forests are the main source of income and even cause land owners to occupy a higher social status. Land size can be one of the factors that affects the economic value obtained from agroforestry systems. The larger the land area, the greater the economic value generated [18]. In addition, the adoption of agroforestry can diversify farmers' livelihoods and increase their income [19]. Moreover, some agroforestry practitioners adjust to an economic focus while keeping an ecologically sound development orientation. However, they occasionally sacrifice the sociocultural aspect [20].

Agroforestry has numerous advantageous effects on the environment, including on how land is used, which leads to ecological, economic, and social benefits [21]. As a result, its sustainability must be preserved. In the case of food security, smallholder farming with commercial agroforestry systems tends to focus on income production, whereas traditional systems concentrate on the benefits of nutritional diversity. Agroforestry benefits the environment and promotes stability [22]. The mixed garden, the most popular agroforestry pattern in Indonesia, also has the highest carbon stock compared to other tree-crops patterns [23].

In Indonesia, subsistence farmers are accustomed to going into debt to obtain the initial capital for farming, which will then be paid off at the time of harvest. When there is asymmetrical information between farmers and creditors about price knowledge and market access, a debt-bondage system to help farmers with financial capital does not appear to be a viable aid to the farmers [24]. Related to income, interactions in factor markets cause price shocks in these markets, which subsequently allow for essential products to reach subsistence producers. Additionally, this lowers wages and land rentals, boosting household subsistence production. As a result, subsistence households' real income decreases [25]. However, under certain circumstances, subsistence farming can operate as a stabilizer and benefit all agriculture [26]. From a macroeconomic perspective, an improvement in semi-subsistence agricultural production might boost economic growth overall, lower the trade deficit, enhance household incomes, and boost government revenue [27].

Despite all these advantages, the adoption of agroforestry systems is still low, and the adoption gap remains largely unexplained [17]. There are disincentives for planting trees among the understory, including a lack of knowledge, upfront costs, long periods of time before there is a return, and reduced short-to-medium-term cash flow and/or household food production [28]. For subsistence farmers, the existence of trees will be detrimental to agricultural crop production if soil tillage is carried out as in conventional agriculture. The ability to integrate trees on agricultural land is strongly influenced by the perception and knowledge of farmers [29], where farmer managerial skills in implementing agroforestry are still low [30]. In a limited treatment of trees, the presence of trees can reduce the growth of commercial food crops [31], while no-tillage can produce higher maize and cassava yields than tillage [32]. As a result, farmers who are unable to produce enough food for their livelihood, and who depend on cash income to meet many expenses, engage in irregular, nonagricultural commercial activities to generate income for the provision of food and other necessities [33].

Likewise, the adoption of such promising land-use practices has been slow in terms of achievement [1]. Agroforestry programs should not be considered a poverty alleviation strategy [1]. This is because smallholders may not be able to cope with the initial production losses resulting from the transition from conventional agriculture to agroforestry. Hence, policy interventions are essential in order to involve smallholders in the promotion of agroforestry. Attention and incentives should be given to traditional smallholder agroforestry farmers who have helped to balance between biodiversity conservation and economic growth [8].

Most agroforestry programs in Indonesia prioritize land rehabilitation and are focused on generating long-term economic benefits from a particular crop. Little is known about whether and how poor farmers behave differently from nonpoor farmers in adopting agroforestry practices. To reap the greatest benefits from agroforestry systems, a fundamental understanding of how and why farmers make long-term land-use decisions is required [34].

It is truly becoming a question of why the high potential advantages of the agroforestry system seem to have no impact on alleviating poverty for farmers in Indonesia. Various references to agroforestry practices indicate that, until recently, they were still at the subsistence level, e.g., Refs. [35–38], even though some are already at either the subsistence or commercial levels, e.g., Refs. [3,20,39]. Studies on subsistence farming in agroforestry practices seem to be scattered in case studies that are probably site-specific to a region and only emphasize certain topics. Therefore, a comprehensive review is needed to understand the subsistence farming phenomenon and formulate possible solutions to obtain an alternative management practice. This paper aims to review the nature of subsistence farming, identify related factors, and offer alternative solutions. We overview the developmental phases of agroforestry in Indonesia, then we highlight subsistence farming and identify the internal and external contributing factors. We also discuss the alternative strategies needed to achieve sustainable farming based on the strengths, weaknesses, opportunities, and treatments of the smallholder farmer side of the equation.

2. Methods

This review was conducted based on agroforestry publications derived from various reputable sources through the thematic proximity approach to capture the nature of agroforestry in Indonesia, especially farming under subsistence conditions. Some keywords in English and Indonesian were used to find relevant issues by employing a search engine. An intensive search for online publications for 2000–2022 was carried out in February–March 2022. The stages of searching and screening the publications are shown in Figure 1. This process resulted in 123 articles, which were then deeply studied because they were relevant to the subdiscussions: (1) existing conditions of agroforestry practices, (2) factors influencing subsistence in agroforestry management (biophysical, social, and economic), and (3) subsistence agroforestry management strategies.

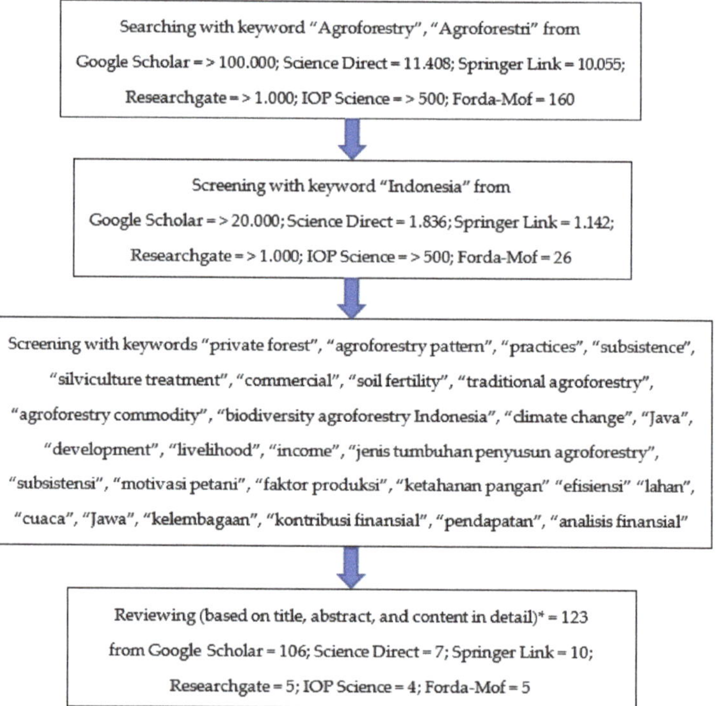

Figure 1. Stages of the literature selection. Note: * The same 14 articles were found in two different sites.

The review was carried out by reading the contents of the literature in detail. The selecting criteria for the literature reviewed in the subdiscussions (that is, of the existing conditions of agroforestry practices) were that they discussed issues related to (1) traditional agroforestry systems in Indonesia, including the types of plants and cropping patterns;

(2) the knowledge and motivation of farmers to develop traditional agroforestry; (3) the management, institutions, and problems of traditional agroforestry; and (4) the benefits of agroforestry, including social, economic, and environmental benefits. The selecting criteria for the literature reviewed in the subdiscussion on the factors influencing subsistence in agroforestry management were that the references discussed issues related to: (1) the explicit or implicit definition of subsistence agroforestry, farmer perceptions of privately owned forest management, agroforestry on limited land, traditional agroforestry, and dry land use; (2) the characteristics of subsistence agroforestry and the characteristics of farmers in subsistence agroforestry; (3) the motivations and factors that influence farmers to manage subsistence agroforestry, factors that influence farmers to adopt agroforestry patterns, and factors that influence the sustainability of agroforestry; and (4) the contribution of agroforestry to farmers, especially with respect to small-scale (household) needs and their problems. In the subdiscussions on subsistence agroforestry management strategies, the selected literature considered external and internal factors to fill in the SWOT framework analysis. To obtain internal factors (strengths and weaknesses), the selected literature focused on smallholder agroforestry management practices in Indonesia. Determination of external factors (opportunities and challenges) was identified from references to general conditions related to agroforestry businesses and agroforestry policies in Indonesia. After finding the external and internal factors, several agroforestry development strategies could be analyzed. Synchronization between the results of the SWOT analysis with the results of previous studies was carried out by searching for related literature.

After obtaining the existing conditions of agroforestry, the factors causing subsistence, and the development strategy, more technical and detailed steps were needed about smallholder agroforestry practices in Indonesia. This was achieved by synthesizing the literature on the selection of agroforestry species; multi-businesses with integrated farming; the intensification of agroforestry, soil, and water conservation on critical lands; and adaptation to climate change. Studies related to social policy and marketing dimensions were needed to support the intensification of agroforestry in order to help us improve agroforestry management practices and provide more welfare for smallholders. The articles reviewed were written in English and Indonesian, such as journal articles, proceedings, reports, and theses, as listed in Appendix A, Table A1.

The results of the literature search were used to present the development and existing conditions of agroforestry in Indonesia. The review then focused on factors causing subsistence in agroforestry practices in Indonesia, as well as the strategies required to develop sustainable agroforestry management for smallholder farmers. Cases that occurred in Indonesia were used as references to seek the causative factors of subsistence in Indonesia. However, given the limited publications regarding external factors, the references referred to were enriched from outside cases in various countries (e.g., Bangladesh, Brazil, India, Kenya) similar to those in Indonesia.

Our strategy to overcome the subsistence condition was analyzed using SWOT analysis, a mapping of external and internal factors that affect smallholder agroforestry businesses. The internal factors were the strengths (S) and weaknesses (W) of smallholder agroforestry businesses in Indonesia, while the external factors were opportunities (O) and threats (T) in developing smallholder agroforestry businesses. In this section, some points will inevitably be similar to the factors causing subsistence e.g., limited landownership, limited financial capital, low education level, and limited knowledge of agroforestry. However, the points in the SWOT analysis are more for smallholder agroforestry in general, in this case, subsistence, semi-commercial, and commercial.

3. Results and Discussion

3.1. Development Phase of Agroforestry Practices

Historically, there were three phases of agroforestry development, namely, classical, premodern, and modern agroforestry. Classical agroforestry was practiced in approximately 700 BC in the form of slash-and-burn, including shifting cultivation, which was a transformation from a hunting lifestyle and food gathering into plant and animal domestication. Premodern agroforestry, recognized as Taungya, was mainly aimed at producing forest products, and in Indonesia, it was introduced by the colonial government in the form of teak forest development at the end of the 19th century. Attention to agricultural components, farmers, and crop production in the Taungya system was low, but it was designed more to benefit the government's agenda. This philosophy was intended to mobilize landless and jobless laborers in rural areas to work in state forests, with compensation granting them the right to utilize space among the trees to plant crops. Therefore, rural communities felt fewer benefits from such systems. Several international organizations were generated to provide policy and programs to improve food productivity and environmental conservation, such as Social Forestry (SF) by the World Bank, Forestry for Rural Development (FRD) by the FAO, and the agroforestry research institute ICRAF by the World Agroforestry Centre.

3.2. Existing Conditions of Agroforestry in Indonesia

The practice of agroforestry in private forests plays a pivotal role in cultivating trees outside the state forest areas, providing a farmer safety net in terms of economic value through production functions [40], and serving as the last bastion of biodiversity conservation for flora and fauna amid the accelerated deforestation rate in natural forests [41–44]. Agroforestry has contributed to strengthening smallholder farmers' resilience to climate change in Indonesia by offering 20% more food variety in the traditional pattern and a fivefold income increase in the commercial pattern [22]. Agroforestry systems also provide environmental service benefits, such as increasing soil organic content; improving soil health through nutrient repair and fertility processes; improving soil biological dynamics; improving soil carbon sequestration and climate change mitigation; and improving water quality, climate change mitigation, and adaptation [39,45,46].

Some of the benefits related to soil nutrients and fertility, as well as soil carbon sequestration, can be obtained from minimal or no-tillage practices, even without trees. However, the presence of trees in agroforestry systems can add more benefits. The presence of trees and perennial plants in agroforestry produces the highest aboveground carbon stocks, including belowground carbon stocks, thereby improving carbon sequestration and mitigating climate change [23,39]. Trees in agroforestry also enhance soil organic content, increase soil nutrients and fertility, and increase soil microbial dynamics, which have a positive effect on soil health [45]. Trees improve soil quality in agroforestry through three main processes, namely, increasing input with tree fertilizers (N2-fixing), increasing soil nutrients from the production and decomposition of tree biomass (falling of leaves, branches, twigs), absorption, and the utilization of nutrients from deep tree roots, thus creating a nutrient cycle in the agroforestry system [46].

Traditional agroforestry systems have long been applied in different parts of Indonesia and have become local wisdom in each area with various vernacular names, as presented in Table 1. A distinctive feature of traditional agroforestry systems is the selection of diverse crops to sustain the resilience of farmer households in accessing sources of food, timber, firewood, and herbal medicines [3,4]. These agroforestry patterns are determined based on the skill of the local farmers and agroecological conditions [11]. The diversity of agroforestry constituents consists of two categories, namely, simple agroforestry and complex agroforestry. In general, simple agroforestry uses one species of commercial tree as the main plant intercropped with one commercial understory, resulting in low plant diversity [47]. Complex agroforestry involves growing two or more species as the main trees along with moderate plant diversity, and the practice is usually found in mixed gardens or home garden patterns [48–51]. However, the diversity of annual crop and tree species in agroforestry systems is higher than that of agricultural and forest crops [38].

Along with the socioeconomic, cultural, and policy development of rural communities, some agroforestry patterns have also improved from traditional subsistence management to business–commercial management with the selection of several types of commercial or industrial crops, such as coffee [52] and porang [53], or integration with ecotourism [54]. The relative profitability of agroforestry business models can be measured based on the associated risks and timeframes [55]. However, the basis of the farmers' considerations in choosing the type of crop is influenced by economic, occupational, cultural, and educational background [56,57]. It is common to have a combination of commercial and noncommercial crops to optimize land use in order to fulfill household needs in terms of subsistence or as a source of daily, monthly, and annual financial income [3,4]. In general, farmer preferences are more biased toward economic benefits and mindfully maintaining the availability of a food supply, but they pay less attention to ecological aspects [19]. Crops that provide instant extra income are more desirable to farmers [58].

Generally, farmers in Java with narrow farmland ownerships tend to cultivate commercial timber trees (such as teak, sengon, mahogany, gmelina, manglid, African wood, and jabon) combined with understory crops that can support daily needs such as food, buildings, and traditional medicines, including coffee, chocolate, cardamom, and tea [59–62]. Some examples of simple agroforestry patterns on smallholder farmer land in Java are displayed in Table 1. Additionally, farmers on islands other than Java choose many types of nontimber trees or fruit trees as the main stand, as well as industrial crops as the understories, as presented in Table 1. Some examples of agroforestry patterns that are widely practiced by the community, especially on the island of Java Island, can be seen in Figure 2.

Figure 2. Agroforestry models in Indonesia: (**a**) *Paracerianthes moluccana* and *Amomum compactum* (picture by Achmad, 2012), (**b**) *Anthocephalus cadamba Miq* and coffee (picture by Achmad, 2020); (**c**) *Tectona grandis* and *Manihot utilissima* (picture by Widiyanto, 2016); (**d**) *Tectona grandis* and *Oryza sativa* L. (picture by Widiyanto, 2015).

The main objective of silviculture is to improve timber productivity by applying tree improvement, site manipulation, and plant protection [63]. With agroforestry, the science of silviculture is expected to encounter issues beyond timber production [64]. Moreover, silviculture treatment on timber-based versus nontimber-based forest products provides a different result for land productivity [22]. Traditional agroforestry practices are usually subsistence and are conducted in a small area of land with minimal silviculture treatments [65,66].

The recognition to apply agroforestry also arises when the benefits of ancient integrated agricultural systems survive after a series of land problems, such as deforestation, soil degradation, and biodiversity decline [46]. This implies that agroforestry contributes to the environment by reducing erosion rates, restoring degraded land, reducing landslide risk, increasing carbon stock, and affecting microclimate and soil moisture [67–69]. The agroforestry systems developed in several locations in West Java are able to decrease soil erosion and restore land degradation, while the highest carbon stock is produced from mixed types with a larger number of trees [23]. Agroforestry, which dominates the foothills of volcanic mountains prone to landslides in Java, can reduce landslide reactivation by selecting species of trees and plants with certain ecological functions to reduce surface runoff, water absorption, and soil moisture without compromising the social and economic value of farms [70]. A mixture of deep-rooted tree species and grasses with smooth and dense rooting can maintain the stability of riverbanks and high hillsides [71]. The diversity of tree root distribution between species can reduce the landslide risk in productive coffee agroforestry systems [71]. The agroforestry system is an easy and affordable way to mitigate microclimate instability and soil moisture degradation, and it can be potentially employed as an adaptive strategy to counter extreme climate impacts through shade cover arrangements [72].

The transition from subsistence to semicommercial or commercial agroforestry is not a simple process. Problems regarding landownership and access, market opportunities, and farmer regeneration always arise from farmers. It is common knowledge that rural farmers are generally villagers over the age of 50 who need guidance on how to maintain their agricultural businesses, especially in regard to utilizing wireless networks. Currently, few among the young generations want to continue working as farmers. Ironically, most parents who work as rural farmers are also unwilling to pass down the profession [73]. The factors that hinder farmer regeneration are the low income generated by the agricultural sector, demanding and laborious work, the perception that farming is only suited to those from a poor background with limited education, and the type of work chosen as the last alternative [74]. This results in only 26.67% of farmers' children having high motivation to become farmers [75]. The education level of farmers, according to Statistics Indonesia (BPS), is still dominated by elementary and junior high school levels. Meanwhile, college graduates and diplomas only account for 0.57%. This low education level can affect farmers' managerial abilities in developing agroforestry. Farmers' managerial capacity in planning, implementing, and evaluating the application of agroforestry systems is a major factor that influences their motivation to apply technology [76]. The low quality of human resources is a drawback in the agricultural sector, which may lower agricultural productivity [77].

In Lampung, many smallholder farmers have the basic skills to turn their subsistence horticultural systems into semicommercial ventures. To facilitate this process, farmers obtain assistance to (1) identify appropriate horticultural species/cultivars that are suitable to the land's biophysical and their own socio-economic conditions; (2) adapt to vegetative propagation and other horticultural management practices; and (3) develop a permanent market relationship. In West Sumatra, 1–2 hectares of agroforestry provide 26–30% of total household income. This percentage range is due to differences in land size, crop selection, level of commercialization, and the intensity of agriculture [78].

The success of agroforestry development needs support from internal and external factors. However, the existing institutions have not yet optimally supported the development of agroforestry [79]. The institutional development of agroforestry can be guaranteed if: (1) incentives are provided for agroforestry farmers or organizations; (2) organizational reinforcement is provided; (3) equality in infrastructure and information asymmetry is guaranteed; (4) assurance about ownership and access to resources is obtained; (5) control over opportunistic behavior is established; and (6) some rules are enforced and obeyed [80]. The key factors of agroforestry development in Indonesia are farmer institutions [81], policy support, the availability of technology packages, the optimization of stakeholder involvement [79], and leadership [82]. Farmer groups are required to accommodate member activities; facilitate farmer-to-farmer relationships; and handle peer group mentoring concerning silvicultural techniques, such as pruning [83]. In addition, farmers gain better interaction and communication skills by joining the farming community [84]. Therefore, the farmer group, as an institution, has benefits and roles to play in improving each member's economic wellbeing as well as sustaining the forest. The implementation of these activities should rely on the personal interests and motivations of smallholder farmers to improve their livelihoods rather than depend on the government's project financing. The improvement and expansion of smallholder horticultural systems can also serve public environmental purposes [85]. Unfortunately, the value of the environmental service of agroforestry is underappreciated and is unreliable as a cash source, at least with respect to today's conditions.

Table 1. Existing agroforestry patterns in Indonesia.

Planting Patterns	Location	Key Commodities	Plant Diversity	Source
Tea-based agroforestry	Tasikmalaya, West Java 850 msl	*Magnolia champaca* and *Camellia sinensis*	18 species Shannon–Wiener Diversity Index (H') = 0.37–0.66 (low)	[86]
Sengon + clove + spices	Magetan, Central Java	*Paraserianthes falcataria*, *Zingiber* sp. *Syzygium aromaticum* and *Curcuma longa*		[87]
Cocoa-based agroforestry	Bulukumba, South Sulawesi	*Theobroma cacao*	29 species H' < 1 (low)	[88]
Coffee-based agroforestry	Bulukumba, South Sulawesi	*Coffea* spp.	17 species H' < 1 (low)	
Coconut-based agroforestry	Bulukumba, South Sulawesi	*Cocos nucifera*	26 species H' < 1 (low)	
Cashews + guava	Bulukumba, South Sulawesi	*Anacardium occidentale*	8 species H < 1 (low)	
Mixed crop home garden	Tana Toraja, Bone & Bulukumba South Sulawesi	*Casuarina junghuhniana*, *Pinus merkusii*, *Elmerillia pubescens*, *Syzigium aromaticum*, *Tectona grandis*, *Aleurites moluccana*, *Artocarpus heterophyllum*, *Coffea arabica*, and *T. cacao*	47 species H' = 1.25–2.18 (moderate)	[56]
Damar-based agroforestry	Lampung	*Shorea javanica*	93 species	[51,89]
Dukuh agroforestry system, South Kalimantan	Banjar Regency, South Kalimantan	*Durio zebethinus*, *Artocarpus integer*, *Lansium parasiticum*, *Hevea brasiliensis*, *C. longa*, *Kaempferia galanga*, *Zingiber officinale*, *Alpinia galanga*, and *Mussa paradisiaca*	9 species	[90]
Sengon-based agroforestry	West Java	*P. falcataria*, *Swietenia macrophylla*, *Maesopsis eminii*, *Cocos nucifera*, and *Amomum cardamomum*	7 species	[58,91]
Bamboo-based agroforestry	Bandung Regency, West Java 700–900 masl	*Gigantochloa verticillate*, *Gigantochloa pseudoarundinacea*, *Gigantochloa apus*, and *Gigantochloa ater*, *Bamboosa vulgaris*, *M. eminii*, *P. falcataria*, *Hibiscus macrophyllus*, *Melia azedarach*, *Toona sureni*, *M. champaca*, *Lansium* spp., *Persea americana*, *Syzygium polycephalum*, *Mangifera odorata*, *Baccaurea racemose*, *Syzygium aromaticum*, and *C. arabica*	76 species	[48]
Mixed garden (home garden)	Central Bengkulu	*H. brasiliensis*	38 species	[92]
Traditional mixed garden (traditional agroforestry)–mixed garden	Central Sulawesi	*A. moluccana*, *Parkia speciosa*, *Artocarpus heterophyllus*, *T. grandis*, *C. nucifera*, and *T. cacao*	43 species in Ta'a Wana, 52 species in Javanese's village, 39 species in Balinese's village	[93]
Talun (mixed forest)	Bogor, West Java	Bamboo	94 species (44 families)	[49]
Mixed garden	Bogor City, West Java	Fruit trees	83 species (41 families)	[49]
Garden farming	Bogor City, West Java	Fruit trees	100 species (45 families)	[49]
Home garden agroforestry	Banyuwangi, East Java 600–800 masl	*C. nucifera*, *Garcinia mangostana*, *D. zibethinus*, *S. aromaticum*, *Coffea liberica*, *Coffea canephora*, and *Nephelium lappaceum*	Trees (39 species), Poles (9 species), Herbs (41 species), Liana (8 species)	[94]
Mindi + sengon + african wood + baros + palawija	Garut, West Java 750–1400 masl	*Melia dubia Cavanilles*, *P. falcataria*, *M. eminii*, and *Manglieta glauca*	14 species	[95]
Cofee-based agroforestry	Banjar, South Kalimantan	*H. brasiliensis*, *Coffea* sp, *Lansium parasiticum*, *Arthocarpus champeden*, *C. nucifera*, and *Musa* sp.	5 species	[96]
Trees + rice + cassava	Bondowoso, East Java 500–900 masl	*Falcataria moluccana*, *T. grandis*, *Gmelina arborea*, *Oryza sativa*, *Manihot esculenta*, and *Zea mays*	35 species	[97]

3.3. Why Subsistence

3.3.1. Subsistence Outlook

Subsistence farming is defined as self-sufficient farming in which farmers focus on cultivating sufficient quantities of food for their families. In addition, subsistence agriculture is characterized by such things as having a wide variety of crops and livestock to eat, and sometimes fiber for clothing and building materials. The decision in determining the type of plant to use usually depends on the type of food that will be consumed in the coming year. It is also determined by market prices, where if the price of a commodity is considered too high, they choose to plant their own [98]. Although they are considered to prioritize self-sufficiency for their families, most subsistence farmers also trade a few of their agricultural products, especially for obtaining goods that cannot be produced from the land, such as salt and kitchen equipment. Most subsistence farmers currently live in developing countries. Numerous subsistence farmers grow alternative crops and have agricultural capabilities that are not found in advanced agricultural methods [99]. Subsistence refers to those who are periodically food insecure, relying on irregular cash income from diversification into a range of sources [100].

Some farmers apply agroforestry systems based on economic considerations rather than social and ecological considerations [30]. This is indicated by the selection of plant species that make up agroforestry with the main objective of utilizing the results to meet the needs of farmer households in the short, medium, and long term. Agroforestry is widely adopted by farmers because this system can increase income while also diversifying their livelihoods [19]. Agroforestry provides income to farmers in the form of weekly, monthly, and annual income [101]. Farmers in Wonogiri earn weekly income from cayenne pepper, monthly income from secondary crops, and annual income from timber plantations [101]. Farmers apply simple agroforestry by intercropping trees with one or more seasonal crop types. Sengon (*Paracerianthes moluccana*) is the main crop because it is considered to possess high economic value and is a form of family savings that can be used for certain urgent needs [102]. Sengon and salak (*Salacca zalacca*) agroforestry can meet the daily needs of families and can support a balanced work structure [103]. The traditional agroforestry system is carried out to meet daily needs, and some of the surpluses are sold to collectors and weekly village markets, such as palm sugar (*Arenga pinnata*), banana (*Musa paradisiaca*), sapodilla (*Manilkara* sp.), mango (*Mangifera indica*), avocado (*Persea americana*), petai (*Parkia speciosa*), and jengkol (*Archidendron pauciflorum*) [104].

Farmers also obtain animal feed from community forests so that they do not need to look for grass in places far from home [101]. Income from the agroforestry system can meet the needs of a family with four to five dependents. Household expenses can be reduced because some foodstuffs can be obtained from the forest, such as vegetables, cayenne pepper (*Capsicum frutescens*), cassava (*Manihot esculenta*), corn (*Zea mays*), turmeric (*Curcuma longa*), ginger (*Zingiber officinale*), and galangal (*Alpinia galanga*).

Benefits in the form of income obtained by farmers from plant cultivation activities have caused farmers to continue cultivating plants, including forest plants. More often than not, farmers' primary motivation to sell wood is to fulfill their urgent needs, for example, wanting to hold a celebration, going on a pilgrimage, paying for children's educational fees, or other needs, which are often referred to as the cutting-and-needed system. More than half of agroforestry income comes from selling sengon wood as the main crop [102]. Agroforestry has increased the security of farmer livelihoods as a safety net that helps households pass through periods of increased vulnerability, for example, due to crop failure and illness [105]. Agroforestry systems are used to support subsistence needs, increase income through the sale of surplus produce, and strengthen the ownership situation of farmers.

3.3.2. Factors Related to Subsistence in Agroforestry

The subsistence level of agroforestry can be seen from the perceived benefits of the agroforestry system, which are limited to meeting the needs of the family, not for commercial purposes. Two factors influence the agroforestry farming system, namely, internal factors such as farmer experience, motivation, landownership area, number, and type of plants, as well as external factors in the form of support from agroforestry extension institutions and community leaders [106]. Likewise, the subsistence of farmers is also influenced by these two factors.

1. Internal factors

The internal factors that influence the subsistence of agroforestry farmers are presented in Table 2.

Table 2. Internal factors that lead to the subsistence of agroforestry farmers.

No	Internal Factor	Source
1.	Limited landownership	[18,107,108]
2.	The character of farmers who are less willing to take risks	[17]
3.	Low education level and poor agroforestry knowledge	[21,38,101]
4.	Limited financial capital	[109,110]
5.	Farmer preferences related to gender and cultural identity	[108,111]

- Limited landownership

Limited land is one of the factors causing farmers to grow only crops to meet family needs. Most farmers own less than 1 ha of land [112] with an average agroforestry area of 0.5 ha [102]. The dominant species are plants that can be consumed in the form of food crops. Farmers who have narrow lands prefer to grow various types of crops to meet subsistence needs and, at the same time, have savings [113], although some farmers replace traditional crops such as rice, corn, and vegetables with valuable commercial crops such as taro, pineapple, banana, papaya, and teak trees [105]. The area of land managed by farmers can come from their own land, a rental system, or profit-sharing. Land fragmentation increases when adult household members marry, create their own families, and manage land separately. Land expansion is difficult if the available land is limited, which results in crop yields that are not sufficient for family needs [105].

Limited landownership may push farmers to choose a management system, whether it is intensive agriculture or agroforestry. One consideration for farmers converting agricultural land into community forests is the management of agricultural land, which, in addition to requiring large capital support, also requires a large amount of labor [101].

- Not willing to take risks

Decisions regarding agroforestry adoption are carried out based on natural and social endowments such as preferences, incentives, and risk, as well as uncertainty assessments across three dimensions, that is, profitability, feasibility, and acceptability [114]. The majority of investments are designed to produce direct production for domestic use. However, any surplus can be sold on occasion. Households with minimal assets may find themselves with less, if any, assets for other occupations after allocating land, labor, cattle, time, and tools to one activity. Escaping food poverty demands investments, but because margins are tight, their willingness to invest in new technologies may give way to cost considerations, resulting in non-adoption if decisions are based as much on past and current conditions as on potential profits [17].

Production hazards in small-scale agriculture include animal and crop illness, drought, flood, climatic unpredictability and change, and/or market shocks, which can affect individuals or entire communities [100]. Even the most attractive investments might come with enormous risks if they fail. Even though they are sometimes used interchangeably, risk and uncertainty "convey different aspects" [115].

Risk minimization is consequently critical, not just profit maximization, especially for food-insecure farmers. While agroforestry adoption is a high-risk endeavor, other investments, such as raising animals, are frequently viewed as insurance [116]. External risk reduction options, unrelated to agroforestry but rather to animal husbandry as a key part of subsistence farming, may include insurance programs or warranties, as well as the provision of animal health support services. In poor households, livestock investments are frequently a top priority for savings [117].

Small farmers tend to prefer continuous yields, even though they are small compared to large yields, which are intermittent. This is because smallholders mostly need certain short-term income rather than uncertain long-term income (both in terms of yield and price) and risk. The experience of tumor rust disease in sengon stands is an example of the risk of loss experienced by Ciamis farmers.

- Low education level and poor agroforestry knowledge

The knowledge and perceptions of farmers influence how they manage their land with agroforestry patterns. Their knowledge causes them to have low motivation in managing agroforestry optimally, so the results obtained are only sufficient to meet the needs of their families [76]. A farmer's economic orientation may not be for commercial purposes but rather to meet daily household food reserves, as well as to use the farm as a vegetable source [118]. In addition, farmers also use wood twigs as fuel [101]. Cultivators continue to carry out agroforestry cooperation programs because they can help meet daily needs and improve farmer welfare if the management is carried out optimally [119].

The knowledge and skills involved in cultivating timber and agriculture are obtained by farmers from their parents and other ancestors from childhood onward [101], and on average, farmers have to manage community forests for more than twenty years. Experience in agroforestry agriculture can support the process of increasing a farmer's capacity [113].

Formal education is significantly correlated with farmers' perceptions of community forest management at a significance level of 5%. The higher their level of formal education, the higher their perceptions of community forest management using agroforestry patterns [101]. They do not perform thinning and assume that the more trees they have, the more results they will obtain [101]. Understanding the perception and knowledge of smallholder farmers regarding the integration of trees on farms is essential for minimizing the barriers to integrating trees on farms [120]. Farmers' capacity to implement agroforestry systems is still categorized as low, especially in terms of farmer managerial capacity [30].

Farmers' knowledge of the types of plants that are resistant to shade is still limited. Some farmers still make decisions in choosing understory crops based on market demand [121]. Ultimately, the expected results are typically not obtained. For the crop types commonly grown on their land, farmers plant based on hereditary knowledge from their ancestors [122]. Some ecological aspects that affect the sustainability of agroforestry include the rate of plant pest and disease attacks and the farmers' level of understanding with respect to soil and water conservation, land conservation measures, the availability of organic fertilizer manufacturing technology, the availability of organic material sources, the productivity of produce, land fertilization, soil processing, and pesticide use [21].

- Limited financial capital

One factor has also become an obstacle for forest farmers is the lack of financial capital to procure seed and fertilizer production facilities [109]. This is very reasonable because, generally, seeds and fertilizers are important production factors in farming [123–126]. In addition, the area of land and the number of plant types cultivated are two production factors affecting the income of agroforestry farmers [127]. The selection of types and cropping patterns indirectly affects the success of agroforestry management. Cropping patterns are physical capital that can be adaptively used to overcome the crisis and determine farmer income; the higher the income of farmers, the higher the ability to save and manage land [128].

The level of farmer income is itself often interlinked with land size. The land factor significantly affects the income level of farmers [129]. The land is the main capital in producing goods/services in agriculture [130] and is closely related to the level of income obtained by forest farmers, so its limitations are often regarded to be an obstacle [107]. The optimization of land production factors is very important so that agroforestry farming is more productive and efficient [131]. Several results from farming research state that the land factor affects productivity both technically and economically [132–136]. The low efficiency in the use of production factors is one of the causes of low farm productivity [137]. However, the production factor is an energy input in agricultural production that also determines the level of agricultural output produced [138].

Limited financial capital is also linked to the lack of alternative income sources for farmer livelihood. Educational qualifications and literacy skills are highly related [139]. Various factors cause the absence of other reliable sources of income, including the low level of education of farmers, which limits their ability to obtain jobs according to their skill competencies.

- Farmer preferences related to gender and cultural identity

Subsistence in agroforestry practices is often caused by the preferences of the farmers themselves. Some farmers actually have the basic skills required to shift from subsistence to semi-commercial enterprises [85]. However, farmers often continue to practice subsistence agroforestry as their economic orientation due to personal beliefs, which may be related to gender in selecting plant species. Gendered species preference is related to tree diversity in agroforestry systems [111]. Therefore, the extent to which this subsistence practice is applied will be influenced by the involvement of female farmers in making decisions about species selection. In addition, farmer preferences, as part of their cultural identity, can also cause agroforestry subsistence practices to persist. The Badui Luar, for instance, practice subsistence agroforestry on shifting land systems as part of their cultural identity [108].

2. External Factors

External factors that affect the subsistence of agroforestry farmers are presented in Table 3.

Table 3. External factors that lead to the subsistence of agroforestry farmers.

No	External Factors	Sources
1.	Government policies that are not responsive to household interests (increases in the price of seeds, fertilizers, and medicines).	[17]
2.	Bad weather/climate change.	[140–142]
3.	Labor shortage.	[101]
4.	Limited market access:	
	- Weak market linkages and poor access to market information;	[143–145]
	- Difficulties imposed related to costs, the management system, and local market constraints;	[146–148]
	- Price instability, poor market information, and poor market infrastructure for the tree products;	[143,149–151]
	- Long period of growth;	[152–154]
	- Preference for the near-term benefits.	[155]
5.	Food import.	[156–159]

- Government policies that are not responsive to household interests

Subsistence farmers need information and input support in developing agroforestry [17]. Therefore, government policies related to capital are necessary. The variables affecting the income of agroforestry farmers are age, plantation area, number of workers, ethnicity, religion, land slope, and credit assistance [160]. Agroforestry farmers develop their businesses by only relying on personal and family capital [113]. They do not rely on capital support from public, private, and/or cooperative financial institutions. Furthermore, they are not interested in obtaining business capital loans due to the high-interest rates and the complicated bureaucracy of financial institutions. Farmers think that agroforestry farming is a gamble. If they are lucky, they will earn large profits; otherwise, if they have bad luck, they will suffer losses. This is due to the uncertainty of product prices.

Four key factors determine the sustainability of agroforestry businesses: the role of extension workers, the availability of technology packages, the existence of farmer groups, and the role of the government [21].

- Bad weather/climate change

Subsistence farmers are most vulnerable to climate change [140]. Climate change increases the severity and frequency of extreme weather [142]. Extreme weather affects the sustainability of agroforestry farming. The forms of weather changes include hurricanes, continuous rain, long dry spells, peaks of pest and disease attacks, and crop failure [141]. Furthermore, in tree-based agroforestry, weather changes cause a decrease in the number of harvests or even crop failures in the given year [141].

- Social commitment to maintaining traditional agricultural practices

Rural communities in Indonesia, which generally live off agriculture, including agroforestry, have several characteristics, including close kinship relations with fellow citizens and still holding strong customs and traditions. Ignoring these traditions may lead to social sanctions that can take the form of social alienation. This situation requires them to dedicate quite a lot of time to meeting the social and cultural demands of their environment. As a result, farmers tend to look for easy cultivation techniques and do not have enough time to implement an intensive cultivation system [161]. Community forests, particularly in the form of agroforestry, are often underdeveloped and managed traditionally [162].

- Labor shortage

Indonesian farmers tend to prioritize agricultural activities other than tree-based farming. This is particularly because agricultural farming requires more intensive management and labor, besides being related to their food security. Smallholder farmers, therefore, prefer to grow trees using traditional agroforestry. The labor required to manage traditional agroforestry is only supplied by family members [101]. Hence, it is only natural that agroforestry only produces products to fulfill their personal and family needs.

- Limited market access

Among the problems faced by farmers is their lack of information about the marketing chain of community timber and an unawareness of the growing demand for community timber [101]. Nontimber agroforestry products, such as fruits, also have fluctuating prices [105]. These situations certainly do not motivate the community to cultivate trees intensively in agroforestry practices. Market access can affect price certainty, which will impact the sustainability of agroforestry businesses. Market certainty is an aspect that affects farmer motivation. Points of access to the market from agroforestry areas may also be considered by farmers [113]. Although agroforestry has various benefits for smallholders, local governments usually prefer oil palm plantations over agroforestry because they possess higher potential income [163].

- Limited access to capital/credit

Limited access to capital causes farmers not to optimally manage their land. Due to a lack of institutional support for agroforestry in agriculture policy, borrowers frequently lack information regarding financing availability [154].

- Food import

The agricultural sector still plays an important role in spurring the national economy through labor absorption in agriculture in order to create food security and foreign exchange earnings through exports and imports [156,157]. In the era of liberalization, the marketing of agricultural products in the agricultural system and the fate of farmers depend entirely on the free market mechanism [158].

The challenges of developing food crop agriculture in the globalization era are getting bigger, so this requires Indonesia to improve production efficiency and product competitiveness to reduce opportunities for import flows [157]. The increasing number of imported food products entering Indonesia indicates that the comparative advantage of domestic food products is lower than that of other countries [159].

In addition to adding new sources of income and livelihoods for communities around forests, agroforestry also represents the government's efforts to improve national food security by adding food sources from forest areas [164]. The impact of food import policies are also felt by agroforestry farmers, whose livelihoods depend on food products from the forest. A policy of import duties on food commodities as a form of social policy sometimes creates a dilemma. It benefits farmers, but on the other hand, also places a burden on the poor [156]. The negative impact will be felt more by small and poor farmers who are unable to compete directly with imported products without assistance and protection from the government [158].

3.4. Sustainable Agroforestry Management Strategies for Smallholder Farmers

Agroforestry has become a common practice for farmers in Indonesia. The contribution of agroforestry to the household economy of farmers is one of the determining factors for the sustainability of the agroforestry system. Improving agroforestry governance from subsistence to commercial ones requires a comprehensive strategy and involves many stakeholders. Internal and external factors from the farmers' side are presented in Table 4.

Table 4. SWOT analysis of smallholder agroforestry in Indonesia.

Strength (S)	Weakness (W)
1. Availability of existing managerial and technical capacity: • Smallholder agroforestry has over 10 years of experience [165]; • Traditional agroforestry practices [166], silviculture practices [65], and soil and water conservation technologies are available to smallholders [167,168]. 2. Recognized social, economic, and ecological benefits of agroforestry practices: • Potentially enhancing the resilience of smallholders in disaster-prone areas [165]; • Strengthening social cohesion when sharing with neighbors [48,169]; • Contributing positively to productivity and smallholder livelihoods in agroforestry and home garden systems [170–173]; • Significantly assisting smallholders with climate change adaptation, improved soil fertility and conservation, pest and disease control, income diversification, and offsetting fuelwood [174–176]; • Agroforests contribute to maintaining on-farm diversity [48]. 3. Availability of institutions at the farmers' level: • The existence of community groups that participate in agroforestry practices, such as farmer groups and soil and water conservation groups.	1. Limited landownership: • The average farm size ranges from less than 0.1 ha to 1 ha [18,165,177,178]. 2. Low education level and poor agroforestry knowledge: • Smallholder farmers have low education [18,165] • Farmers' knowledge about agroforestry is still lacking [101]; • Farmers are reluctant to implement agroforestry due to culture and capacity [178]. 3. Limited financial capital: • Limited capital for the provision of seed and fertilizer production facilities [109]; • Low access to credit/financial assistance [179–181]. 4. Poor management practices: • A lack of awareness regarding the positive influence of proper silvicultural management and limited technical capacity [182]; • Agroforestry is regarded as the second livelihood for some farmers [18,165]; • Some farmers do not fertilize trees or only do so during the intensive maintenance of crops [165]. *Kebun campuran* (mixed garden) is not intensively managed [18]. 5. Poor perception of agroforestry benefits: • Farmers' understanding of the benefits of agroforestry (income, food, and conservation) is still low [165]; • Farmers' perceptions of the benefit of agroforestry practices show varying results (high in Central Java [101] and East Java [183,184] but low in Lampung [185] and Ciamis West Java [76]); • Smallholders regard trees as competitors for annual crops/smallholders consider trees to make gardens difficult to be cleared periodically [48]. 6. Limited market linkage and farmer bargaining: • Low volume of trade, limited access to information, and weak market linkage information [182]; • Low bargaining position of smallholder farmers [186,187].

Table 4. *Cont.*

Opportunity (O)	Threat (T)
1. Increasing global interest in local agroforestry practices: • Local practices in various parts of the world, especially in the tropics, have become a global interest that is increasingly relevant because of their sustainability on a local-to-global scale [188]. 2. Increasing global demand for agroforestry products and services: • Adaptation to high population growth and a rise in the market economy [48]; • Globalization of food and increases in demand for food security [189]; • Mixed gardens have become international tourist destinations and sources of fresh organic produce for traditional food [18]; • An important measure to obtain multidimensional benefits as pointed out in the Sustainable Development Goals [190] and biodiversity conservation strategy [18]. 3. Availability of financial support schemes: • Financial support from central and local governments. Central and regional agencies whose main tasks are related to the management of forestry, agriculture, and environmental resources (RHL program with agroforestry and community development [191] and social forestry) [192,193]; • Finding alternatives to swidden agriculture [178,194]; • CDM/carbon trade becomes part of rural development and the availability of incentives that allow for small land and direct payments [195]. 4. Legal support: • The availability of regulations on food security [196], the sustainable land protection of food agriculture, conservation, environmental, community empowerment, etc.	1. Decreasing tree-based land availability: • Population growth and economic pressure lead to increased demand for food production, conventional agricultural land [189], housing, and other land uses [18,189]; on the other hand, it decreases agroforestry land [189,197]; • Volcano eruption reduces land area [165]. 2. Forest and land degradation: • Forest and land degradation or conventional farming practices that are not environmentally friendly [198–200]. 3. Climate change: • Climate change affects agroforestry to keep food production [140,156]. 4. Labor shortage: • A lack of laborers or family members to do the work and a rise in the wages of agricultural workers [48]. 5. Market uncertainty: • Market uncertainty and low-profit margins of agroforestry products [186,187]; • The increasing price of seeds, fertilizers, and medicines; high social burden; high land rent; rising wages for agricultural workers; and imported agricultural products [156]; • The increasing number of imported food products entering Indonesia [156,159]. 6. Poor coordination and technical assistance from the government: • Weak coordination between central and local agencies, as well as between regional and cross-regional governments [201]; • Limited extension services [105].

The mapping of strengths, weaknesses, opportunities, and challenges above was used as the basis for designing an agroforestry business development strategy. The strategy was prepared based on the interaction between internal and external factors (SO, ST, WO, and WT). It takes into account the importance of agroforestry business development (Table 5).

Table 5. Strategies of agroforestry business development in Indonesia.

SO Strategy	ST Strategy
• Improving the performance of existing stakeholders (GO, NGO, local community) for supporting the agroforestry business using the community development/RHL program [202,203]; • Utilizing the existing social capital and labor support effectively and efficiently in maintaining agroforestry practices [204]; • Transferring agroforestry silviculture and soil conservation technology through smallholder farmer groups, environmental groups, communities, and so forth [65,203]; • Fulfilling food needs due to population growth through increased agroforestry practices [22,164] • Strengthening of farmer groups (institutional and capacity, skill and capital) [202].	• Increasing the success of agroforestry business practices for addressing livelihood, smallholder income, food security, land conservation, population growth pressure, lands rehabilitations; and obtaining benefits from climate change resilience [22,205,206] • Controlling land-use change in maintaining and increasing agroforestry areas for SDGs [206,207] • Improving agroforestry silviculture techniques, improving soil and water conservation practices for improving land productivity, and reducing land degradation [208]

Table 5. Cont.

WO Strategy	WT Strategy
• Improving the effectiveness of community development to improve agroforestry practices [164,191]; • Improving land availability for farmers with social forestry programs/RHL program [192]; • Improving market access for agroforestry products; • Financial support from regional governments and NGOs for smallholder farmer groups with respect to agroforestry practices [194,209]; • Capital and production tools form incentives that allow for small land and direct payments from CDM/carbon trading [195], developing integrated agroforestry management policy with respect to the SDGs [210].	• Maintaining local knowledge in agroforestry practices [166]; • Keeping traditional practices/tree diversity from conventional agriculture [37]; • Enhancing community awareness in food security, land degradation, climate change, and environmental services [211]; • Improving synergy between stakeholder/cross-sectoral institutions at both the central and regional levels [202,203,209].

3.5. Enabling Smallholder of Agroforestry Practices in Indonesia

3.5.1. Selection of Agroforestry Plant Species

The existence of agroforestry outside the state forest area (private forest) is proven to provide great benefits for landowners. Cultivated tree species provide yields for building wood, firewood, animal feed, species, and medicinal plants [212]. The use of native species with multiple benefits needs to be encouraged with research and policy interventions because native trees can maintain biodiversity [213]. The greater the plant diversity in agroforestry, the greater the value of the forest from the yield variety and the environmental benefits. In rubber plantations, more shrubs will increase soil C and N as well as the infiltration rate from soil pores [214].

3.5.2. Integrated Farming (Agrosilvopasture/Livestock and Plants)

Product diversification from agriculture–livestock–forestry (agrosilvopasture), with market access and management efficiency, will result in income stability [215,216]. In agrosilvopasture, the use of livestock manure (fertilizer) and crop waste (animal feed) residues is an efficiency that can reduce tradeoffs between agriculture and the environment [217]. The integration of livestock and crops can increase economic efficiency by 38.4% [218,219]. The existence of plants for forage can supply animal feed needs during the dry season [220]. This system can also maintain the N content in the soil, the bacterial community cycle, and the N cycle process [221].

3.5.3. Soil and Water Conservation

The use of trees, especially legume trees, plays a pivotal role in maintaining soil fertility. Tree roots increase the rate of soil infiltration and the availability of water as a reserve during the dry season, thereby reducing plant stress [222]. The management of agroforestry plant biomass is a key factor in maintaining soil fertility. Cocoa leaf waste can be decomposed quickly when mixed with gliricidia leaves so that it can maintain soil nutrient input [208]. Farmers in mountainous areas need to apply alley cropping as a soil conservation technique. The productivity of annual crops in the tree aisles hanging 48 m and 96 m is still quite high even despite tree canopy cover; however, it still possesses high environmental value [223].

3.5.4. Improved Access to Land and Markets

Population growth and economic pressures have caused the availability of agricultural land to be increasingly limited. This has become an obstacle in the development of the agroforestry business in Java, Indonesia. Governments can provide access to forest land, community participation-based planning, and agricultural diversification [105]. The Indonesian government has made efforts to distribute access to land resources for the community by granting state forest land management permits through the social forestry program.

One of the factors affecting the level of commodity prices produced by private forests is land access to markets. The price of wood in private forests can be influenced by the location's accessibility to transportation. Good roads increase the efficiency of transport to market, thereby promoting more commercial agricultural cultivation [224]. The existence of village funds can be allocated to improve road access to private forest land. The government can also provide information on market access through extension workers [225]. In general, private forest harvests are sold to local middlemen and then to larger middlemen; thus, to increase farmer incomes, partnerships between industry and farmers are needed and incentives are provided to empower private forest farmers [226].

3.5.5. Climate Change Adaptation

Adaptation to climate change is required to reduce the risk of failure in crop cultivation. Farmers adapting to climate change will be influenced by farmer institutions, access to finance, information on climate, and extension [211]. Farmer adaptation practices to increase land productivity include the application of soil and water conservation techniques, the use of varieties that are resistant to high temperatures/shade, and the use of mulch [227]. The use of shade trees in agroforestry patterns is needed to deal with climate change that causes an increase in temperature [228]. Agroforestry has proven to be a technique that is suitable for farmers to apply in the era of climate change.

3.5.6. Intensification of Smallholder Agroforestry

Previous research shows that intensive agroforestry practices (with environmental manipulations) result in higher productivity than traditional practices [229–231]. Meanwhile, the agroforestry patterns in smallholder forests vary, including border trees, alley cropping, mix patterns, and alternate rows [232]. Yard agroforestry and complex agroforestry forms are still focused on experimental forms of intensive silvicultural applications such as fertilization, spacing, and pruning [233,234]. Both agroforestry forms are associated with the development of private forests with various characteristics on a household scale with limited lands, such as in Indonesia today. As a matter of course, many findings are not necessarily suitable for general application and produce trial/experimental results that are either not suitable or error-prone. Discretion in selecting research recommendations to suit the specific characteristics of private forest development is the key to success.

In agroforestry development, there are two main activities: (1) species selection and (2) management [235]. Species selection is determined by land conditions [236], latitude [237], light intensity [237,238], the social culture of the community [239], and the market [187]. Treatment in agroforestry will regulate the availability of growth resources [240,241], such as light intensity [237], water [242], and nutrients [214]. This is achieved by sharing growth resources [232] or minimizing competition among species [243] and additional growth factor inputs [244,245]. The components of the treatment carried out are spacing or density [239,246,247], thinning [248] and pruning [249], rotation [250], and fertilization [244]. The factors for implementing agroforestry technology are influenced by the feasibility of financial analysis [22,251], policy support [252], market networks [187], collective action to solve limited landownership [253], adoption power [114,216], and extension [254,255]. The picture below tries to illustrate how the implementation of agroforestry works (Figure 3).

So far, the implementation of agroforestry has relied more on knowledge passed down from ancestors and the use of subsistence (fulfillment of daily needs) or market demand from cultivated plant commodities. The types of seasonal food crops, which are usually understory and perennial fruit trees, are timber plants that have been developed for a long time and have contributed to the local food supply. The forms of environmental manipulation or land management that can be carried out based on studies so far are (1) the use of tolerant varieties that have been created; (2) the implementation of thinning/pruning/tree-spacing to make room for incoming light; (3) fertilization, manure being relatively safe for environmental sustainability, as is the use of inorganic fertilizers, especially N types; and (4) optimizing the benefits of current trees in the system. In this effort, policies established by the government and other stakeholders collaborate to empower the community to achieve food security, such as in (1) the creation of new shade-resistant varieties; (2) converging research results that produce agroforestry models that adapt to the characteristics of private forests (biophysical variations, household scale/areas of 0.25 ha), and management collective action in farmer groups to achieve commercial areas; (3) the extension of intensive silvicultural concepts in multifunctional agroforestry entry for farmers; and (4) the provision of production facilities and infrastructure (subsidized seeds and fertilizers).

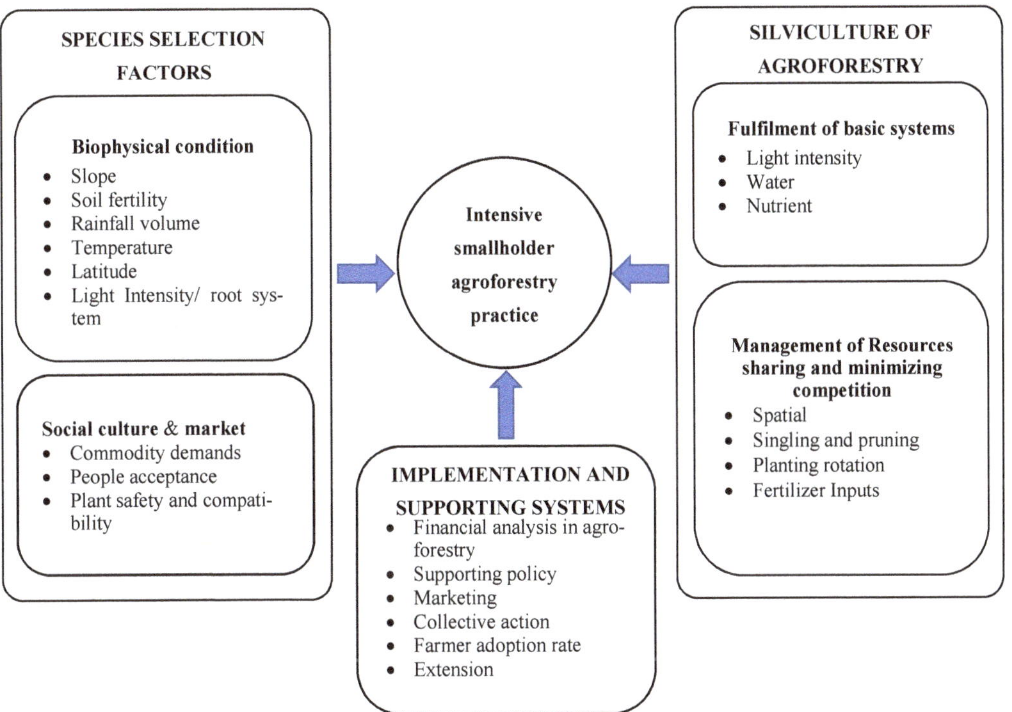

Figure 3. Framework for enabling intensification of smallholder agroforestry practices. Source: compiled from [22,114,187,214,216,232,236–246,248–255].

Agroforestry development during the private forest cycle is not a static environment; agroforestry development for any kind of seasonal crop cannot be separated from other dimensions, such as social policies and marketing dimensions. If agroforestry in Indonesia is assessed according to the characteristics of success required by [256], namely, productivity, yield sustainability, and adoption, then it only fulfills the productivity aspects, while the latter two are still challenging to achieve [257]. There are several obstacles in agroforestry development that are especially faced by farmers: technical, financial, market, and social doubts [203]. In addition, there is no binding legal provision and less (half-hearted) partisanship in supporting agroforestry development from relevant parties.

The support and synergy of all parties for the development of agroforestry can include various factors [202,203]: (1) support from government policies and programs that are pro-agroforestry (technical, financial, legality); (2) capacity building from the community, assistants, governments, and related institutions; (3) support from science and technology packages and community facilitation; and (4) market support. Furthermore, synergy and support from all parties for the development of agroforestry, with respect to optimizing land use in order to support food security and farmer investment, are expected to increase Indonesian farmer welfare and purchasing power. Agroforestry is a solution to many problems, and it has the potential to become the future of agriculture by achieving the environmental benefits and services it promises [5].

4. Conclusions and Recommendations

Based on this review, it can be concluded that the combination of limited landownership, low average literacy, and limited financial capital are the main factors causing smallholders to be unable to increase their income from agroforestry practices. A feasible strategic effort is to intensify forest management using agroforestry patterns by optimizing the use of growing space and maintaining plants by applying cultivation technology. However, in reality, the technology has failed to be adopted by farmers because of limited capital and low knowledge due to low average literacy levels. This means that assistance in the form of technology alone has not been able to touch the subsistence problem of small farmers. Given this situation, government policies that are oriented toward providing other incentives aimed at encouraging an increase in the number and diversity of income sources need to be stimulated. Until recently, forest services have not been utilized as a source of income for farmers, although the government often calls for the importance of maintaining the ecological function of forests. Therefore, the upstream–downstream compensation model should be employed as an entry point for initiating the appreciation of other forest services, such as the carbon sequestration capacity of agroforestry and the control of soil erosion or landslides. This upstream–downstream compensation mechanism is expected to have positive implications for the reforestation of bare hills, which are widespread in several areas.

In the end, we need to consider that each region has different characteristics related to biophysical and socioeconomic statuses of farmers. The proposed strategies to improve the sustainability of farming are not necessarily applicable to all local and regional conditions. Further research is needed to review successful cases of intensive farming and commercially oriented agroforestry practices, as well as to evaluate how these cases can be applied to other areas. On the other hand, it is too early to assume that subsistence farming is tantamount to failure. We realize some potential benefits may be obtained from subsistence farming, such as conservative farming due to non-tillage practices, high biodiversity spots due to the cultivation of noncommercial tree species, and higher potential carbon sequestration due to the longer harvesting cycle of the trees. Sustainability seems to be the key to further agroforestry development for smallholder farmers as an alternative to other intensive, modern, and commercial practices. Therefore, a more comprehensive study is needed to compare the productivity, socioeconomic, and environmental aspects between traditional and modern agroforestry, subsistence and commercial agroforestry, and non-intensive and intensive agroforestry.

Author Contributions: All authors made equal contributions to resources, conceptualization, methodology, analysis, validation, writing, review, and editing of the manuscript. All authors have read and agreed to the published version of the manuscript.

Funding: This research received no external funding.

Institutional Review Board Statement: Not applicable.

Informed Consent Statement: Not applicable.

Data Availability Statement: No additional datasets have been used.

Acknowledgments: We thank the anonymous reviewers for their expert assessments, detailed comments, and valuable inputs.

Conflicts of Interest: The authors declare no conflict of interest.

Appendix A

Table A1. Full list of reviewed studies.

No.	Bibliography	Search Site	Sub-Discussions
	International Literature		
1.	[202]	GS	3
2.	[191]	GS	3
3.	[55]	GS, SP	1
4.	[152]	SD	3
5.	[39]	GS	1
6.	[11]	GS	1
7.	[57]	GS	1
8.	[22]	GS, SP	1
9.	[71]	GS	1
10.	[94]	GS	1
11.	[54]	GS	1
12.	[42]	GS, FM	1
13.	[53]	GS	1
14.	[52]	GS, RG	1
15.	[44]	GS	1
16.	[108]	GS	2
17.	[104]	GS	1, 2
18.	[37]	GS	3
19.	[90]	GS	1
20.	[153]	GS	2
21.	[60]	GS, IS	1
22.	[61]	GS, SP	1
23.	[177]	GS, SD	3
24.	[186]	GS	3
25.	[204]	GS	3
26.	[162]	GS	2
27.	[254]	SP	3
28.	[66]	GS	1
29.	[114]	SP	2
30.	[19]	GS	1
31.	[207]	GS	3
32.	[201]	GS	3
33.	[48]	GS	1, 3
34.	[56]	GS, IS	1
35.	[50]	RG	1
36.	[18]	GS	3

Table A1. *Cont.*

No.	Bibliography	Search Site	Sub-Discussions
37.	[205]	GS	3
38.	[93]	GS	1
39.	[49]	GS	1
40.	[70]	GS	1
41.	[251]	GS	3
42.	[253]	GS	3
43.	[178]	SP	3
44.	[194]	GS	3
45.	[105]	SD	2, 3
46.	[193]	SD	3
47.	[95]	GS, IS	1
48.	[83]	GS, SP	1
49.	[241]	GS	3
50.	[187]	GS	3
51.	[151]	RG	2
52.	[85]	IS	1
53.	[180]	GS, SD	2, 3
54.	[62]	GS	1, 3
55.	[182]	GS	2, 3
56.	[165]	GS	3
57.	[65]	GS, SP	1
58.	[163]	SD	2
59.	[111]	GS	2
60.	[59]	GS	1, 3
61.	[23]	GS	1
62.	[67]	GS	1
63.	[112]	SD	2
64.	[97]	GS	1
65.	[78]	GS	1
66.	[38]	GS, SP	1, 2, 3
67.	[113]	GS	2
68.	[103]	SD	2
69.	[181]	GS	3
70.	[110]	GS	2
71.	[92]	GS	1
72.	[43]	GS	1
73.	[3]	GS	1, 3
74.	[4]	GS	1
75.	[209]	SP	3

Table A1. *Cont.*

No.	Bibliography	Search Site	Sub-Discussions
	Local Literature		
76.	[189]	GS	3
77.	[107]	GS	2
78.	[109]	GS	2, 3
79.	[225]	GS	3
80.	[128]	GS	2
81.	[134]	GS	2
82.	[58]	GS	1
83.	[82]	FM	1
84.	[255]	GS	3
85.	[89]	GS	1
86.	[87]	RG	1
87.	[130]	GS	2
88.	[5]	GS	3
89.	[47]	GS	1
90.	[125]	GS	2
91.	[184]	GS	3
92.	[244]	GS	3
93.	[101]	GS	2, 3
94.	[91]	GS	1
95.	[119]	GS	2
96.	[96]	GS	1
97.	[122]	GS	2
98.	[102]	GS	2
99.	[161]	GS	2
100.	[141]	GS	2
101.	[164]	GS	2
102.	[84]	RG	1
103.	[129]	GS	2
104.	[74]	GS	1
105.	[160]	GS	2
106.	[183]	GS	3
107.	[21]	GS	2
108.	[76]	GS	1, 2, 3
109.	[79]	GS, FM	1
110.	[30]	GS, FM	2
111.	[106]	GS	2
112.	[88]	GS	1
113.	[131]	GS	2

Table A1. *Cont.*

No.	Bibliography	Search Site	Sub-Discussions
114.	[226]	GS	3
115.	[232]	GS	3
116.	[185]	GS	3
117.	[118]	GS	2
118.	[121]	GS	2
119.	[81]	FM	1
120.	[51]	GS	1, 3
121.	[86]	GS	1
122.	[41]	GS	1
123.	[127]	GS	2

Note—Search sites: GS (Google Scholar), SD (Science Direct), SP (Springer), RG (Researchgate), IS (IOP Science), FM (Forda-Mof). Subdiscussions: 1 (existing agroforestry), 2 (subsistence factor of agroforestry), 3 (strategy of agroforestry).

References

1. Dhakal, A.; Rai, R.K. Who adopts agroforestry in a subsistence economy?—Lessons from the Terai of Nepal. *Forests* **2020**, *11*, 565. [CrossRef]
2. Pandey, D.K.; De, H.K.; Geetarani, L.; Hazarika, B.N. Problems and prospects of agricultural diversification in shifting cultivation area of north east India: An empirical study. *J. Pharmacogn. Phytochem.* **2019**, *8*, 69–73.
3. Wiyanti, D.T. Agroforestry and mobile local "bank" system (Bank Kuriling) in Cianjur, West Java. In Proceedings of the International Conference of Integrated Microfinance Management (IMM-16), Bandung, Indonesia, 19–22 September 2016; Atlantis Press: Paris, France, 2016; pp. 131–135.
4. Wiyanti, D.T. People and agroforestry (the use of non-commercial product of agroforestry). In Proceedings of the International Social Science, Humanity and Education Research Congress (SSHERC-16), Bali, Indonesia, 20–21 July 2016; pp. 41–44.
5. Hairiah, K.; Ashari, S. Pertanian masa depan: Agroforestri, manfaat dan layanan lingkungan. In Proceedings of the Prosiding Seminar Nasional Agroforestri 2013; Kuswantoro, D.P., Widyaningsih, T.S., Fauziyah, E., Rachmawati, R., Eds.; BPTA-Faperta UB-ICRAF-MAFI: Malang, Indonesia, 2013; pp. 23–35.
6. Mutua, J.; Muriuki, J.; Gachie, P.; Bourne, M.; Capis, J. *Conservation Agriculture with Trees: Principles and Practice a Simplified Guide for Extension Staff and Farmers*; World Agroforestry Centre: Nairobi, Kenya, 2014.
7. Kabir, M.H. Effects of Organic and Inorganic Fertilizers on Growth and Yield of Garlic. Master's Thesis, Department of Horticulture, BAU, Mymensingh, Bangladesh, 2004.
8. Khan, M.L.; Anukachalam, K. Balancing traditional jhum cultivation with modern agroforestry in eastern Himalaya—A biodiversity hot spot. *Curr. Sci.* **2002**, *83*, 117.
9. Sileshi, G.W.; Mafongoya, P.L.; Akinnifesi, F.K.; Phiri, E.; Chirwa, P.; Beedy, T.; Makumba, W.; Nyamadzawo, G.; Njoloma, J.; Wuta, M.; et al. Agroforestry: Fertilizer trees. In *Encyclopedia of Agriculture and Food Systems*; Van Alfen, N., Ed.; Elsevier: San Diego, CA, USA, 2014; Volume 1, pp. 222–234.
10. Jose, S. Agroforestry for ecosystem services and environmental benefits: An overview. *Agrofor. Syst.* **2009**, *76*, 1–10. [CrossRef]
11. Budiadi; Jihad, A.N.; Lestari, L.D. An overview and future outlook of Indonesian agroforestry: A bibliographic and literature review. In *Proceedings of the E3S Web of Conferences*; EDP Sciences: Les Ulis, France, 2021; Volume 305, p. 7002.
12. Dhakal, A.; Cockfield, G.; Maraseni, T.N. Evolution of agroforestry based farming systems: A study of Dhanusha District, Nepal. *Agrofor. Syst.* **2012**, *86*, 17–33. [CrossRef]
13. Neupane, R.P.; Thapa, G.B. Impact of agroforestry intervention on soil fertility and farm income under the subsistence farming system of the middle hills, Nepal. *Agric. Ecosyst. Environ.* **2001**, *84*, 157–167. [CrossRef]
14. Farage, P.K.; Ardö, J.; Olsson, L.; Rienzi, E.A.; Ball, A.; Pretty, J.N. The potential for soil carbon sequestration in three tropical dryland farming systems of Africa and Latin America: A modelling approach. *Soil Tillage Res.* **2007**, *94*, 457–472. [CrossRef]
15. Córdova, R.; Hogarth, N.J.; Kanninen, M. Sustainability of smallholder livelihoods in the Ecuadorian Highlands: A comparison of agroforestry and conventional agriculture systems in the indigenous territory of Kayambi People. *Land* **2018**, *7*, 45. [CrossRef]
16. Quinion, A.; Chirwa, P.; Akinnifesi, F.K.; Ajayi, O.O. Do agroforestry technologies improve the livelihoods of the resource poor farmers? Evidence from Kasungu and Machinga districts of Malawi. *Agrofor. Syst.* **2010**, *80*, 457–465. [CrossRef]
17. Jerneck, A.; Olsson, L. More than trees! Understanding the agroforestry adoption gap in subsistence agriculture: Insights from narrative walks in Kenya. *J. Rural Stud.* **2013**, *32*, 114–125. [CrossRef]
18. Parikesit; Withaningsih, S.; Rozi, F. Socio-ecological dimensions of agroforestry called kebun campuran in tropical karst ecosystem of West Java, Indonesia. *Biodiversitas* **2021**, *22*, 122–131. [CrossRef]

19. Nöldeke, B.; Winter, E.; Laumonier, Y.; Simamora, T. Simulating agroforestry adoption in rural Indonesia: The potential of trees on farms for livelihoods and environment. *Land* **2021**, *10*, 385. [CrossRef]
20. Ngaji, A.U.K.; Baiquni, M.; Suryatmojo, H.; Haryono, E. Assessing the sustainability of traditional agroforestry practices: A case of Mamar agroforestry in Kupang-Indonesia. *For. Soc.* **2021**, *5*, 438–457. [CrossRef]
21. Ruhimat, I.S. Status keberlanjutan usahatani agroforestry pada lahan masyarakat: Studi kasus di Kecamatan Rancah, Kabupaten Ciamis, Provinsi Jawa Barat. *J. Penelit. Sos. Dan Ekon. Kehutan.* **2015**, *12*, 99–110. [CrossRef]
22. Duffy, C.; Toth, G.G.; Hagan, R.P.O.; McKeown, P.C.; Rahman, S.A.; Widyaningsih, Y.; Sunderland, T.C.H.; Spillane, C. Agroforestry contributions to smallholder farmer food security in Indonesia. *Agrofor. Syst.* **2021**, *95*, 1109–1124. [CrossRef]
23. Siarudin, M.; Rahman, S.A.; Artati, Y.; Indrajaya, Y.; Narulita, S.; Ardha, M.J.; Larjavaara, M. Carbon sequestration potential of agroforestry systems in degraded landscapes in West Java, Indonesia. *Forests* **2021**, *12*, 714. [CrossRef]
24. Salampessy, M.L.; Febryano, I.G.; Zulfiani, D. Principal agent in tree mortgage system on traditional agroforestry management in Moluccas Indonesia. In *IOP Conference Series: Earth and Environmental Science*; IOP Publishing: Bristol, UK, 2019; Volume 285, p. 12013.
25. Dyer, G.A.; Boucher, S.; Taylor, J.E. Subsistence response to market shocks. *Am. J. Agric. Econ.* **2006**, *88*, 279–291. [CrossRef]
26. Kostov, P.; Lingard, J. Subsistence agriculture in transition economies: Its roles and determinants. *J. Agric. Econ.* **2004**, *55*, 565–579. [CrossRef]
27. Amoussouga Gero, A.; Egbendewe, A.Y.G. Macroeconomic effects of semi-subsistence agricultural productivity growth: Evidence from Benin and extension to the WAEMU countries. *Sci. Afr.* **2020**, *7*, e00222. [CrossRef]
28. Toth, G.G.; Nair, P.K.R.; Duffy, C.P.; Franzel, S.C. Constraints to the adoption of fodder tree technology in Malawi. *Sustain. Sci.* **2017**, *12*, 641–656. [CrossRef]
29. Buyinza, J.; Nuberg, I.K.; Muthuri, C.W.; Denton, M.D. Psychological factors influencing farmers' intention to adopt agroforestry: A structural equation modeling approach. *J. Sustain. For.* **2020**, *39*, 854–865. [CrossRef]
30. Ruhimat, I.S. Analisis pengambilan keputusan petani dalam pemilihan jenis tanaman bawah pada kebun campuran. *J. Agroforestri Indones.* **2020**, *3*, 111–122.
31. Deiss, L.; de Moraes, A.; Dieckow, J.; Franzluebbers, A.J.; Gatiboni, L.C.; Lanzi Sassaki, G.; Carvalho, P.C.F. Soil phosphorus compounds in integrated crop-livestock systems of subtropical Brazil. *Geoderma* **2016**, *274*, 88–96. [CrossRef]
32. Hulugalle, N.R.; Ndi, J.N. Effects of no-tillage and alley cropping on soil properties and crop yields in a Typic Kandiudult of southern Cameroon. *Agrofor. Syst.* **1993**, *22*, 207–220. [CrossRef]
33. Gabrielsson, S.; Ramasar, V. Widows: Agents of change in a climate of water uncertainty. *J. Clean. Prod.* **2013**, *60*, 34–42. [CrossRef]
34. Pratiwi, A.; Suzuki, A. Reducing agricultural income vulnerabilities through agroforestry training: Evidence from a randomised field experiment in Indonesia. *Bull. Indones. Econ. Stud.* **2019**, *55*, 83–116. [CrossRef]
35. Baka, W.K.; Rianse, I.S.; Rianse, U.; Abdullah, W.G. Zulfikar Pattern of palm-based agroforestry the Bugis etnic community in the Regency of Kolaka Indonesia. In *IOP Conference Series: Earth and Environmental Science*; IOP Publishing: Bristol, UK, 2019; Volume 314, p. 12041.
36. Afentina; McShane, P.; Wright, W. Ethnobotany, rattan agroforestry, and conservation of ecosystem services in Central Kalimantan, Indonesia. *Agrofor. Syst.* **2020**, *94*, 639–650. [CrossRef]
37. Iskandar, J.; Iskandar, B.S. Local knowledge of the Baduy Community of South Banten (Indonesia) on the traditional landscapes. *Biodiversitas* **2017**, *18*, 928–938. [CrossRef]
38. Suryanto, P.; Widyastuti, S.; Sartohadi, J.; Awang, S.A. Budi Traditional knowledge of homegarden-dry field agroforestry as a tool for revitalization management of smallholder land use in Kulon Progo, Java, Indonesia. *Int. J. Biol.* **2012**, *4*, 173–183. [CrossRef]
39. Baliton, R.S.; Wulandari, C.; Landicho, L.D.; Cabahug, R.E.D.; Paelmo, R.F.; Comia, R.A.; Visco, R.G.; Budiono, P.; Herwanti, S.; Rusita, R.; et al. Ecological services of agroforestry landscapes in selected watershed areas in the Philippines and Indonesia. *Biotropia* **2017**, *24*, 71–84. [CrossRef]
40. Suryanto, P.; Budiadi; Sabarnudin, M.S. Silvikultur agroforestri dan masa depan hutan rakyat. In *Hutan Rakyat di Simpang Jalan*; Maryudi, A., Nawir, A.A., Eds.; Gadjah Mada University Press: Yogyakarta, Indonesia, 2017; pp. 45–64. ISBN 9786023862580.
41. Winara, A.; Widiyanto, A. Keanekaragaman Jenis Pohon pada Sistem Agroforestri di DAS Balangtieng. In *Jasa Lingkungan Agroforestri: Belajar Dari Masyarakat DAS Balangtieng Sulawesi Selatan*; Rahayu, S., Hasanah, N., Lusiana, B., Eds.; IPB Press: Bogor, Indonesia, 2019; pp. 89–107.
42. Hani, A.; Suhaendah, E. Diversity of soil macro fauna and its role on soil fertility in manglid agroforestry. *Indones. J. For. Res.* **2019**, *6*, 61–68. [CrossRef]
43. Withaningsih, S.; Parikesit; Alham, R.F. Diversity of bird species in the coffee agroforestry landscape: Case study in the Pangalengan Sub-District, Bandung District, West Java, Indonesia. *Biodiversitas* **2020**, *21*, 2467–2480. [CrossRef]
44. Imron, M.A.; Campera, M.; Al Bihad, D.; Rachmawati, F.D.; Nugroho, F.E.; Budiadi, B.; Wianti, K.F.; Suprapto, E.; Nijman, V.; Nekaris, K.A.I. Bird assemblages in coffee agroforestry systems and other human modified habitats in Indonesia. *Biology* **2022**, *11*, 310. [CrossRef]
45. Dollinger, J.; Jose, S. Agroforestry for soil health. *Agrofor. Syst.* **2018**, *92*, 213–219. [CrossRef]
46. Nair, P.K.R. Agroforestry systems and environmental quality: Introduction. *J. Environ. Qual.* **2011**, *40*, 784–790. [CrossRef]
47. Hartoyo, A.P.P.; Wijayanto, N.; Karimatunnisa, T.; Ikhfan, A. Keanekaragaman hayati vegetasi pada praktik agroforestri dan kaitannya terhadap fungsi ekosistem di Taman Nasional Meru Betiri, Jawa Timur. *J. Hutan Trop.* **2019**, *7*, 145–157.

48. Okubo, S.; Parikesit, P.; Harashina, K.; Muhamad, D.; Abdoellah, O.S.; Takeuchi, K. Traditional perennial crop-based agroforestry in West Java: The tradeoff between on-farm biodiversity and income. *Agrofor. Syst.* **2010**, *80*, 17–31. [CrossRef]
49. Prastiyo, Y.B.; Kaswanto, R.L.; Arifin, H.S. Plants diversity of agroforestry system in Ciliwung riparian landscape, Bogor Municipality. In *IOP Conference Series: Earth and Environmental Science*; IOP Publishing Ltd.: Bristol, UK, 2020; Volume 477.
50. Parikesit; Takeuchi, K.; Tsunekawa, A.; Abdoellah, O.S. Kebon tatangkalan: A disappearing agroforest in the Upper Citarum Watershed, West Java, Indonesia. *Agrofor. Syst.* **2005**, *63*, 171–182. [CrossRef]
51. Wijayanto, N.; Hartoyo, A.P.P. Biodiversitas berbasiskan agroforestry. In Proceedings of the Prosiding Seminar Nasional Masyarakat Biodiversitas Indonesia, Makassar, Indonesia, 5 October 2015; Volume 1, pp. 242–246.
52. Herwanti, S.; Febryano, I.G.; Zulfiani, D. Economic value analysis of community forest food products in Ngarip Village, Ulu Belu Subdistrict, Tanggamus Regency (A case from Indonesia). *For. Ideas* **2019**, *25*, 314–328.
53. Hermudananto, H.; Permadi, D.B.; Septiana, R.M.; Riyanto, S.; Pratama, A.A. Adoption of agroforestry-porang model for land utilization under teak stands. *J. Pengabdi. Kpd. Masy. Indonesian J. Community Engag.* **2019**, *5*, 416–436. [CrossRef]
54. Hakim, L.; Rahardi, B.; Guntoro, D.A.; Mukhoyyaroh, N.I. Coffee Landscape of Banyuwangi Geopark: Ecology, Conservation, and Sustainable Tourism Development. *J. Trop. Life Sci.* **2022**, *12*, 107–116. [CrossRef]
55. Applegate, G.; Freeman, B.; Tular, B.; Sitadevi, L.; Jessup, T.C. Application of agroforestry business models to tropical peatland restoration. *Ambio* **2022**, *51*, 863–874. [CrossRef]
56. Paembonan, S.A.; Millang, S.; Dassir, M.; Ridwan, M. Species variation in home garden agroforestry system in South Sulawesi, Indonesia and its contribution to farmers' income. In *IOP Conference Series: Earth and Environmental Science*; IOP Publishing: Bristol, UK, 2018; Volume 157, pp. 1–6.
57. Desmiwati, D.; Veriasa, T.O.; Aminah, A.; Safitri, A.D.; Hendarto, K.A.; Wisudayati, T.A.; Royani, H.; Dewi, K.H.; Raharjo, S.N.I.; Sari, D.R. Contribution of agroforestry systems to farmer income in state forest areas: A case study of Parungpanjang, Indonesia. *For. Soc.* **2021**, *5*, 109–119. [CrossRef]
58. Diniyati, D.; Achmad, B.; Santoso, H.B. Analisis finansial agroforestry sengon di Kabupaten Ciamis (Studi kasus di Desa Ciomas Kecamatan Panjalu). *J. Penelit. Agrofor.* **2013**, *1*, 13–30.
59. Siarudin, M.; Awang, S.A.; Sadono, R.; Suryanto, P. The pattern recognition of small-scale privately-owned forest in Ciamis Regency, West Java, Indonesia. *For. Soc.* **2022**, *6*, 104–120. [CrossRef]
60. Karim, A.; Hifnalisa; Jufri, Y.; Fazlina, Y.D.; Megawati. Distribution of organic carbon in Ultisol soils with citronella and pine vegetation, at Gayo Highlands, Aceh. In *IOP Conference Series: Earth and Environmental Science*; IOP Publishing Ltd.: Bristol, UK, 2022; Volume 951, pp. 1–9.
61. Khasanah, N.; Perdana, A.; Rahmanullah, A.; Manurung, G.; Roshetko, J.M.; van Noordwijk, M. Intercropping teak (*Tectona grandis*) and maize (*Zea mays*): Bioeconomic trade-off analysis of agroforestry management practices in Gunungkidul, West Java. *Agrofor. Syst.* **2015**, *89*, 1019–1033. [CrossRef]
62. Roshetko, J.M.; Rohadi, D.; Perdana, A.; Sabastian, G.; Nuryartono, N.; Pramono, A.A.; Widyani, N.; Manalu, P.; Fauzi, M.A.; Sumardamto, P.; et al. Teak agroforestry systems for livelihood enhancement, industrial timber production, and environmental rehabilitation. *For. Trees Livelihoods* **2013**, *22*, 241–256. [CrossRef]
63. Soekotjo. *Teknik Silvikultur Intensif (SILIN)*, 1st ed.; Gadjah Mada University Press: Yogyakarta, Indonesia, 2009; ISBN 979-420-694-6.
64. Achim, A.; Moreau, G.; Coops, N.C.; Axelson, J.N.; Barrette, J.; Bédard, S.; Byrne, K.E.; Caspersen, J.; Dick, A.R.; D'Orangeville, L.; et al. The changing culture of silviculture. *Forestry* **2022**, *95*, 143–152. [CrossRef]
65. Sabastian, G.E.; Yumn, A.; Roshetko, J.M.; Manalu, P.; Martini, E.; Perdana, A. Adoption of silvicultural practices in smallholder timber and NTFPs production systems in Indonesia. *Agrofor. Syst.* **2019**, *93*, 607–620. [CrossRef]
66. Maryudi, A.; Nawir, A.A.; Sekartaji, D.A.; Sumardamto, P.; Purwanto, R.H.; Sadono, R.; Suryanto, P.; Soraya, E.; Soeprijadi, D.; Affianto, A.; et al. Smallholder farmers' knowledge of regulations governing the sale of timber and supply chains in Gunungkidul District, Indonesia. *Small-Scale For.* **2017**, *16*, 119–131. [CrossRef]
67. Silvianingsih, Y.A.; Hairiah, K.; Suprayogo, D.; van Noordwijk, M. Kaleka agroforest in Central Kalimantan (Indonesia): Soil quality, hydrological protection of adjacent peatlands, and sustainability. *Land* **2021**, *10*, 856. [CrossRef]
68. Suárez, L.R.; Suárez Salazar, J.C.; Casanoves, F.; Ngo Bieng, M.A. Cacao agroforestry systems improve soil fertility: Comparison of soil properties between forest, cacao agroforestry systems, and pasture in the Colombian Amazon. *Agric. Ecosyst. Environ.* **2021**, *314*, 107349. [CrossRef]
69. Masebo, N.; Menamo, M. A review paper on: The role of agroforestry for rehabilitation of degraded soil. *J. Biol. Agric. Healthc.* **2016**, *6*, 128–135.
70. Purwaningsih, R.; Sartohadi, J.; Setiawan, M.A. Trees and crops arrangement in the agroforestry system based on slope units to control landslide reactivation on volcanic foot slopes in Java, Indonesia. *Land* **2020**, *9*, 327. [CrossRef]
71. Hairiah, K.; Widianto, W.; Suprayogo, D.; Van Noordwijk, M. Tree roots anchoring and binding soil: Reducing landslide risk in Indonesian agroforestry. *Land* **2020**, *9*, 256. [CrossRef]
72. Lin, B.B. Agroforestry management as an adaptive strategy against potential microclimate extremes in coffee agriculture. *Agric. For. Meteorol.* **2007**, *144*, 85–94. [CrossRef]
73. Susilowati, S.H. Fenomena penuaan petani dan berkurangnya tenaga kerja muda serta implikasinya bagi kebijakan pembangunan pertanian. *Forum Penelit. Agro Ekon.* **2016**, *34*, 35–55. [CrossRef]

74. Nugroho, A.D.; Waluyati, L.R.; Jamhari, J. Upaya Memikat Generasi Muda Bekerja Pada Sektor Pertanian di Daerah Istimewa Yogyakarta. *JPPUMA J. Ilmu Pemerintah. dan Sos. Polit. Univ. Medan Area* **2018**, *6*, 76. [CrossRef]
75. Prawesti, N.; Witjaksono, R.; Raya, A.B. Motivasi anak petani menjadi petani. *Agro Ekon.* **2010**, *17*, 11–18. [CrossRef]
76. Ruhimat, I.S. Tingkat motivasi petani dalam penerapan sistem agroforestry. *J. Penelit. Sos. Dan Ekon. Kehutan.* **2015**, *12*, 131–147. [CrossRef]
77. Sandy BPS Sebut Tingkat Pendidikan Petani Masih Rendah. Available online: https://www.idxchannel.com/economics/bps-sebut-tingkat-pendidikan-petani-masih-rendah (accessed on 15 March 2022).
78. Suryana, A.; Simatupang, P. Socio-economic aspects of pekarangan land in Java: A literature review. *Indones. Agric. Res. Dev. J.* **1992**, *14*, 7–16.
79. Ruhimat, I.S. Faktor kunci dalam pengembangan kelembagaan agroforestry pada lahan masyarakat. *J. Penelit. Sos. Dan Ekon. Kehutan.* **2016**, *13*, 73–84. [CrossRef]
80. Djogo, T.; Sunaryo; Suharjito, D.; Sirait, M. *Bahan Ajaran Agroforestry 8: Kelembagaan Dan Kebijakan Dalam Pengembangan Agroforestry*; World Agroforestry Center (ICRAF): Bogor, Indonesia, 2003.
81. Widiyanto, A.; Hani, A. The role and key success of agroforestry (A review). *J. Agroforestri Indones.* **2021**, *4*, 69–80.
82. Diniyati, D.; Achmad, B. Identifikasi dan strategi pengembangan kelembagaan petani hutan rakyat. *J. Penelit. Agrofor.* **2013**, *1*, 83–100.
83. Reid, R. Developing farmer and community capacity in Agroforestry: Is the Australian Master TreeGrower program transferable to other countries? *Agrofor. Syst.* **2017**, *91*, 847–865. [CrossRef]
84. Nikoyan, A.; Kasim, S.; Uslinawaty, Z.; Yani, R. Peran dan manfaat kelembagaan kelompok tani pelestari hutan dalam pengelolaan Taman Hutan Raya Nipa-Nipa. *Perennial* **2020**, *16*, 34–39.
85. Roshetko, J.M.; Purnomosidhi, P. Smallholder agroforestry fruit production in Lampung, Indonesia: Horticultural strategies for smallholder livelihood enhancement. In Proceedings of the 4th International Symposium on Tropical and Subtropical Fruits, Bogor, Indonesia, 3–7 November 2008; Palupi, E.R., Warrington, K.I.J., Eds.; ISHS: Leuven, Belgium, 2013; pp. 671–678.
86. Winara, A.; Suhaendah, E. Keragaman jenis tumbuhan pada agroforestri Manglid. *Pros. Semin. Nas. Penelit. Dan PKM Sains Dan Teknol.* **2016**, *6*, 80–87.
87. Dwijo, S.; Supardi, S.; Marwanti, S. Analisis kelayakan finansial pengusahaan kayu sengon (Paraserianthes falcataria) dengan pola tanam agroforestry di Kabupaten Magetan. *J. Penelit. Sos. Dan Ekon. Kehutan.* **2020**, *17*, 29–37. [CrossRef]
88. Siarudin, M.; Winara, A.; Indrajaya, Y.; Badrunasar, A.; Rahayu, S.; Roshetko, J.M. *Seri Agroforestri Dan Kehutanan di Sulawesi: Keanekaragaman Hayati Jenis Pohon Pada Hutan Rakyat Agroforestri di DAS Balangtieng, Sulawesi Selatan*; World Agroforestry Centre (ICRAF); Southeast Asia Regional Program dan Balai Penelitian dan Pengembangan Teknologi Agroforestry (BPTA): Bogor, Indonesia, 2017.
89. Dwi Susanti, A.; Wijayanto, N.; Hikmat, A. Keanekaragaman jenis tumbuhan obat di Agroforestri Repong Damar Krui, Provinsi Lampung. *Media Konserv.* **2018**, *23*, 162–168.
90. Itta, D.; Hafizianor, R.H.; Rampun, E.L.A.; Elma, M. Strategy for developing Dukuh agroforestry system in Ati'im Village, Pengaron Sub-District, Banjar Regency, South Kalimantan Province, Indonesia. *J. Hunan Univ. Sci.* **2021**, *48*, 1–9.
91. Indrajaya, Y.; Sudomo, A. Analisis finansial agroforestry sengon dan kapulaga di Desa Payungagung Kecamatan Panumbangan Ciamis. *J. Penelit. Agrofor.* **2013**, *1*, 123–132.
92. Wiryono, W.; Puteri, V.N.U.; Senoaji, G. The diversity of plant species, the types of plant uses and the estimate of carbon stock in agroforestry system in Harapan Makmur Village, Bengkulu, Indonesia. *Biodiversitas* **2016**, *17*, 249–255. [CrossRef]
93. Pitopang, R.; Atmoko, A.T.; Yusran; Wardah; Mertosono, S.R.; Ramawangsa, P.A. Plant diversity in agroforestry system and its traditional use by three different ethnics in Central Sulawesi Indonesia. In *IOP Conference Series: Earth and Environmental Science*; IOP Publishing Ltd.: Bristol, UK, 2021; Volume 886.
94. Hakim, L.; Pamungkas, N.R.; Wicaksono, K.P.; Soemarno, S. The conservation of osingnese traditional home garden agroforestry in Banyuwangi, East Java, Indonesia. *Agrivita* **2018**, *40*, 506–514. [CrossRef]
95. Rambey, R.; Wijayanto, N.; Siregar, I.Z.; Onrizal; Susilowati, A. Study of agroforestry mindi planting pattern (*Melia dubia cavanilles*) in Selaawi Village, Garut District, West Java Province. In *IOP Conference Series: Earth and Environmental Science*; IOP Publishing Ltd.: Bristol, UK, 2019.
96. Kissinger; Pitri, R.M.N. Bioekologi agroforestry kopi: Tutupan vegetasi dan pola tumbuhan penyusun agroforestry kopi (*Coffea* sp.) di Kecamatan Pengaron Kabupaten Banjar Kalimantan Selatan. *EnviroScienteae* **2017**, *13*, 150–156. [CrossRef]
97. Danarto, S.A. Cropping patterns and plants diversity in agroforests in Wringin village subdistrict of Wringin Bondowoso East Java. In *IOP Conference Series: Earth and Environmental Science*; IOP Publishing Ltd.: Bristol, UK, 2020; Volume 456.
98. Waters, T. *The Persistence of Subsistence Agriculture: Life Beneath the Level of the Marketplace*; Lexington Books: Lanham, MD, USA, 2007; ISBN 978-0739107683.
99. Miracle, M.P. "Subsistence agriculture": Analytical problems and alternative concepts. *Am. J. Agric. Econ.* **1968**, *50*, 292–310. [CrossRef]
100. Morton, J.F. *The Impact of Climate Change on Smallholder and Subsistence Agriculture*; National Academy of Sciences of the United States of America: Washington, DC, USA, 2007; Volume 104, pp. 19680–19685.
101. Hudiyani, I.; Purnaningsih, N.; Asngari, P.S. Hardjanto Persepsi Petani terhadap Hutan Rakyat Pola Agroforestri di Kabupaten Wonogiri, Provinsi Jawa Tengah. *J. Penyul.* **2017**, *13*, 64–78. [CrossRef]

102. Madyantoro, A.; Muttaqin, Z.; Lidiawati, I. Kajian kontribusi sistem agroforestri terhadap pendapatan petani (Studi kasus: Desa Cibatok Dua, Kecamatan Cibungbulang, Kabupaten Bogor, Provinsi Jawa Barat). *J. Nusa Sylva* **2015**, *15*, 11–16.
103. Szulecka, J.; Obidzinski, K.; Dermawan, A. Corporate-society engagement in plantation forestry in Indonesia: Evolving approaches and their implications. *For. Policy Econ.* **2016**, *62*, 19–29. [CrossRef]
104. Iskandar, J.; Iskandar, B.S.; Partasasmita, R. Responses to environmental and socio-economic changes in the Karangwangi traditional agroforestry system, South Cianjur, West Java. *Biodiversitas* **2016**, *17*, 332–341. [CrossRef]
105. Rahman, S.A.; Sunderland, T.; Roshetko, J.M.; Healey, J.R. Facilitating smallholder tree farming in fragmented tropical landscapes: Challenges and potentials for sustainable land management. *J. Environ. Manag.* **2017**, *198*, 110–121. [CrossRef]
106. Saputra, A.; Itta, D. Asysyifa Peran agroforestri karet terhadap pendapatan petani karet di Desa Maburai Provinsi Kalimantan Selatan. *J. Hutan Trop.* **2021**, *9*, 244–251. [CrossRef]
107. Abdurachman, D.; Andung Rokhmat, H.; Setiawan, H.I. Hubungan luas garapan hutan rakyat dengan pendapatan petani (Kasus pada Kelompok Tani Alam Raya Desa Pamedaran Kecamatan Ketanggungan, Kabupaten Brebes). *J. Agrijati* **2015**, *28*, 35–44.
108. Iskandar, B.S.; Iskandar, J.; Partasasmita, R. Strategy of the Outer Baduy community of South Banten (Indonesia) to sustain their swidden farming traditions by temporary migration to non-Baduy areas. *Biodiversitas* **2018**, *19*, 453–464. [CrossRef]
109. Achmad, B.; Simon, H.; Diniyati, D.; Widyaningsih, T.S. Persepsi petani terhadap pengelolaan dan fungsi hutan rakyat di Kabupaten Ciamis. *J. Bumi Lestari* **2012**, *12*, 123–136.
110. Wijaya, K.; Budidarsono, S.; Roshetko, J. *Socioeconomic Baseline Studies: Agro-Forestry and Sustainable Vegetables Production in Southeast Asian Watershed;* ICRAF Southeast Asia: Bogor, Indonesia, 2007.
111. Sari, R.R.; Saputra, D.D.; Hairiah, K.; Rozendaal, D.M.A.; Roshetko, J.M.; Noordwijk, M. Van Gendered species preferences link tree diversity and carbon stocks in Cacao agroforest in Southeast Sulawesi, Indonesia. *Land* **2020**, *9*, 108. [CrossRef]
112. Siregar, U.J.; Rachmi, A.; Massijaya, M.Y.; Ishibashi, N.; Ando, K. Economic analysis of sengon (*Paraserianthes falcataria*) community forest plantation, a fast growing species in East Java, Indonesia. *For. Policy Econ.* **2007**, *9*, 822–829. [CrossRef]
113. Suyadi, S.; Sumardjo, S.; Uchrowi, Z.; Tjitropranoto, P. Factors affecting agroforestry farmers' capacity surrounding national park. *Indones. J. For. Res.* **2019**, *6*, 27–41. [CrossRef]
114. Mercer, D.E. Adoption of agroforestry innovations in the tropics: A review. *Agrofor. Syst.* **2004**, *61*, 311–328. [CrossRef]
115. Kostov, P.; Lingard, J. Risk management: A general framework for rural development. *J. Rural Stud.* **2003**, *19*, 463–476. [CrossRef]
116. Reubens, B.; Moeremans, C.; Poesen, J.; Nyssen, J.; Tewoldeberhan, S.; Franzel, S.; Deckers, J.; Orwa, C.; Muys, B. Tree species selection for land rehabilitation in Ethiopia: From fragmented knowledge to an integrated multi-criteria decision approach. *Agrofor. Syst.* **2011**, *82*, 303–330. [CrossRef]
117. Hardy, M.; Dziba, L.; Kilian, W.; Tolmay, J. Rainfed farming systems in South Africa. In *Rainfed Farming Systems;* Tow, P., Cooper, I., Partridge, I., Birch, C., Eds.; Springer: Dordrecht, The Netherlands, 2011; pp. 395–432, ISBN 978-1-4020-9132-2.
118. Triwanto, J.; Syarifuddin, A.; Mutaqin, T. Aplikasi agroforestry di Desa Mentaraman Kecamatan Donomulyo Kabupaten Malang. *Dedikasi* **2012**, *9*, 13–21.
119. Isnaini, D.N. Kelayakan Usaha Agroforestry di Hutan Pendidikan Gunung Walat Sukabumi Jawa Barat. Undergraduate Thesis, Institut Pertanian Bogor: Bogor, Indonesia, 2006.
120. Buyinza, J.; Nuberg, I.K.; Muthuri, C.W.; Denton, M.D. Assessing smallholder farmers' motivation to adopt agroforestry using a multi-group structural equation modeling approach. *Agrofor. Syst.* **2020**, *94*, 2199–2211. [CrossRef]
121. Wahyu, I.; Pranoto, H.; Supriyanto, B. Kajian produktivitas tanaman semusim pada sistem agroforestri di Kecamatan Samboja Kabupaten Kutai Kartanegara. *J. Agroekoteknologi Trop. Lembab* **2018**, *1*, 24–33. [CrossRef]
122. Kusumedi, P.; Jariyah, N.A. Analisis finansial pengelolaan agroforestri dengan sengon kapulaga di Desa Tirip, Kecamatan Wadaslintang, Kabupaten Wonosobo. *J. Penelit. Sos. Dan Ekon. Kehutan.* **2010**, *7*, 93–100. [CrossRef]
123. Onibala, A.G.; Sondakh, M.L.; Kaunang, R.; Mandei, J. Analisis faktor-faktor yang mempengaruhi produksi padi sawah di Kelurahan Koya, Kecamatan Tondano Selatan. *Agri-SosioEkonomi Unsratkonomi* **2017**, *13*, 237–242. [CrossRef]
124. Sarina, S.; Silamat, E.; Puspitasari, D. Analisis faktor-faktor yang mempengaruhi produksi cabai merah di Desa Kampung Melayu Kecamatan Bermani Ulu Kabupaten Rejang Lebong. *J. Agroqua* **2015**, *13*, 57–67.
125. Heriyanto, H.; Darus, D. Analisis efisiensi faktor produksi karet di Kabupaten Kampar Provinsi Riau. *J. Din. Pertan.* **2017**, *XXXIII*, 121–128. [CrossRef]
126. Zwingly, K.R.A.; Lolowang, T.F.; Pangemanan, L.R.J. Analisis faktor produksi yang mempengaruhi produksi kopra di Kecamatan Tomohon Barat. *Agri-SosioEkonomi Unsrat* **2018**, *14*, 17–32. [CrossRef]
127. Winarni, S.; Yuwono, S.B.; Herwanti, S. Struktur pendapatan, tingkat kesejahteraan dan faktor produksi agroforestri kopi pada Kesatuan Pengelolaan Hutan Lindung Batutegi (Studi di Gabungan Kelompok Tani Karya Tani Mandiri). *J. Sylva Lestari* **2016**, *4*, 1–10. [CrossRef]
128. Armas, A.A.; Dassir, M.; Millang, S. Peranan berbagai pola agroforestri terhadap tingkat resiliensi petani di sub DAS Minraleng Hulu. *J. Hutan Dan Masy.* **2020**, *12*, 120–135. [CrossRef]
129. Nopitasari, R.; Nurlaila, A.; Deni, D. Kontribusi agroforestri terhadap tingkat pendapatan rumah tangga petani Desa Cibinuang Kuningan Jawa Barat. *Wanaraksa* **2019**, *13*. [CrossRef]

130. Dyah, P.S. Manajemen usahatani pada lahan kering di Kabupaten Gunung Kidul Daerah Istimewa Yogyakarta. In Proceedings of the Prosiding Interdisciplinary Postgraduate Student Conference 3rd. Program Pascasarjana Universitas Muhammadiyah Yogyakarta (PPs UMY); Program Pascasarjana Universitas Muhammadiyah Yogyakarta (PPs UMY): Yogyakarta, Indonesia, 2017; pp. 274–278.
131. Suhartono, S.; Widiyanto, A. Optimalisasi penggunaan faktor produksi usahatani kopi di bawah tegakan pinus (*Pinus merkusii*). *J. Penelit. Sos. Dan Ekon. Kehutan.* **2020**, *17*, 39–47. [CrossRef]
132. Berliantara, B.; Zakaria, W.A.; Lestari, D.A.H. Analisis efisiensi produksi dan keuntungan usahatani tomat dataran rendah di Kabupaten Lampung Selatan. *J. Ilmu-Ilmu Agribisnis* **2016**, *4*, 342–350.
133. Suroso, S.; Watemin, W.; Utami, P. Efisiensi ekonomi usahatani padi semi organik di Desa Sawangan Kecamatan Kebasen Kabupaten Banyumas. *Agritech* **2016**, *XVIII*, 60–72.
134. Dewi, N.L.P.R.; Utama, M.S.; Yuliarmi, N.N. Faktor-faktor yang mempengaruhi produktivitas usaha tani dan keberhasilan program Simantri di Kabupaten Klungkung. *E-J. Ekon. Dan Bisnis Univ. Udayana* **2017**, *6*, 701–728.
135. Panjaitan, E.; Paman, U. Darus Analisis pengaruh faktor produksi terhadap produktivitas usahatani kelapa sawit pola swadaya di Desa Sungai Buluh Kecamatan Kuantan Singingi Hilir, Kabupaten Kuantan Singingi. *J. Din. Pertan.* **2020**, *XXXVI*, 61–68. [CrossRef]
136. Nazeb, A.; Darwanto, D.H.; Suryantini, A. Efisiensi alokatif usahatani padi pada lahan gambut di Kecamatan Pelalawan, Kabupaten Pelalawan, Riau. *J. Ekon. Pertan. Dan Agribisnis* **2019**, *3*, 267–277. [CrossRef]
137. Isyanto, A.Y. Faktor-faktor yang berpengaruh terhadap produksi pada usahatani padi di Kabupaten Ciamis. *Cakrawala Galuh* **2012**, *I*, 1–8.
138. Adekanye, T.; Oni, K.C. Comparative energy use in cassava production under different farming technologies in Kwara State of Nigeria. *Environ. Sustain. Indic.* **2022**, *14*, 100173. [CrossRef]
139. Massing, N.; Schneider, S.L. Degrees of competency: The relationship between educational qualifications and adult skills across countries. *Large-Scale Assess. Educ.* **2017**, *5*, 434. [CrossRef]
140. Ghosh-Jerath, S.; Kapoor, R.; Ghosh, U.; Singh, A.; Downs, S.; Fanzo, J. Pathways of climate change impact on agroforestry, food consumption pattern, and dietary diversity among indigenous subsistence farmers of Sauria Paharia tribal community of India: A mixed methods study. *Front. Sustain. Food Syst.* **2021**, *5*, 1–16. [CrossRef] [PubMed]
141. Martini, E.; Dewi, S.; Janudianto; Setiawan, A.; Roshetko, J. Strategi penghidupan petani agroforest dalam menghadapi perubahan cuaca yang tidak menentu: Contoh kasus di Sulawesi Selatan dan Sulawesi Tenggara. In *Proseding Seminar Nasional Agroforestri*; Kuswantoro, D.P., Widyaningsih, T.S., Fauziyah, E., Rachmawati, R., Eds.; BPTA-Faperta UB-ICRAF-MAFI: Malang, Indonesian, 2013; pp. 604–609.
142. Thorlakson, T.; Neufeldt, H. Reducing subsistence farmers' vulnerability to climate change: Evaluating the potential contributions of agroforestry in western Kenya. *Agric. Food Secur.* **2012**, *1*, 15. [CrossRef]
143. Rahman, S.A.; Rahman, M.F.; Sunderland, T. Causes and consequences of shifting cultivation and its alternative in the hill tracts of eastern Bangladesh. *Agrofor. Syst.* **2012**, *84*, 141–155. [CrossRef]
144. Arocena-Francisco, H.; de Jong, W.; Doanh, L.; de Guzmn, R.S.; Koffa, S.; Kuswanda, M.; Lawrence, A.; Pagulon, A.; Rantan, D.; Widawati, E. External factors affecting the domestication of agroforestry trees (economics and policy). In *Domestication of Agroforestry Trees in Southeast Asia*; Forest, Farm and Community Tree Research Reports; Winrock International: Morrilton, AR, USA, 1999.
145. Hammett, A.L. Developing community-based market information systems. In Proceedings of the Marketing Multipurpose Tree Species in Asia, Baguio City, Philippines, 6–9 December 1993; Raintree, J.B., Francisco, H.A., Eds.; Winrock International: Bangkok, Thailand, 1994; pp. 289–300.
146. Cortner, O.; Garrett, R.D.; Valentim, J.F.; Ferreira, J.; Niles, M.T.; Reis, J.; Gil, J. Perceptions of integrated crop-livestock systems for sustainable intensification in the Brazilian Amazon. *Land Use Policy* **2019**, *82*, 841–853. [CrossRef]
147. Garrett, R.D.; Niles, M.T.; Gil, J.D.B.; Gaudin, A.; Chaplin-Kramer, R.; Assmann, A.; Assmann, T.S.; Brewer, K.; de Faccio Carvalho, P.C.; Cortner, O.; et al. Social and ecological analysis of commercial integrated crop livestock systems: Current knowledge and remaining uncertainty. *Agric. Syst.* **2017**, *155*, 136–146. [CrossRef]
148. Gil, J.D.B.; Garrett, R.; Berger, T. Determinants of crop-livestock integration in Brazil: Evidence from the household and regional levels. *Land Use Policy* **2016**, *59*, 557–568. [CrossRef]
149. Asian Development Bank and Government of Bangladesh. *Second Chittagong Hill Tracts Rural Development Project (CHTRDP-2)*; Technical Assistance Consultant's Report; Asian Development Bank: Mandaluyong, Philippines, 2016.
150. Perdana, A.; Roshetko, J.M. Survival strategy: Traders of smallholder teak in Indonesia. *Int. For. Rev.* **2015**, *17*, 461–468. [CrossRef]
151. Roshetko, J.M.; Kurniawan, I.; Budidarsono, S. *Smallholder Cultivation of Katuk (Sauropus androgynus) and Kucai (Allium odorum): Challenges in Sustaining Commercial Production and Market Linkage*; Susila, A.D., Purwoko, B.S., Roshetko, J.M., Palada, M.C., Kartika, J.G., Dahlia, L., Wijaya, K., Rahmanulloh, A., Raimadoya, M., Koesoemaningtyas, T., et al., Eds.; World Association of Soil and Water Conservation (WASWAC): Bangkok, Thailand, 2012; pp. 215–230. ISBN 978-974-350-655-0.
152. Arvola, A.; Brockhaus, M.; Kallio, M.; Pham, T.T.; Chi, D.T.L.; Long, H.T.; Nawir, A.A.; Phimmavong, S.; Mwamakimbullah, R.; Jacovelli, P. What drives smallholder tree growing? Enabling conditions in a changing policy environment. *For. Policy Econ.* **2020**, *116*, 102173. [CrossRef]

153. Kang, B.T.; Akinnifesi, F.K. Agroforestry as alternative land-use production systems for the tropics. *Nat. Resour. Forum* **2000**, *24*, 137–151. [CrossRef]
154. Rahman, S.A.; de Groot, W.T.; Snelder, D.J. Exploring the agroforestry adoption gap: Financial and socioeconomics of litchi-based agroforestry by smallholders in Rajshahi (Bangladesh). In *Smallholder Tree Growing for Rural Development and Environmental Services: Lessons from Asia, Advances in Agroforestry*; Snelder, D.J., Lasco, R.D., Eds.; Springer: Dordrecht, The Netherlands, 2008; pp. 227–243, ISBN 978-1-4020-8261-0.
155. Hosier, R.H. The economics of smallholder agroforestry: Two case studies. *World Dev.* **1989**, *17*, 1827–1839. [CrossRef]
156. Abidin, M.Z. Dampak kebijakan impor beras dan ketahanan pangan dalam perspektif kesejahteraan sosial. *Sosio Inf.* **2015**, *1*, 213–230. [CrossRef]
157. Utomo, S. Dampak impor dan ekspor jagung terhadap produktivitas jagung di Indonesia. *J. Etikonomi* **2012**, *11*, 158–179. [CrossRef]
158. Hasibuan, A.I.R.S. Kebijakan pangan pasca ratifikasi Agreement on Agriculture (AoA)-WTO. *J. Kaji. Polit. Dan Masal. Pembang.* **2015**, *11*, 1633–1644.
159. Maramis, A.Y. Dampak impor cabai dari Tiongkok terhadap perekonomian Indonesia tahun 2010–2015. *Jom Fisip* **2018**, *5*, 1–15.
160. Olivi, R.; Qurniati, R.; Firdasari, F. Kontribusi agroforestri terhadap pendapatan petani di Desa Sukoharjo Kabupaten Pringsewu. *J. Sylva Lestari* **2015**, *3*, 1–12. [CrossRef]
161. Martin, E.; Suharjito, D.; Darusman, D.; Sunito, S.; Winarno, B. Etika subsistensi petani kopi: Memahami dinamika pengembangan agroforestri di dataran tinggi Sumatera Selatan. *Sodality J. Sosiol. Pedesaan* **2016**, *4*, 92–102. [CrossRef]
162. Martial, T. The need of tree tenure security to maintenance the traditional agroforestry (Parak) in Koto Malintang Village, West Sumatera. In Proceedings of the 2nd Annual International Conference Syiah Kuala University 2012 & the 8th IMT-GT Uninet Biosciences Conference, Banda Aceh, Indonesia, 22–24 November 2012; Volume 2, pp. 181–186.
163. Sahide, M.A.K.; Giessen, L. The fragmented land use administration in Indonesia—Analysing bureaucratic responsibilities influencing tropical rainforest transformation systems. *Land Use Policy* **2015**, *43*, 96–110. [CrossRef]
164. Mayrowani, H.; Ashari, N. Pengembangan agroforestry untuk mendukung ketahanan pangan dan pemberdayaan petani sekitar hutan. *Forum Penelit. Agro Ekon.* **2011**, *29*, 83–98. [CrossRef]
165. Rozaki, Z.; Rahmawati, N.; Wijaya, O.; Mubarok, A.F.; Senge, M.; Kamarudin, M.F. A case study of agroforestry practices and challenges in Mt. Merapi risk and hazard prone area of Indonesia. *Biodiversitas* **2021**, *22*, 2511–2518. [CrossRef]
166. Sinclair, F.L.; Walker, D.H. A utilitarian approach to the incorporation of local knowledge in agroforestry research and extension. In *Agroforestry in Sustainable Agricultural Systems*; Buck, L.E., Lassoie, J., Fernandes, E.C.M., Eds.; Lewis Publishers Inc.: Boca Raton, FL, USA, 1999; pp. 245–275. ISBN 1566702941.
167. Nugroho, H.Y.S.H.; Basuki, T.M.; Pramono, I.B.; Savitri, E.; Purwanto; Indrawati, D.R.; Wahyuningrum, N.; Adi, R.N.; Indrajaya, Y.; Supangat, A.B.; et al. Forty years of soil and water conservation policy, implementation, research and development in Indonesia: A review. *Sustainability* **2022**, *14*, 2972. [CrossRef]
168. de Graaff, J.; Aklilu, H.A.; Bodnar, F.; Posthumus, H.; Tenge, A.J.M. Adoption of soil and water conservation measures. In Proceedings of the EFARD Conference, Zurich, Switzerland, 27–29 April 2005.
169. Rahman, S.A.; Foli, S.; Al Pavel, M.A.; Al Mamun, M.A.; Sunderland, T. Forest, trees and agroforestry: Better livelihoods and ecosystem services from multifunctional landscapes. *Int. J. Dev. Sustain.* **2015**, *4*, 479–491.
170. Acheampong, E.; Insaidoo, T.F.G.; Ros-Tonen, M.A.F. Management of Ghana's modified taungya system: Challenges and strategies for improvement. *Agrofor. Syst.* **2016**, *90*, 659–674. [CrossRef]
171. Boffa, J.-M.; Kindt, R.; Katumba, B.; Jourget, J.-G.; Turyomurugyendo, L. Management of tree diversity in agricultural landscapes around Mabira Forest Reserve, Uganda. *Afr. J. Ecol.* **2008**, *46*, 24–32. [CrossRef]
172. Kabir, M.E.; Webb, E.L. Can homegardens conserve biodiversity in Bangladesh? *Biotropica* **2008**, *40*, 95–103. [CrossRef]
173. Moreno-Calles, A.I.; Casas, A.; García-Frapolli, E.; Torres-García, I. Traditional agroforestry systems of multi-crop "milpa" and "chichipera" cactus forest in the arid Tehuacán Valley, Mexico: Their management and role in people's subsistence. *Agrofor. Syst.* **2012**, *84*, 207–226. [CrossRef]
174. Carlson, M.J.; Mitchell, R.; Rodriguez, L. Scenario analysis to identify viable conservation strategies in Paraguay's imperiled Atlantic forest. *Ecol. Soc.* **2011**, *16*, 8. [CrossRef]
175. Lasco, R.D.; Delfino, R.J.P.; Espaldon, M.L.O. Agroforestry systems: Helping smallholders adapt to climate risks while mitigating climate change. *Wiley Interdiscip. Rev. Clim. Chang.* **2014**, *5*, 825–833. [CrossRef]
176. Mbow, C.; Smith, P.; Skole, D.; Duguma, L.; Bustamante, M. Achieving mitigation and adaptation to climate change through sustainable agroforestry practices in Africa. *Curr. Opin. Environ. Sustain.* **2014**, *6*, 8–14. [CrossRef]
177. Kurniasih, H.; Ford, R.M.; Keenan, R.J.; King, B. The evolution of community forestry through the growth of interlinked community institutions in Java, Indonesia. *World Dev.* **2021**, *139*, 105319. [CrossRef]
178. Rahman, S.A.; Sunderland, T.; Roshetko, J.M.; Basuki, I.; Healey, J.R. Tree culture of smallholder farmers practicing agroforestry in Gunung Salak Valley, West Java, Indonesia. *Small-Scale For.* **2016**, *15*, 433–442. [CrossRef]
179. Rahman, S.A.; Rahman, M.F.; Sunderland, T. Increasing tree cover in degrading landscapes:'Integration'and 'Intensification'of smallholder forest culture in the Alutilla Valley, Matiranga, Bangladesh. *Small-Scale For.* **2014**, *13*, 237–249. [CrossRef]
180. Roshetko, J.M.; Nugraha, E.; Tukan, J.C.M.; Manurung, G.; Fay, C.; van Noordwijk, M. Agroforestry for Livelihood Enhancement and Enterprise Development. 2007. Available online: https://www.worldagroforestry.org/ (accessed on 13 June 2022).

181. Tomich, T.P.; de Foresta, H.; Dennis, R.; Ketterings, Q.M.; Murdiyarso, D.; Palm, C.; Stolle, F.; Suyanto, S.; van Noordwijk, M. Carbon offsets for conservation and development in Indonesia? *Am. J. Altern. Agric.* **2002**, *17*, 125–137. [CrossRef]
182. Roshetko, J.M. Smallholder Tree Farming Systems for Livelihood Enhancement and Carbon Storage. Ph.D. Thesis, Department of Geosciences and Natural Resource Management, University of Copenhagen, Copenhagen, Denmark, 2013.
183. Purnomo, A.S.; Laili, S.; Zayadi, H. Persepsi masyarakat tentang agroforestri di Desa Sumberejo Poncokusumo Malang. *Biosaintropis (Biosci.-Trop.)* **2022**, *7*, 9–14. [CrossRef]
184. Hidayat, A.S.; Laili, S.; Zayadi, H. Studi persepsi masyarakat tentang agroforestri tanaman kopi di Desa Patokpicis Kecamatan Wajak, Kabupaten Malang. *Biosaintropis (Biosci.-Trop.)* **2021**, *6*, 1–7.
185. Susanti, Y.; Wulandari, C.; Kaskoyo, H.; Safe'i, R.; Yuwono, S.B. Persepsi masyarakat terhadap pengelolaan agroforestri di Tahura Wan Abdul Rachman, Bandarlampung. *J. Hutan Trop.* **2021**, *9*, 472–487. [CrossRef]
186. Kurniawan, I.; Roshetko, J.M. Market's risk and production uncertainty: Drivers of agroforestry land use diversification of smallholder teak grower in Gunungkidul District, Indonesia. In Proceedings of the 2 World Congress of Agroforestry: Agroforestry-The Future of Global Land Use, Nairobi, Kenya, 24–28 August 2009.
187. Rohadi, D.; Herawati, T. Agroforestry research in Indonesia: Where to go for the next. In Proceedings of the World Congress on Agroforestry, New Delhi, India, 10–14 February 2014; pp. 1–13.
188. Sood, K.K.; Mitchell, C.P. Identifying important biophysical and social determinants of on-farm tree growing in subsistence-based traditional agroforestry systems. *Agrofor. Syst.* **2009**, *75*, 175–187. [CrossRef]
189. Abdurachman, A.; Dariah, A.; Mulyani, A. Strategi dan teknologi pengelolaan lahan kering mendukung pengadaan pangan nasional. *J. Litbang Pertan.* **2008**, *27*, 43–49.
190. Anderson, V. Deep Adaptation: A Framework for Climate Resilience, Decarbonization, and Planetary Health. Ph.D. Thesis, University of Toronto, Toronto, ON, Canada, 2018.
191. Anisykurlillah, R.; Domai, T.; Hermawan, R. Community empowerment efforts in private forest development of Malang Regency, Indonesia. *Russ. J. Agric. Socio-Econ. Sci.* **2019**, *11*, 252–256. [CrossRef]
192. Pambudi, A.S. The development of social forestry in Indonesia. *JISDeP J. Indones. Sustain. Dev. Plan.* **2020**, *1*, 57–66. [CrossRef]
193. Rakatama, A.; Pandit, R. Reviewing social forestry schemes in Indonesia: Opportunities and challenges. *For. Policy Econ.* **2020**, *111*, 102052. [CrossRef]
194. Rahman, S.A.; Jacobsen, J.B.; Healey, J.R.; Roshetko, J.M.; Sunderland, T. Finding alternatives to swidden agriculture: Does agroforestry improve livelihood options and reduce pressure on existing forest? *Agrofor. Syst.* **2017**, *91*, 185–199. [CrossRef]
195. Hadi, Y.S.; Saleh, M.B. Strategy for the implementation of CDM and carbon trade in Indonesia. In Proceedings of the Workshop Forests for Poverty Reduction: Opportunities with Clean Development Mechanism, Environmental Services and Biodiversity, Seoul, Korea, 27–29 August 2003; Sim, H.C., Appanah, S., Youn, Y.C., Eds.; FAO of the UN Regional Office for Asia and the Pacific: Bangkok, Thailand, 2004; pp. 89–98.
196. Undang-Undang Republik Indonesia Nomor 18 Tahun 2012 Tentang Pangan 2012. Available online: https://luk.staff.ugm.ac.id/atur/UU18-2012Pangan.pdf (accessed on 13 June 2022).
197. Tan, Z.; Lal, R.; Wiebe, K.D. Global soil nutrient depletion and yield reduction. *J. Sustain. Agric.* **2005**, *26*, 123–146. [CrossRef]
198. Kubitza, C.; Krishna, V.V.; Urban, K.; Alamsyah, Z.; Qaim, M. Land property rights, agricultural intensification, and deforestation in Indonesia. *Ecol. Econ.* **2018**, *147*, 312–321. [CrossRef]
199. Olson, J.M.; Berry, L. *Land Degradation in Java, Indonesia: Its Extent and Impact*; World Bank: Washington, DC, USA, 2001.
200. Sitorus, S.R.P.; Pravitasari, A.E. Land degradation and landslide in Indonesia. *Sumatra J. Disaster Geogr. Geogr. Educ.* **2017**, *1*, 61–71.
201. Nurrochmat, D.R.; Pribadi, R.; Siregar, H.; Justianto, A.; Park, M.S. Transformation of agro-forest management policy under the dynamic circumstances of a two-decade regional autonomy in Indonesia. *Forests* **2021**, *12*, 419. [CrossRef]
202. Ajayi, O.C.; Place, F. Policy support for large-scale adoption of agroforestry practices: Experience from Africa and Asia. In *Agroforestry-The Future of Global Land Use*; Nair, P., Garrity, D., Eds.; Springer: Dordrecht, The Netherlands, 2012; pp. 175–201. ISBN 978-94-007-4675-6.
203. Santoso, H. Peran sektor kehutanan dalam mendukung akses pangan. In Proceedings of the Bahan Presentasi pada Seminar Nasional Hari Pangan Sedunia XXI, Jakarta, Indonesia, 29 September 2011.
204. Lee, Y.; Rianti, I.P.; Park, M.S. Measuring social capital in Indonesian community forest management. *Forest Sci. Technol.* **2017**, *13*, 133–141. [CrossRef]
205. Perdana, A.; Roshetko, J.M.; Kurniawan, I. Forces of competition: Smallholding teak producers in Indonesia. *Int. For. Rev.* **2012**, *14*, 238–248. [CrossRef]
206. Waldron, A.; Garrity, D.; Malhi, Y.; Girardin, C.; Miller, D.C.; Seddon, N. Agroforestry can enhance food security while meeting other sustainable development goals. *Trop. Conserv. Sci.* **2017**, *10*, 1–6. [CrossRef]
207. van Noordwijk, M.; Speelman, E.; Hofstede, G.J.; Farida, A.; Abdurrahim, A.Y.; Miccolis, A.; Hakim, A.L.; Wamucii, C.N.; Lagneaux, E.; Andreooti, F.; et al. Sustainable agroforestry landscape management: Changing the game. *Land* **2020**, *9*, 243. [CrossRef]
208. Kaba, J.S.; Yamoah, F.A.; Acquaye, A. Towards sustainable agroforestry management: Harnessing the nutritional soil value through cocoa mix waste. *Waste Manag.* **2021**, *124*, 264–272. [CrossRef]

209. Zinngrebe, Y.; Borasino, E.; Chiputwa, B.; Dobie, P.; Garcia, E.; Gassner, A.; Kihumuro, P.; Komarudin, H.; Liswanti, N.; Makui, P.; et al. Agroforestry governance for operationalising the landscape approach: Connecting conservation and farming actors. *Sustain. Sci.* **2020**, *15*, 1417–1434. [CrossRef]
210. Kidd, C.V.; Pimentel, D. (Eds.) *Integrated Resource Management: Agroforestry for Development*; Elsevier Science Publishing Co. Inc.: San Diego, CA, USA, 1992; ISBN 9780124064102.
211. Antwi-Agyei, P.; Wiafe, E.A.; Amanor, K.; Baffour-Ata, F.; Codjoe, S.N.A. Determinants of choice of climate change adaptation practices by smallholder pineapple farmers in the semi-deciduous forest zone of Ghana. *Environ. Sustain. Indic.* **2021**, *12*, 100140. [CrossRef]
212. Legesse, A.; Negash, M. Species diversity, composition, structure and management in agroforestry systems: The case of Kachabira district, Southern Ethiopia. *Heliyon* **2021**, *7*, e06477. [CrossRef]
213. Lelamo, L.L. A review on the indigenous multipurpose agroforestry tree species in Ethiopia: Management, their productive and service roles and constraints. *Heliyon* **2021**, *7*, e07874. [CrossRef]
214. Wu, J.; Zeng, H.; Zhao, F.; Chen, C.; Liu, W.; Yang, B.; Zhang, W. Recognizing the role of plant species composition in the modification of soil nutrients and water in rubber agroforestry systems. *Sci. Total Environ.* **2020**, *723*, 138042. [CrossRef]
215. Asante, B.O.; Ennin, S.A.; Osei-Adu, J.; Asumadu, H.; Adegbidi, A.; Saho, M.; Nantoume, H. Performance of integrated crop-small ruminant production systems in West Africa. *Agrofor. Syst.* **2019**, *93*, 989–999. [CrossRef]
216. Atangana, A.; Khasa, D.; Chang, S.X.; Degrande, A. Definitions and classification of agroforestry systems. In *Tropical Agroforestry*; Springer: Berlin/Heidelberg, Germany, 2014; pp. 35–47.
217. Dene, A.B.N.; Diogo, R.V.C.; Houehanou, T.D.; Paul, B.K. Improving forage productivity for increased livestock production using biochar and green manure amendments. In Proceedings of the 20th Australian Agronomy Conference, Toowomba, Australia, 18–22 September 2022; pp. 1–5.
218. Hairiah, K.; van Noordwijk, M.; Suprayogo, D. *Interaksi Antara Pohon—Tanah—Tanaman Semusim: Kunci Keberhasilan atau Kegagalan Dalam Sistem Agroforestri*; World Agroforestry Centre (ICRAF) Southeast Asia: Bogor, Indonesia, 2016; pp. 19–42, ISBN 978-979-493-415-9.
219. Tesema, T. Determinants of allocative and economic efficiency in crop-livestock integration in western part of Ethiopia evidence from Horro district: Data envelopment approach. *Heliyon* **2021**, *7*, e07390. [CrossRef]
220. Duan, C.; Shi, P.; Zong, N.; Wang, J.; Song, M.; Zhang, X. Feeding solution: Crop-livestock integration via crop-forage rotation in the southern Tibetan Plateau. *Agric. Ecosyst. Environ.* **2019**, *284*, 106589. [CrossRef]
221. Portilho, I.I.R.; Savin, M.C.; Borges, C.D.; Tsai, S.M.; Mercante, F.M.; Roscoe, R.; de Carvalho, L.A. Maintenance of N cycling gene communities with crop-livestock integration management in tropical agriculture systems. *Agric. Ecosyst. Environ.* **2018**, *267*, 52–62. [CrossRef]
222. Udawatta, R.P.; Gantzer, C. Soil and water ecosystem services of agroforestry. *J. Soil Water Conserv.* **2022**, *77*, 5A–11A. [CrossRef]
223. Swieter, A.; Langhof, M.; Lamerre, J.; Greef, J.M. Long-term yields of oilseed rape and winter wheat in a short rotation alley cropping agroforestry system. *Agrofor. Syst.* **2019**, *93*, 1853–1864. [CrossRef]
224. Acheampong, E.O.; Sayer, J.; Macgregor, C. Road improvement enhances smallholder productivity and reduces forest encroachment in Ghana. *Environ. Sci. Policy* **2018**, *85*, 64–71. [CrossRef]
225. Anatika, E.; Kaskoyo, H.; Febryano, I.G.; Banuwa, I.S. Pengelolaan hutan rakyat di Kabupaten Tulang Bawang Barat. *J. Sylva Lestari* **2019**, *7*, 42–51. [CrossRef]
226. Sukwika, T.; Darusman, D.; Kusmana, C.; Nurrochmat, D.R. Skenario kebijakan pengelolaan hutan rakyat berkelanjutan di Kabupaten Bogor. *J. Pengelolaan Sumberd. Alam dan Lingkung* **2018**, *8*, 207–215. [CrossRef]
227. Adeagbo, O.A.; Ojo, T.O.; Adetoro, A.A. Understanding the determinants of climate change adaptation strategies among smallholder maize farmers in South-west, Nigeria. *Heliyon* **2021**, *7*, e06231. [CrossRef]
228. Ranjitkar, S.; Lawley, B.; Tannock, G.; Engberg, R.M. Bacterial succession in the broiler gastrointestinal tract. *Appl. Environ. Microbiol.* **2016**, *82*, 2399–2410. [CrossRef]
229. Cardinael, R.; Mao, Z.; Prieto, I.; Stokes, A.; Dupraz, C.; Kim, J.H.; Jourdan, C. Competition with winter crops induces deeper rooting of walnut trees in a Mediterranean alley cropping agroforestry system. *Plant Soil* **2015**, *391*, 219–235. [CrossRef]
230. Dufour, L.; Gosme, M.; Le Bec, J.; Dupraz, C. Does pollarding trees improve the crop yield in a mature alley-cropping agroforestry system? *J. Agron. Crop Sci.* **2020**, *206*, 640–649. [CrossRef]
231. Fletcher, E.H.; Thetford, M.; Sharma, J.; Jose, S. Effect of root competition and shade on survival and growth of nine woody plant taxa within a pecan [*Carya illinoinensis* (Wangenh.) C. Koch] alley cropping system. *Agrofor. Syst.* **2012**, *86*, 49–60. [CrossRef]
232. Suryanto, P.; Tohari; Sabarnurdin, M.S. Dinamika sistem berbagi sumberdaya (resources sharing) dalam agroforestri: Dasar pertimbangan penyusunan strategi silvikultur. *Ilmu Pertan.* **2005**, *12*, 165–178.
233. Bazié, H.R.; Bayala, J.; Zombré, G.; Sanou, J.; Ilstedt, U. Separating competition-related factors limiting crop performance in an agroforestry parkland system in Burkina Faso. *Agrofor. Syst.* **2012**, *84*, 377–388. [CrossRef]
234. Dilla, A.M.; Smethurst, P.J.; Huth, N.I.; Barry, K.M. Plot-scale agroforestry modeling explores tree pruning and fertilizer interactions for maize production in a Faidherbia parkland. *Forests* **2020**, *11*, 1175. [CrossRef]
235. Schroth, G.; Teixeira, W.G.; Seixas, R.; da Silva, L.F.; Schaller, M.; Macêdo, J.L.V.; Zech, W. Effect of five tree crops and a cover crop in multi-strata agroforestry at two fertilization levels on soil fertility and soil solution chemistry in central Amazonia. *Plant Soil* **2000**, *221*, 143–156. [CrossRef]

236. Moreno, G.; Obrador, J.J.; García, A. Impact of evergreen oaks on soil fertility and crop production in intercropped dehesas. *Agric. Ecosyst. Environ.* **2007**, *119*, 270–280. [CrossRef]
237. Dupraz, C.; Blitz-Frayret, C.; Lecomte, I.; Molto, Q.; Reyes, F.; Gosme, M. Influence of latitude on the light availability for intercrops in an agroforestry alley-cropping system. *Agrofor. Syst.* **2018**, *92*, 1019–1033. [CrossRef]
238. Suárez Salazar, J.C.; Melgarejo, L.M.; Casanoves, F.; Di Rienzo, J.A.; DaMatta, F.M.; Armas, C. Photosynthesis limitations in cacao leaves under different agroforestry systems in the Colombian Amazon. *PLoS ONE* **2018**, *13*, e0206149. [CrossRef]
239. Atangana, A.; Khasa, D.; Chang, S.; Degrande, A. Socio-cultural aspects of agroforestry and adoption. In *Tropical Agroforestry*; Springer: Dordrecht, The Netherlands, 2014; pp. 323–332. ISBN 978-94-007-7723-1.
240. Mantino, A.; Tozzini, C.; Bonari, E.; Mele, M.; Ragaglini, G. Competition for light affects alfalfa biomass production more than its nutritive value in an olive-based alley-cropping system. *Forests* **2021**, *12*, 233. [CrossRef]
241. Rizwan, M.; Rauf, A.; Rahmawaty; Akub, E.N. Physiology response of soybean variety to various types of shading in agroforestry system. In Proceedings of the The 7th International Conference on Multidisciplinary Research, Sumatera Utara, Indonesia, 5–6 September 2018; SCITEPRESS—Science and Technology Publications: Setúbal, Portugal, 2018; pp. 225–230.
242. Livesley, S.J.; Gregory, P.J.; Buresh, R.J. Competition in tree row agroforestry systems. 3. Soil water distribution and dynamics. *Plant Soil* **2004**, *264*, 129–139. [CrossRef]
243. Wu, J.; Zeng, H.; Zhao, F.; Chen, C.; Jiang, X.; Zhu, X.; Wang, P.; Wu, Z.; Liu, W. The nutrient status of plant roots reveals competition intensities in rubber agroforestry systems. *Forests* **2020**, *11*, 1163. [CrossRef]
244. Hilmanto, R. Local Ecological Knowledge Dalam Teknik Pengelolaan Lahan Pada Sistem Agroforestri. Master's Thesis, Universitas Lampung, Bandar Lampung, Indonesia, 2009.
245. Huxley, P.A. *Tropical Agroforestry*, 1st ed.; Blackwell Science: Hoboken, NJ, USA, 1999; ISBN 978-0-632-04047-6.
246. Belsky, A.J.; Mwonga, S.M.; Duxbury, J.M. Effects of widely spaced trees and livestock grazing on understory environments in tropical savannas. *Agrofor. Syst.* **1993**, *24*, 1–20. [CrossRef]
247. Sudomo, A.; Hani, A. The influence of agroforestry silviculture on productivity and quality of Canna edulis Ker on private forest land. *J. Degrad. Min. Lands Manag.* **2014**, *1*, 137–142.
248. Tsai, H.-C.; Chiang, J.-M.; McEwan, R.; Lin, T.-C. Decadal effects of thinning on understory light environments and plant community structure in a subtropical forest. *Ecosphere* **2018**, *9*, 1–14. [CrossRef]
249. Nicodemo, M.L.F.; Castiglioni, P.P.; Pezzopane, J.R.M.; Tholon, P.; Carpanezzi, A.A. Reducing competition in agroforestry by pruning native trees. *Rev. Árvore* **2016**, *40*, 509–518. [CrossRef]
250. de Mazancourt, C.; Isbell, F.; Larocque, A.; Berendse, F.; de Luca, E.; Grace, J.B.; Haegeman, B.; Polley, H.W.; Roscher, C.; Schmid, B.; et al. Predicting ecosystem stability from community composition and biodiversity. *Ecol. Lett.* **2013**, *16*, 617–625. [CrossRef]
251. Purwoko, A.; Ahmad, A.G.; Saragih, J.M. Management model and feasibility of agroforestry practices in Sipolha Horison village, Simalungun regency, North Sumatra province, Indonesia. In Proceedings of the E3S Web of Conferences; EDP Sciences: Les Ulis, France, 2021; Volume 332, p. 9001.
252. Buck, L.E. Agroforestry policy issues and research directions in the US and less developed countries: Insights and challenges from recent experience. *Agrofor. Syst.* **1995**, *30*, 57–73. [CrossRef]
253. Qurniati, R.; Febryano, I.G.; Zulfiani, D. How trust influence social capital to support collective action in agroforestry development? *Biodiversitas J. Biol. Divers.* **2017**, *18*, 1201–1206. [CrossRef]
254. Martini, E.; Roshetko, J.M.; Paramita, E. Can farmer-to-farmer communication boost the dissemination of agroforestry innovations? A case study from Sulawesi, Indonesia. *Agrofor. Syst.* **2017**, *91*, 811–824. [CrossRef]
255. Diniyati, D.; Achmad, B. Pengaruh penyuluhan terhadap pengembangan kapulaga di hutan rakyat: Kasus di Kabupaten Ciamis dan Tasikmalaya, Jawa Barat. *J. Penelit. Sos. Dan Ekon. Kehutan.* **2016**, *13*, 25–36. [CrossRef]
256. Nair, P.K.R. *An Introduction to Agroforestry*; Kluwer Academic Publishers: Dordrecht, The Netherlands, 1993; ISBN 0-7923-2134-0.
257. Sabarnurdin, M.; Budiadi; Suryanto, P. *Agroforestri Untuk Indonesia: Strategi Kelestarian Hutan Dan Kemakmuran*; Cakrawala Media: Yogyakarta, Indonesia, 2011; ISBN 9789791841355.

Review

Mainstreaming Smart Agroforestry for Social Forestry Implementation to Support Sustainable Development Goals in Indonesia: A Review

Dona Octavia [1,*], Sri Suharti [1], Murniati [1], I Wayan Susi Dharmawan [1], Hunggul Yudono Setio Hadi Nugroho [1,*], Bambang Supriyanto [2], Dede Rohadi [3], Gerson Ndawa Njurumana [1], Irma Yeny [1], Aditya Hani [1], Nina Mindawati [1], Suratman [4], Yelin Adalina [5], Diana Prameswari [6], Etik Erna Wati Hadi [1] and Sulistya Ekawati [7]

1 Research Center for Ecology and Ethnobiology, National Research and Innovation Agency (BRIN), Jl. Raya Jakarta-Bogor Km. 46, Cibinong 16911, Jawa Barat, Indonesia; sris021@brin.go.id (S.S.); murn006@brin.go.id (M.); iway028@brin.go.id (I.W.S.D.); gers001@brin.go.id (G.N.N.); irma015@brin.go.id (I.Y.); adit029@brin.go.id (A.H.); nina.mindawati@brin.go.id (N.M.); etik004@brin.go.id (E.E.W.H.)

2 Directorate General of Social Forestry and Environmental Partnership, Ministry of Environment and Forestry, Jl. Gatot Subroto, Senayan 10270, Jakarta, Indonesia; bambang_halimun@yahoo.com

3 Strengthening of Social Forestry in Indonesia Project, Directorate General of Social Forestry and Environmental Partnership, Ministry of Environment and Forestry, Jl. Gatot Subroto, Senayan 10270, Jakarta, Indonesia; drohadi22@yahoo.co.id

4 Research Center for Geospatial, National Research and Innovation Agency (BRIN), Jl. Raya Jakarta-Bogor Km. 46, Cibinong 16911, Jawa Barat, Indonesia; sura025@brin.go.id

5 Research Center for Biomass and Bioproducts, National Research and Innovation Agency (BRIN), Jl. Raya Jakarta-Bogor Km. 46, Cibinong 16911, Jawa Barat, Indonesia; yeli001@brin.go.id

6 Research Center for Plant Conservation, Botanic Gardens and Forestry, National Research and Innovation Agency (BRIN), Kusnoto Building, Jl. Ir. H. Juanda No. 18, Bogor 16122, Jawa Barat, Indonesia; dian074@brin.go.id

7 Research Center for Society and Culture (BRIN), Jl. Gatot Subroto No. 10, Mampang Prapatan 12710, Jakarta, Indonesia; sulistya.ekawati@brin.go.id

* Correspondence: donasyifa@gmail.com or dona002@brin.go.id (D.O.); hunggul.yudono.setio.hadinugroho@brin.go.id (H.Y.S.H.N.)

Citation: Octavia, D.; Suharti, S.; Murniati; Dharmawan, I.W.S.; Nugroho, H.Y.S.H.; Supriyanto, B.; Rohadi, D.; Njurumana, G.N.; Yeny, I.; Hani, A.; et al. Mainstreaming Smart Agroforestry for Social Forestry Implementation to Support Sustainable Development Goals in Indonesia: A Review. *Sustainability* **2022**, *14*, 9313. https://doi.org/10.3390/su14159313

Academic Editor: Victor Rolo

Received: 9 June 2022
Accepted: 27 July 2022
Published: 29 July 2022

Abstract: The increasing need for forest resources and cultivated land requires a solution in forest management to realize sustainable land use. Smart agroforestry (SAF) is a set of agriculture and silviculture knowledge and practices that is aimed at not only increasing profits and resilience for farmers but also improving environmental parameters, including climate change mitigation and adaptation, biodiversity enhancement, and soil and water conservation, while assuring sustainable landscape management. SAF, a solution for land management systems to reduce the rate of deforestation, is a smart effort to overcome the food crisis and mitigate climate change that is prospectively applied mainly in the social forestry area. Optimized forest land utilization could be achieved by implementing SAF and applying silvicultural and crop cultivation techniques to optimize productivity and meet sustainability and adaptability goals. This paper reviews the existing conditions, opportunities, and challenges in the mainstreaming of SAF in social forestry implementation to support the Sustainable Development Goals in Indonesia. Mainstreaming SAF should include policy innovation and regulation implementation, the use of appropriate technology, and compromises or trade-offs among benefits, risks, and resources. SAF is a strategy to revive the rural economy and community prosperity through the optimal use of local resources as well as a form of smart landscape and land-use management that has significant roles in soil and water conservation, bioenergy, climate change responses, and enhanced biodiversity conservation.

Keywords: agrisilviculture; climate change; food security; livelihoods; tropical forest management

1. Introduction

Indonesia has the third largest area of tropical forest [1] as well as the second largest biodiversity and the second highest number of indigenous medicinal plants in the world [2]. It covers 10% of the global tropical forest with 50% of the world's biodiversity, flora, mammals, amphibians, reptiles, primates, and birds, and provides 25% of the medicinal plants for human health. Blessed with high rainfall throughout the year, Indonesia tropical forest regulates water supply for agriculture, domestic needs, and industries. Meanwhile, more than 25 thousand villages are close to the forest area, with a total population of about 9.2 million households, of which 1.7 million are classified as poor. Tropical forests store about 30% of the world's carbon and Indonesia's forest has the most significant tracts of rainforest and has become a pillar toward Agenda 2030 for Sustainable Development Goals (SDGs) [1]. In spite of its potential, Indonesia's tropical forest management encounters some challenges due to the high community demand for forest resources, including the need for land in state-owned forest areas. Meanwhile, [3] states that conversion to agricultural land is the main cause of deforestation, therefore, it faces a high challenge in preserving the existence of forests. In addition, the demand to meet community needs requires proper strategies to manage forest resources. Several efforts have been developed by the government of Indonesia for these two interests through several forest management policy changes including the Social Forestry (SF) Program.

Along with policy changes in forest management, the deforestation rate in 2017–2018 has decreased by 0.49 million ha [4,5] as forest management policies became increasingly popular in the Social Forestry scheme. Deforestation decreased by 75.03% in the 2019–2020 period, covering an area of 115.46 thousand ha. This achievement is the lowest deforestation rate in recent history. Subsequently, to provide more opportunities for the community to gain access in forest management and get direct benefit from it, the Indonesian government issued the Minister of Environment and Forestry (MoEF) Regulation 9/2021 concerning Social Forestry Management. It is a derivative of Law 11/2020 concerning job creation and Government Regulation 23/2021 concerning forestry management. In this regard, Regulation of MoEF No. 8 of 2021 concerning forest management and preparation of forest management plans as well as forest utilization in protection forests and production forests encourage the increased productivity of forest land through the application of agroforestry and forestry multi-business. Utilization of state forests, especially protection forests and production forests, can be carried out through program activities in the Social Forestry schemes.

SF is a sustainable forest management system implemented in state forests or private/customary forests by local or traditional indigenous communities as the main actors to improve well-being, environmental balance, and sociocultural dynamics in the form of village forests, community forestry, community plantation forests, traditional forests, and forestry partnerships [6–8]. A review of the Indonesian literature showed that social and economic perspectives on SF development received more attention than environmental perspectives. Economic opportunities are deemed to be the main benefit of social forestry, while social and environmental challenges seem to be the major barriers to implementation [8]. Three main keys in the SF program are how to improve the institutional governance, forest governance, and business governance. SF management needs innovation, technology, and collaboration to provide broader benefits for communities in terms of forest land and the use of forest products.

Agroforestry is a silvicultural plan that responds to the challenges of sustainable forest management, especially adjacent to community settlements. Agroforestry is a prospective solution to reduce the rate of deforestation and overcome the food crisis problem [3,9]. This is an integrated approach to a sustainable land-use system (traditional and modern) in which there are interactions between ecological and economic components (timber/forestry plants with seasonal/perennial tree crops, livestock or fisheries inside or outside forest areas). Agroforestry provides ecosystem services, including climate change mitigation and benefits for smallholders, as well as prospects for sustainable food production [10–12].

In the context of Indonesia, the legal framework for agroforestry is laid out in Government Regulation 23/2021 and MoEF Regulations 8 and 9/2021. These regulations state that agroforestry involves optimizing the utilization of forest land through a combination of planting patterns of forest trees and agricultural tree crops and/or animals to increase the productivity of forest land without changing its main function. This is a suitable utilization of forest areas and products in SF management to increase forest land productivity and support meeting the needs for food and energy, medicine, and/or fodder. It is also an alternative solution to resolving social and land conflicts and/or increasing local community income [6,7,13]. Agroforestry offers a great opportunity to support the SDGs for synergy in the agriculture and forestry sectors in the areas of food, energy, water, and income [14,15]. Agroforestry technologies implemented according to an SF scheme have many advantages in both ecological and economic aspects that support the achievement of several of the SDGs, specifically #1 (no poverty), #2 (zero hunger), #3 (good health and well-being), #5 (gender equality), #6 (clean water and proper sanitation), #7 (affordable and clean energy), #13 (climate action), and #15 (life on land).

Another benefit of agroforestry development is the availability of new off-farm jobs in rural areas, such as drying crop yields, wood cutting, and making furniture. In addition, new jobs can also benefit women, as it opens up opportunities for them to be involved in production activities, which will increase gender equality and ultimately contribute to improving the rural economy. Furthermore, agroforestry can also contribute by encouraging community sociocultural development and participation through collective action for people to learn together, rediscover the power of traditional wisdom and knowledge, and then integrate all of that with the development of new knowledge and technology.

Smart agroforestry (SAF) is actually AF "plus" or good AF practices, a set of agriculture and silviculture knowledge and practices aimed not only at increasing profits and resilience for farmers but also at improving environmental parameters, including climate change mitigation and adaptation, biodiversity enhancement, and soil and water conservation, while assuring sustainable landscape management. SAF has subsequently evolved into a science-based pathway for both traditional and modern agroforestry to achieve important goals for natural resource management and socioeconomic benefits. The SAF model has been recommended for development in social forestry areas [7,13], with the aim of providing access for communities around forests in the form of forest farmer groups, facilitated with an extension to enhance farmer capacity and links to capital and the market. The use of SAF technologies has proven to have the potential to improve productivity and livelihood in compliance with environmental guidelines while enabling subsistence and small-scale communities to improve their resources. From the economic perspective, SAF technologies can increase economic resilience through product diversification, especially by increasing the profitability of agroforestry.

The involvement of the community in collaboration with government agencies and the private sector is also important to achieve SAF goals. This is mainly to lower the risk of overharvesting activities on common pool resources. In the context of Indonesia, several legal frameworks are beneficial for the government and private sector to formulate regulations, programs, and strategies to help the community manage its natural resources with support from those institutions. Thus, well-structured formal and informal institutions provide a better collaborative attitude, good management, and natural resource protection, as well as better livelihoods.

Mainstreaming SAF within the SF program becomes a promising alternative to accommodate the interests of good management and natural resource protection as well as better livelihoods. Furthermore, tree planting in the SAF model has become the latest trend in overcoming the climate crisis, motivating legions of people around the world to utilize the incredible carbon-absorbing potential of trees. In this regard, Indonesia signed the Paris Agreement (PA) in April 2016 and ratified it through Law 16/2016 in October 2016, followed by enacting its first nationally determined contribution. As climate change intensifies, Indonesia continues to seek a balance between current and future development and poverty reduction priorities.

We aimed to review and provide information on the existing conditions, opportunities, and challenges in implementing SAF within SF to support the SDGs in Indonesia. The paper was based on a synoptic review approach to SAF- and SF-related publications and nationwide experiences. The reviewed materials come from international and national research papers, regulations, technical reports, and relevant books. The scope of the review covers the historical context of regulations, institutions, and policies of SAF and SF management, and discusses the implementation of SAF in relation to the SDGs, including its benefits in terms of increasing community prosperity, soil and water conservation, climate change adaptation and mitigation, biodiversity conservation, landscape-based resource management, and best practices for agroforestry in some areas.

2. Agroforestry and Social Forestry: Historical Backgrounds

2.1. History of Agroforestry and Social Forestry Development

Agroforestry and social forestry may be two inseparable terms. While agroforestry refers to land-use techniques by which woody plants are combined with agricultural crops or livestock to form ecological and economic interactions between various existing components [16], social forestry emphasizes the strategy of strengthening the tenure of forest management by granting access and management rights over forest areas to local communities (MOEF regulation 9/2021). Social forestry prioritizes the application of agroforestry techniques in land management in order to achieve two main objectives, community welfare and forest resource sustainability. The implementation of agroforestry in SF has been highly encouraged, as it is believed to contribute to the achievement of the Sustainable Development Goals [16].

Although research on agroforestry began to be popular around the 1980s [17], agroforestry has been practiced in Indonesia for a very long time. Farming systems that are known by local names, such as *parak, pelak, repong damar, tembawang, simpukng, talun, wono, tenganan*, and *amarasi* in various regions, basically reflect agroforestry practices that have become part of community culture in land management [16]. Local community wisdom has viewed the agroforestry system as a land management approach that can meet daily needs while proving to be able to conserve natural resources, including forests.

The concept of social forestry, on the other hand, was introduced in Indonesia in the 19th century. The Dutch colonial government in the late 1890s introduced the *taungya* or *tumpangsari* system in Java [17]. The taungya system was a model of granting access (cultivation rights) to farmers over teak forest areas in Java. Farmers were allowed to cultivate food crops between young teak plants up to a certain plant age. Farmers could enjoy the results of their cultivation but were obligated to look after the young teak plants. More recently, initiated by the 8th World Forestry Congress in Jakarta, with the theme "Forests for People", community rights to forest resources have received greater attention among policy makers, bureaucrats, scholars, and forestry practitioners at the national level [18]. The attention of these various groups was also triggered by the increasing tenurial conflict between local communities and the government and forest area permit holders in land uses.

Various initiatives to give the community a greater role in managing forest areas then emerged. Perum Perhutani, with the assistance of various donors in 1985, built 13 pilot models of social forestry in several areas in Java, while other pilot models were built around the early 1990s in South Kalimantan, South Sulawesi, and West Irian [19]. Analyses of these pilot practices indicated that social forestry can potentially improve the welfare of forest-surrounding communities and resolve tenurial conflicts, but there are weaknesses in terms of equal rights (between community groups and Perum Perhutani) and a lack of community interest in investing in the long term, due to uncertain continuation of land management rights [20,21].

With the establishment of the Directorate General of Land Rehabilitation and Social Forestry under the Ministry of Forestry (MoF) in the early 1990s, community involvement in forest area management began to receive better attention in terms of the policy and regulatory aspects. However, as stated in [22], at that time, the government had not given forest management rights to the community. Forest management was still dominated by the "forest first" approach, where community involvement was more directed toward forest rehabilitation programs [23]. The limited budget and human resources for the implementation of forest rehabilitation were the main considerations for community involvement in the management of the forest area, and as a result, the community's tenurial rights to manage the forest area have not been fully granted.

Community involvement in the early era of social forestry programs can be seen from several introduced policies, such as HPH Bina Desa, Community Forestry, and People Plantation Forest [24]. In 1995, the government introduced the Community Forestry (CF) program. In this program, the community was actively involved in forest management activities and obtained the right to use non-timber forest products (NTFPs). The CF management permit was determined based on a contract agreement between the applicant (individual, community group, or cooperative) and the local Provincial Environment and Forestry Service. Although the policy still focused on community activities in forest rehabilitation, the program was successfully implemented through the granting of CF concessions in the Nusa Tenggara region, with an area of approximately 92,000 ha [25].

The concept of social forestry that is close to the current practice might have existed since 2003. At the International Conference on Livelihoods, Forests, and Biodiversity, at a series of Conference of Parties meetings in Bonn in 2003, the official representative of Indonesia introduced a social forestry program with a more comprehensive approach that included ideology, strategy, and implementation in the context of community empowerment in forest resource management. The program provided access for local communities to manage forest areas in order to improve their welfare and conserve forests at the same time [24]. In 2006, the government launched the Community Plantation Forest (Hutan Tanaman Rakyat or HTR) program, which was linked to its efforts to alleviate poverty (propoor), create new jobs (projob), and increase economic growth (progrowth) [26].

Subsequently, in 2015, the development of social forestry entered a new phase. The 2014–2019 Working Cabinet, led by President Joko Widodo, merged the Ministry of Forestry with the Ministry of Environment to become the Ministry of Environment and Forestry (MoEF), and one of the Directorate Generals formed under the MoEF is the Directorate General of Social Forestry and Environmental Partnerships (DG PSKL). The establishment of DG PSKL in 2015 significantly accelerated SF development. Under the Nawa Cita program promoted by the Presidential Cabinet, the DG PSKL set a target for developing social forestry in Indonesia, covering an area of 12.7 million ha, through five SF schemes: community forestry, village forest, people plantation forest, private/customary forest, and forestry partnership [27]. From the emerging ideas of SF up to 2014, the total established SF area was only around 0.4 million ha, while the current total established area reached 5.004 million ha [28].

Similar to SF, agroforestry policies ideally should be well-organized and institutionalized. Many successful collective actions on forest and land management in Indonesia were associated with the implementation of good agroforestry practices by communities [29]. In Indonesia, there are at least two ministries that have been pioneers in the development of agroforestry: the Ministry of Environment and Forestry (MoEF) and the Ministry of Agriculture (MoA). Although the mandate of the MoEF is within the forest area and the MoA is more focused on areas outside forests, there are many overlapping areas on the ground that require close coordination between the two ministries.

Agroforestry has a potential role in increasing farmers' income and sustainable landscape management [30]. Forest ecosystem services significantly support community livelihoods and provide very important sources of livelihood for indigenous people [31]. In other places, such as Paru Village Forest, West Sumatra Province, community plantations of rattan within the agroforestry model show promise as a source of economic income [32]. The application of best practices of agroforestry in SF areas seems promising to reduce people's dependence on expanding the opening of forest lands, and thus could be a better alternative for reducing deforestation as well [33,34]. Last but not least, the integration of the market chain is also among the crucial priorities in developing best practices of agroforestry [35].

2.2. *Rules and Regulation*

The legal system in Indonesia has a clear hierarchical structure for its rules and regulations, including the Constitution, laws, government regulations, ministerial regulations, and regulations of the Director Generals. The regulatory hierarchy covers everything from philosophical principles to tactical and technical aspects of the implementation of government policies and programs. The system of rules and regulations applied to the agroforestry and social forestry sectors also follows this legal system.

Four main laws are used as reference rules for implementing agroforestry and SF: Law 41/1999 on forestry [36], Law 32/2009 on environmental protection and management [37], Law 37/2014 on soil and water conservation [38], and Law 11/2020 on job creation [38]. Observing the substance of the four laws, what is interesting is that three of them, Laws 41/1999, 32/2009, and 37/2014, seem to be in a diametrically opposite position to Law 11/2021. On the one hand, the three laws regulate how forests and land are managed, with a strong sense of ecological protection, while on the other hand, the law on job creation encourages investment for economic purposes. Based on this situation, implementing SAF becomes important and strategic as a counterbalance to achieving the mandate of the four laws. On the one hand, it ensures that the mandates of Laws 41/1999, 32/2009, and 37/2014 can be fulfilled, while at the same time ensuring that the objectives of Law 11/2021 regarding investment and job creation can also be fulfilled. What then becomes important is how the derivative regulations of the four laws can regulate and provide a legal umbrella for the implementation of SAF as a balance between the two poles of interest.

Under these laws, there are various government regulations (PP) that are the source of various derivative regulations focusing on agroforestry and SF. Some of the most relevant rules and regulations around agroforestry and social forestry are presented in Figure 1. They include PP 26/2020 on forest rehabilitation and reclamation and PP 23/2021, concerning the implementation of forestry, which states that agroforestry is a form of reforestation (rehabilitation in forest areas) and afforestation (land rehabilitation outside forest areas) [39]. AF is one of the recommended activities in the business of utilizing protected and production forests, as well as SF management [40].

The job creation law changes a number of regulations in the forestry sector. PP 23/2021 states that the management of social forestry is carried out by planting patterns in agroforestry, silvopasture, silvofishery, and agrosilvopasture. The regulation also states that village forest, community forestry, and people plantation forest approval holders are prohibited from planting oil palm trees in the SF concession area. PP 24/2021 regulates the term for managing oil palm plantations (*jangka benah sawit*) in forest areas that do not have permits in the forestry sector. The previous monoculture oil palm cultivation system must be gradually changed to an agroforestry system to increase land productivity and maintain biodiversity. Oil palm plantations are obliged to return the business part of the forest area to the state 25 years after planting. These two government regulations prove that the Indonesian government prevents forest conversion and encourages the application of agroforestry as a means of conflict resolution.

At the ministerial regulation level, PERMEN Kehutanan P.37/2007, which accommodates agroforestry practice in community forestry, seems to be the first ministerial regulation that mentions wanatani/agroforestry directly. Other regulations on agroforestry include P.11/2020 concerning people plantation forest and P.9/2021 on SF management. PERMEN LHK P.11/2020 regulates the application of agroforestry in cultivated areas based on the principle of sustainability. P.9/2021, concerning SF management, states that it can be carried out with agroforestry, silvopasture, silvofishery, and agrosilvopasture according to forest function and type of space [7]. At the policy level, the two PPs and three PERMENs mentioned above, although some were enacted before Law 11/2021, provide a fairly strong legal basis for the implementation of SAF in Indonesia.

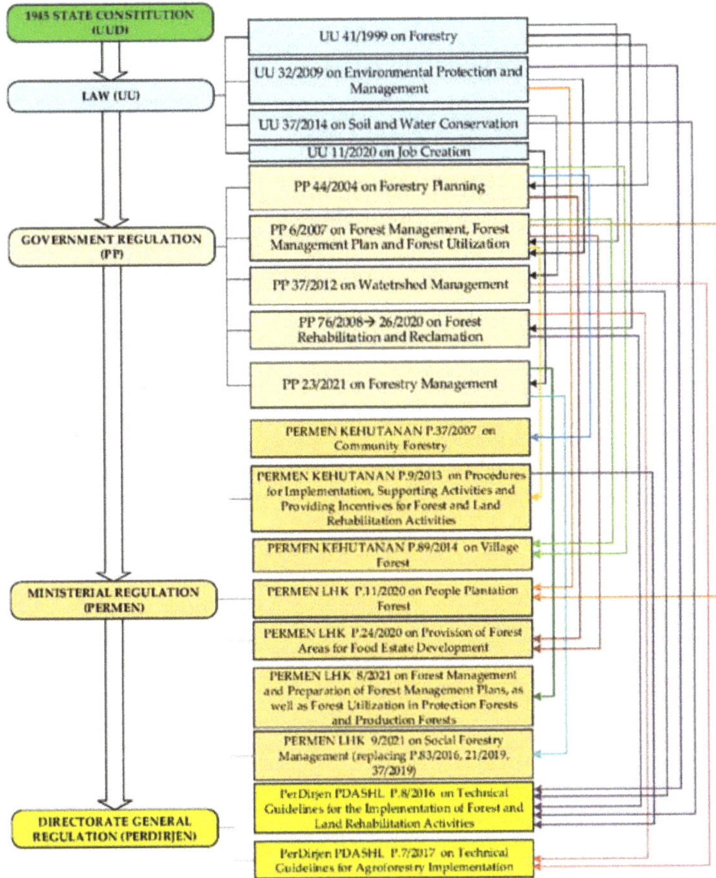

Figure 1. Regulations related to agroforestry and SF at the operational level and their umbrella rules (box colors indicate different rule levels and arrow colors indicate derivative rules).

At the operational level, there are two regulations governing agroforestry, Director General of Watershed Control and Protected Forests (PerDirjen PDASHL) Regulation P.7/2017 concerning technical guidelines for agroforestry implementation, and P.8/2016, concerning technical guidelines for the implementation of forest and land rehabilitation activities. Perdirjen PDASHL P.8/2016 stipulates that agroforestry is a form of forest and land rehabilitation activities on cultivated land and forest areas. The technical guidance is further regulated in PERDIRJEN PDASHL P.7/2017 for implementing agroforestry in protected forests and outside forest areas, as a technical direction for agroforestry activities. The goal is to realize sustainable forest management and increase the carrying capacity of watersheds through agroforestry, for the welfare of the community in terms of economic, ecological, and social aspects [41].

The support of two existing PERDIRJEN as operational guidelines for the implementation of SAF as a derivative of the ministerial regulations, PPs, and the above laws, could become the basis for implementing SAF to improve community welfare while assuring the continuation of the ecological functions of forests and land.

3. SAF Implementation

SAF practices have numerous benefits for many aspects of human life and the environment. They have a substantial function in providing a community's daily needs [42–44], enhancing the biodiversity of flora and fauna [45–48], moderating microclimates, mitigating and adapting to climate change [14,49], promoting soil and water conservation and soil health [50–52], and restoring the landscape [53–55]. The SAF paradigm should drive people, especially policy makers, to create rules and regulations that promote a larger scale of agroforestry implementation, in areas of either forest land or private land. It was suggested in [43] to strengthen the appreciation of smallholder agroforestry farmers as significant contributors to worldwide environmental targets and community economic objectives. Agroforestry is a prospective form of land use that can improve soil health and productivity in a sustainable manner, which makes it a viable strategic option for agricultural communities and policy makers to restore and maintain soil health [52].

Implementing agroforestry systems using different forest management schemes in Indonesia is in line with shifting the paradigm on agroforestry as a smart tool for sustainable land-use management. We distinguish smart agroforestry from agroforestry based on several variables, which are presented in Table 1.

Table 1. Differences between SAF and AF.

Variable	Agroforestry (AF)	Smart Agroforestry (SAF)
Impact on economy	• Use of forest land for subsistence farming with low external input [56,57]	• Use of forest land for subsistence farming • Attention paid to quality assurance and production sustainability [58] • Efficient system of resource use to achieve environmental benefits and high economic profits, produce new products for industry [59]
Impact on ecology	• Adaptive to climate change • Relies on natural regeneration [60] • Extensification tendency [61]	• Adaptive to climate change • Effective and efficient agricultural practices and contributes to reducing deforestation rate [62] • Intensive management to produce sustainable products on degraded land to reduce extensification [63] • Enhanced by assisted natural regeneration (ANR) and intensive artificial regeneration (IAR) * • Attention paid to quality assurance and environmental sustainability • Intensification tendency
Impact on social	• No innovation and integration between sectors [64]	• Contributes to achievement of SDGs 1 and 2 [65]
Knowledge based	• Passed down based on experience across generations in one community group [66,67]	• Based on local resources, use of knowledge (both modern and indigenous), technology, and advance management [68]

* In SAF implementation, plant regeneration can be carried out by using assisted natural regeneration (ANR) and intensive artificial regeneration (IAR). ANR is an accelerated technique focusing on the ecological aspect, while IAR is artificial planting carried out in heavily degraded forest areas with hard-to-find natural seedlings, integrating economic and social aspects of the community, along with the ecological aspect [69].

In order to provide broader insight to the SAF paradigm, we conducted a review of AF system practices that contribute to achieving the SDGs, in particular, those related to increased rural income and prosperity, soil and water conservation, bioenergy, climate change response, biodiversity conservation, and landscape management, from AF practices in Indonesia and other countries. The schematic linkage of SAF activities with SDG objectives is presented in Figure 2.

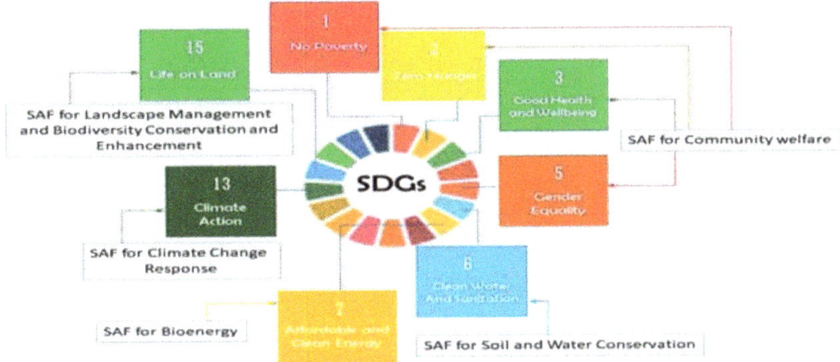

Figure 2. Linkage of SAF activities with SDG objectives.

3.1. *SAF for Community Welfare*

The promulgation of Law 6/2014 concerning rural development programs, with the main objective of fostering community prosperity, is in line with the target for achieving Sustainable Development Goals (SDGs) 1, 2, 3, and 15: alleviating poverty and hunger, promoting good health and well-being, and conserving limited terrestrial resources. In its implementation, Law 6/2014 recommends the optimum use of local resources through the implementation of applicable technology, especially locally based innovations. According to [70], the potency of AF practices in supporting community livelihoods will make it the most approachable strategy.

As a land management system that has long been recognized, AF provides many benefits for farmers [12,71] and will be significant for a long time because of its capacity to support livelihoods, as well as provide environmental benefits and revive the rural economy [70]. As a system that is eco-friendly [72] and can satisfy the socioeconomic needs of the people [73–75], AF can also open up new jobs for gender-friendly rural economic development. Apart from generating income, agroforestry can also play an important role in sociocultural stimulation among communities that starts by studying problems together, rediscovering existing knowledge and traditional wisdom, and integrating new knowledge. In this way, the community becomes more involved and better educated through peer-to-peer discussion and group participation.

The SAF approach is a solution to forest management practices and the development of livelihoods for local communities that depend on forest resources. SAF is expected to accelerate the increase of land cover by increasing the diversity of wood plant species and non-timber forest products in forest areas. This will have implications in terms of increasing the production of forest products and improving the ecological functions and ecosystem services of the forest area. In addition, SAF is an effort to anticipate changes in the sustainability status of the traditional AF system, which is at a moderate level [76–78]. The application of SAF will strengthen the function of forest areas, especially in providing benefits and added value to sociocultural, economic, and ecological aspects, as well as their ecosystem services [79]. Implementing SAF is in line with the vision of realizing sustainable forest management and prosperous communities in Indonesia.

A study in East Nusa Tenggara Province was conducted to examine the relevance of agroforestry management to rural prosperity [70]. The AF practice in the area is dominated by coffee and cashews, and it was found that these two commodities could potentially be promoted to increase the community's welfare. AF practices can support the community livelihoods (income and food security) of smallholder farmers, produce raw materials, and strengthen biodiversity while maintaining productivity and sustainability of soil and water resources [80,81]. Research on AF development in Wan Abdul Rachman Grand Forest Park (WAR GFP), Indonesia, showed that farmers' income increased by 75.63%. There was also increased food crop production, possession of luxury goods, and access to information and financial loans, as well as strengthened farmer institutions [42,54]. AF can also improve livelihoods (income and food security) of smallholder farmers through diversified commodities [80]. As shown by the SAF characteristics presented in Table 1, agroforestry practices in East Nusa Tenggara Province and WAR Grand Forest Park are examples of SAF where the community already uses its knowledge for effective and efficient agricultural practices to gain environmental benefits and high economic profits.

Studies of agroforestry practices in Kenya and other sub-Saharan Africa countries have also confirmed that AF has the potential to provide social and economic benefits that can address household incomes, livelihoods, and food insecurity and provide fuelwood [82,83]. A systematic review of 22 peer-reviewed articles published between 2000 and 2019 assessing the contributions of AF to food security in Indonesia, mostly in Java and Sumatra, found that there are differences between AF and SAF in their contributions to food security [44]. SAF can contribute up to five times more revenue, which enables farmers to ensure their food security. The contribution of agrisilviculture lies along the continuum of timber versus non-timber forest products, showing similar trade-offs between diversity and income. One study highlighted that AF options have up to 98% and 65% greater net present value (for periods over 30 years) compared to traditional agriculture [84,85].

Similarly, AF also contributes to the aspect of public health. AF development increases the diversity of foodstuffs by introducing newly developed fruit and vegetable species, cultivating animal feed for livestock systems (providing more milk and other livestock-derived products), or increasing incomes, enabling the purchase of a wider variety of foods and medicines [43,83,86].

Learning from many studies of existing AF practices in Indonesia and other countries, it has been proven that SAF makes a significant contribution to people's welfare, especially by increasing family income, ensuring food availability and family nutritional intake, increasing access to financial loans, and strengthening farmers' institutions at the local level. To further develop SAF on a broader scale, full support from relevant stakeholders is needed, especially for the development of local mainstay commodities, which will ultimately be able to realize the welfare of rural communities.

3.2. *SAF for Soil, Water Conservation, and Bioenergy*

Globally, agriculture plays a major role in 70% of fresh water use and is a driving factor in various environmental problems. It also became an important factor related to the current global problem of water resources and climate change [81]. Converting from forest to agricultural land, which drastically reduces tree cover, can lead to soil degradation and reduced soil organic carbon, and decreases water accessibility in terms of proper quality and quantity [87–90]. Worldwide, out of a total of approximately 75 tons/ha/year of soil lost due to erosion, agricultural land contributes 13 to 40 tons/ha/year, which causes production losses of 33.7 million tons/year and increases global food prices by 3.5% [51].

AF is one way to maintain a hydrological balance in accordance with efforts to maintain productivity and achieve sustainable soil and water resources. Through AF, a model of sustainable agriculture and landscape management, soil fertility, and water security can be maintained by increasing the infiltration rate, suppressing the erosion rate, and regulating water flows, reducing the amount of water lost as surface runoff, and improving water quality [81,91,92]. This is in line with the target for achieving SDG 6: clean water and sanitation. AF/mixed tree-based farming systems maintain the soil's physical properties better than conventional agriculture [87,91]. Multistrata tree crop systems make up an important category of agroforestry systems, in which 25% of the total biomass production that goes to the roots can be retained in the soil for a longer time than in an annual cropping system [88].

An AF system is structurally and functionally more complex, and spatially and temporally more heterogeneous than a monoculture system of plants or trees, hence it is more efficient in the utilization of nutrients, sunlight, and water [88]. At least 20% tree canopy cover is required to maintain organic matter in the system as a whole [91]. The important element in AF patterns is the presence of deep-rooted plants that have the ability to regulate the water status in the root system for the benefit of shallow rooted crops [91,93]. Based on the rate of infiltration and runoff, the capacity of AF systems to maintain the hydrological cycle, although still under forest, empirically and scientifically is much better than conventional agriculture [94]. Tree–crop combinations reduce the destructive power of rainfall, reduce runoff, and increase the infiltration rate, which in turn reduce erosion [90]. Rubber-based AF systems can increase nitrogen supply and water-use efficiency in each soil layer [95], and can enable rubber trees to acquire shallow soil water up to 24% [96].

One of the good AF models in soil and water conservation is the intercropped contour hedgerow, which is characterized by multiple hedges along the contour at intervals of 4–6 m [97], with agricultural and commercial crops in between [90,97]. In a contour hedgerow system, the eroded soil from the top is blocked by hedgerows at the bottom, which not only functions to block runoff and store trimmed biomass piles, but also to retain and settle sedimentation [90]. This accumulation of biomass can increase the organic C content in the soil. An increase of 1 g kg^{-1} (0.1%) of soil organic carbon was shown to increase available soil water capacity by 6% [98]. Nitrogen-fixing plants have been widely applied in contour hedgerow systems to minimize soil erosion, restore soil fertility, and improve crop productivity [97], due to their high nutrition content [90] and ability to increase soil nitrogen content [91].

SAF also contributes to providing bioenergy through the application of bioenergy-based AF. This system simultaneously functions to mitigate climate change, improve food security and soil quality, increase energy access, and alleviate rural poverty [99]. A study found that short rotation woody crops in shelterbelts that are typically designed to optimize soil protection and enhance conditions for crop growth can produce a number of environmental benefits, including increased biodiversity and carbon sequestration, reduced GHG emissions, and increased soil organic matter and water conservation [100]. Thus, in the context of bioenergy development, the role of agroforestry in the rural economy is to generate bioenergy crop production, which offers increased market access and income diversification strategies for rural populations [101]. AF systems using coppiced tree belts, which combine the advantages of traditional alley cropping with the short rotation coppice, is a new strategy for bioenergy [102] that supports the achievement of SDG 7 (clean and affordable energy).

3.3. SAF for Climate Change Response

The climate change phenomenon and its problems are being faced by all nations around the world [103], triggering a global mitigation commitment to keep the temperature from going up 1.5 °C [104]. Plants can absorb carbon dioxide (CO_2), which results in carbohydrate accumulation in plant biomass. Plant constituents and the age of the land determine the amount of CO_2 absorption. CO_2 gas is the main GHG, with the largest amount originating from human activities that produce high rates of emission, so it is used as a standard or reference for changes in atmospheric composition and global climate change. AF practices such as carbon sequestration have indirect effects in terms of reducing deforestation and forest degradation in primary forests. Apart from that, the AF system that is encouraged in the social forestry scheme in Indonesia is also a solution for revamping areas within forests that have already been used by the community for non-forestry activities, especially agriculture, called the Jangka Benah policy. The Jangka Benah policy is an effort to improve the function of forest areas by encouraging communities to implement AF systems instead of monoculture farming. This policy is also intended to ensure that illegal agriculture or agroforestry activities in forested areas can be reduced slowly and gradually. This system can conserve carbon in the soil to support carbon stock enhancement in both soil and vegetation [105]. One effort to ensure that AF practices can support sustainable forest management was to certify legal and sustainable AF products [106]. One of the sustainable forest management certification bodies in Indonesia has provided certification services for small farmers, hence it is hoped that farmers who practice AF will still pay attention to conservation and environmental principles. The government can also apply a regional priority scale for AF practices based on geological criteria, pedology, land slope, level of anthropogenic factors, forest fragments, potential land-use capacity, and legal reserves [107], so that SAF development is based on the forest condition, land, and environmental carrying capacity.

AF contributes significantly to supporting actions for mitigating [106] and adapting to climate change [107]. Hence, the practices of AF systems and technologies support the achievement of SDG 13 (addressing climate change). In mitigating climate change, AF practices can increase carbon stock in the above- and below-ground biomass and soil carbon pool. AF systems that are integrated into cropping and livestock can enhance carbon sequestration significantly [104]. AF has more carbon stock than land-use systems without trees, such as pastures and fields with annual crops [108], but less than forested areas. There is an estimated 12–228 Mg ha^{-1} of carbon stock [108,109] in tropical humid AF systems, and 68–81 Mg C ha^{-1} in tropical dry lowland areas [108]. Meanwhile, the estimation of carbon stock in AF systems across the world amounts to 0.29–15.21 Mg ha^{-1} yr^{-1} in above-ground biomass and 30–300 Mg ha^{-1} yr^{-1} in 1 m soil depth [110].

The increasing risk of climate change related to drought, flooding, and crop pests will affect farming systems. In this regard, AF management requires conservation oriented toward adapting to climate change [111,112]. Farmers can implement AF systems to continue living on their land as a form of climate adaptation at the farm level. AF can create a resilient microclimate for crop cultivation and livestock activity [113]. In addition, trees as a component of AF can support environmental services to increase resilience to climate change and reduce the vulnerability of local people [11] in terms of water, soil, climate, and hydrology disaster control. AF allows important improvements in microclimates in terms of climate buffering and reduced climate variability [105].

Technology in AF practices for climate change mitigation and adaptation (defined here as SAF for climate change) must meet specific criteria such as availability, applicability, low cost, and environmental friendliness. The Technology Needs Assessment (TNA) for Climate Change Adaptations in Indonesia indicates that “site species matching” is the selected technology and relevant to the mitigation aspect [114]. The smart site species matching technology considering land suitability and local community preferences is indispensable in the development of AF systems to obtain optimal results for climate change mitigation and adaptation [115].

3.4. *SAF for Landscape Management*

SAF in sustainable landscape management is the main strategy for implementing sustainable development [116]. This supports the achievement of SDG 15 (especially for sustainable use of terrestrial ecosystems, halting and reversing land degradation). Various kinds of landscapes, land uses, and land covers in Indonesia are changing very quickly as a result of the global crisis, which has caused several crises in the country, with the impacts felt by urban and rural communities [117,118].

Implementing SAF in integrated landscape management must be in accordance with local land conditions, topography, climate, and community culture. It is a more cost-effective land utilization strategy and production input to enhance multipurpose production and optimal revenue for each unit area, which relates to the rule of sustainable production [42,119]. The productivity of AF patterns by using silviculture techniques has been optimized by means of tree species selection, land suitability and topography, and environmental manipulation, as well as innovation in maintenance techniques, including plant protection from pests, diseases, and weeds [119–123].

Developing SAF toward sustainable landscape management requires the application of several silvicultural and crop cultivation techniques, as follows.

3.4.1. Planting Pattern Arrangement and Species Selection

The planting pattern of an AF system should be arranged to optimize the use of space and time. Plants should be arrayed in such a way as to maximize the growth space. The interface area will be maximized if there are many plant species planted in a hexagonal pattern [124]. AF systems can be classified into agrisilvicultural (crops, tree/shrub crops, and trees), silvopastoral (pasture/animals and trees), and agrosilvopastoral (crops, pasture/animals, and trees) or another specialized system [125]. AF systems were classified in [126,127] into two patterns based on the component number and the canopy strata: simple and complex or multistoried. An along cycle or permanent AF system was introduced in [128], which referred to an intercropping system between some tree species and annual crops, where the crops can be cultivated continuously during the tree cycle. In the intermediate and advanced phases of AF, where the tree crowns are close to each other and the relative sunlight intensity reaching the ground surface is lower, shade tolerant crop species should be chosen. This is because the shade from trees has a negative effect on crop growth and yield [129].

3.4.2. Use of Legumes and Cover Crops in Agroforestry

Intercropping legumes can increase input and reduce the use of external nitrogen fertilizers, hence they can be part of a low-emission farming system. In addition, such intercropping provides legume yield [130] and can improve monoculture cropping when shade trees of legume species are planted [131]. It can also control weeds without reducing crop yields, thereby reducing maintenance costs for farmers [132], and has a function in the cycle and availability of P in the soil [133]. Moreover, as reported in [134], using *Centrosema macrocarpum* (5 kg seeds per ha) as a cover crop could cover 100% of an area in 3 months, and it also controls weeds. In 8 months, the total average *Centrosema* biomass was 8.12 T/ha, with total nitrogen accumulation of 232 kg/ha, soil compaction was reduced at 20 cm depth, and organic matter and potassium levels were increased.

3.4.3. Use of Biofertilizers in Agroforestry

The use of biofertilizers in AF has an ameliorating effect on soil, and biofertilizers multiply easily [135]. Biofertilizer acts as a component of or instrument for soil fertility management in fluctuating weather [136]. Several research results show that applying biofertilizer along with inorganic fertilizer could increase the soil biota and the production of various types of cultivated plants [137]. Nitrogen-fixation plants provide an environmentally friendly and renewable source of low-cost nutrients as well as fertilization efficiency, improve soil health, and maintain environmental sustainability. Biofertilizers can also increase plant resistance to adversarial environmental stresses [138]. Integrated nutrient management, which involves applying organic material, compost, manure, biofertilizer, and inorganic fertilizer together, can increase tree and crop yields, improve the nutrient quality, and minimize losses to the environment [139].

In addition, there are some advantages of AF development, such as its potential to stimulate the production of important secondary metabolites, especially in dry areas where water, light intensity, nutrition, and shade stresses occur [140]. Branches, twigs, and leaves of trees can be sources of secondary metabolites or potential bioactive compounds. The content of primary and secondary metabolites in plants is an important factor that determines their nutrition and health and promotes the plants' value [141]. At a certain level of growth and/or stress, it was found that plants produce secondary metabolites in certain cells and taxonomic groups [142]. The intensity of light received by plants in the shade affects photosynthesis, including the yield of secondary metabolites [143,144].

Another advantage of AF systems is the low risk of pest and disease attacks. It was confirmed in [145] that plant diversity in an AF can decrease pests and diseases. Furthermore, as revealed by [146], complex landscapes in agricultural areas and AF systems by their nature allow multitrophic interactions, which presents opportunities to develop natural pest control. Meanwhile, it was reported in [147] that pest and disease attacks could be reduced by 73% with mixed-crop planting. A more recent study found that an intercropping system with legumes in western Kenya suppressed brown line disease in cassava [148]. The intensity of pest damage in AF systems is lower than in monoculture agriculture [149], because AF planting patterns affect pests and diseases not only based on the plant species, but also other factors, including pest types, pest preferences, and microclimate [150,151]. The diversity of plant species composition in an AF system can cause different types and intensity of pest and disease attacks [151].

3.5. SAF for Biodiversity Conservation and Enhancement

As a mixed land utilization system that integrates tree and crop species based on their designation and use plan, AF increases plant diversity [152,153] and directly supports the achievement of SDG 15 (halting biodiversity loss). Farmers with larger cultivated areas tend to practice multistrata AF with larger amounts of plant species [54]. A multistrata cacao-based AF associated with some shade tree species in the buffer zone of Lore Lindu National Park, Central Sulawesi, Indonesia, succeeded at inviting pollinating insects to help pollinate the cacao plants, so the yield increased to the optimum level [154]. Besides economical purposes, AF practices are also beneficial for the conservation of local native plant species, such as *Acacia seyal* and *Balanitras aegyptiaca* in Ethiopia, which are used for fuelwood, food (the fruit), medicine, charcoal, poles, bee feed, fodder, etc. [155].

An AF system contributes to plant protection by decreasing the pest, disease, and weed populations, increasing natural predators of pests, and increasing plant and land productivity [151,156]. An alley cropping cacao AF practice showed a higher abundance of arthropoda, which has an important role as a decomposer and natural predator, compared to monoculture oil palm plantations [157]. AF systems can regenerate multiple fauna populations and diversity compared to monoculture systems because they have habitats for the provision of food and good food chain cycles [158,159].

Many studies have reported an increase in floral, faunal, and microbial diversity induced by AF in temperate and tropical regions compared to monocropping [47,48,52,160,161] due to favorable soil–plant–water–microclimate conditions [47] and the availability of nutrients from shade trees [162]. Silvopastoral and agrosilvicultural systems have the potential to conserve floral and faunal species [163]. A simple modelling demonstration of the habitat amount hypothesis proved that AF is important in reducing biodiversity loss at farming sites [48].

AF for shade-grown coffee has been practiced worldwide to increase biodiversity conservation [164]. It can preserve biodiversity [47,158,165] and improve the quality of the surrounding environment and ecosystems [166]. Coffee in agroforestry practice requires shade, thereby increasing biodiversity and building multistrata SAF. Furthermore, coffee and cocoa AF systems were found to increase the diversity of fauna such as birds, mammals, and other species [167]. Rubber agroforests in Thailand have greater butterfly richness compared to rubber monoculture [45]. Coffee AF could provide habitats for slow lorises. The trees provide numerous sources of foraging, social activity, and movement for slow lorises [168]. This is in accordance with the findings of [158], showing that coffee AF carried out in the Kemuning Forest of Central Java has an important role in preserving the Javan slow loris (*Nycticebus javanicus*), an endemic primate species that is endangered.

On the southern and eastern slopes of Guatemala, coffee, cardamom, and fallow AF systems significantly increased the connectivity of forest biodiversity in a fragmented landscape and overall landscape biodiversity [169]. AF sites in Pearl Lagoon Basin, Nicaragua, tended to have comparable plant diversity relative to neighboring secondary forest [170].

The research findings of [171] highlighted that the biodiversity of AF systems should be maintained to guarantee the availability of all three types of ecosystem services (provisioning, regulating, and cultural), food security, and community livelihoods. AF practices lead to increased biodiversity and can be linked to food, shelter, habitat, auspicious microclimate, better soil–plant–water relations, and other benefits produced by mixed tree crops [47]. Based on the above findings regarding the numerous contributions of AF to biodiversity (flora, fauna, micro-organisms), we concluded that AF systems represent smart land-use management with a significant role in biodiversity conservation and enhancement.

3.6. Best Practices of SAF

Many studies have found that the main practitioners and innovators of AF are indigenous people with local wisdom. They have practiced AF for thousands of years and have passed on their knowledge of developing, modifying, and adapting AF systems through learning and adjusting [172–174], including developing variations in plant structure and composition according to environmental and socioeconomic factors [175]. One key aspect of their best practice that the world can learn is that they always apply a sustainable and integrated approach to managing ecosystems in a multifunctional way and the innovation decision processes are common to any sector [55,176,177]. A good understanding of the soil conditions, as the main consideration that leads to careful selection of tree planting sites, coupled with good tree maintenance efforts, results in high growth rates, which is widely recognized as a sustainable land management practice, particularly in the tropics [172,178,179].

In deciding on a tree–crop combination, farmers mainly consider the availability of land and the potential for income diversification [180]. Several best practices of AF (SAF) in Indonesia have been widely developed by farmers throughout the country. In Java, people's local wisdom regarding land use is based on biophysical conditions, experience, and cultural heritage, specifically sacred forests, mixed gardens, rice fields, residential areas, and home gardens [181]. The main timber commodities developed are sengon (*Falcataria mollucana*), with a 6-year rotation, and teak (*Tectona grandis*), with a 20-year rotation, which generate economic values of USD 1015/ha^{-1} and USD 2815/ha^{-1}, respectively [182].

Indigenous AF in dryland areas can be found in the eastern part of Indonesia, including Mamar on Timor Island [76,77,183] and Kaliwu on Sumba Island [79]. The socioeconomic contribution of sedentary gardens, silvopasture, and Mamar on Timor Island ranges from 77.74 to 78.99% [76], while Kaliwu agroforests contribute 46.88% to farmers' income on Sumba Island [184]. In terms of ecological functions, in the Mamar and Kaliwu ecosystems, there are 112 plant species structured in multiple strata and 146 plant species [185,186], respectively. Community initiatives to regenerate and conserve NTFPs by implementing AF also need to be enhanced, especially to improve the rate of plant growth [187].

On the easternmost island of Indonesia, the Papuan community has several examples of best AF practices aiming to meet substantive needs. The area for AF development was obtained from degraded forest with an average land size of 0.25–1.0 ha undergoing a shift to the next plantation location within 2–3 years [188]. They applied a traditional AF system through trial and error to meet their basic needs and ensure food security first [189,190]. The most commonly used cropping pattern for the top canopy stratum is timber, fruit, and plantation tree crops, while the lower crown stratum consists of tubers, vegetables, and herbs [191–193]. The selected plants are species that do not need intensive care or attention, and are local mainstays [188]. The most commonly developed species are multipurpose tree species (MPTS) such as *Pometia* sp., *Artocarpus communis*, *Durio zibethinus*, *Nephelium lappaceum*, and *Mangifera indica* [191]. The AF practice in Papua makes a significant contribution (53.5%) to the farmers' income [192]. SAF can reduce degraded forest areas in the future through intensive planting with three pillars to increase land productivity: the use of superior seeds from quality sources, environmental modification, and integrated pest and disease control (intensive silviculture) [194].

In the western part of Indonesia, best AF practices for benzoin (*Styrax benzoin*) can be found in a customary management system adopted by a local community in North

Sumatra which has traditionally owned benzoin forests for generations. They collect seeds or seedlings around benzoin-producing trees and then plant them in areas where there are Styrax trees [195]. Beginning at 8–15 years, farmers start to tap the resin of benzoin trees, and if they do it right, they can extract the resin for about 60 years, or across generations. Study [196] revealed that benzoin trees help in forest conservation and promote local economic and cultural values. Another interesting best AF practice is found in WAR GFP, Lampung, where farmers with limited land availability have successfully developed a cacao-based AF and ensured the fulfilment of their basic needs by gaining access to food, fodder, and fuelwood [42], while restoring the previously degraded conservation forest [54]. Benzoin agroforestry and AF practices in WAR GFP are other examples of SAF being successfully practiced by local communities, which provide multiple benefits, as depicted in Table 1.

Best practices of SAF are also developed in many countries, including northern Bangladesh and some African countries. Farmers improved their livelihoods significantly by practicing AF, as it provides multiple benefits for them [11,75,197]. AF practices were found to ensure economic returns and sustain farmers' livelihoods [75,198] as well as increase plant species diversity, since high-canopy trees enable the underneath crops to capture optimal sunlight, while deep-rooted plants can ensure optimal nutrient absorption [124,199,200]. In addition, indigenous AF also plays a role in land and water conservation, especially landscape conservation around springs and hilly areas [183,184]. Examples of successful SAF practices in other countries include agroforestry for the sustainable management of sloping land areas in Southeast Asia and the Pacific [201], agroforestry practices in the form of biomass transfer of nitrogen-fixing trees combined with rainwater harvesting techniques and integrated nutrient management in a semi-arid region of Zimbabwe [202]. The use of smart agroforestry technologies in Mali is also an example of SAF to address the constraints to adopting proven agroforestry techniques for improved market access to increase food and nutritional security and build resilience of farming systems [203]

Best practices of AF systems, both traditional and modern, that have a significant impact on the economy, ecology, and social aspects can be categorized as smart agroforestry (SAF) or agroforestry plus (AFP). Other examples of SAF practices in Indonesia include Repong Damar in the Krui coastal area of Lampung Province, Tembawang, in West Kalimantan, Parak in Maninjau, West Sumatera Province, and Mixed Garden in West Java [204]. Another SAF practice is the AF Smart Farming initiative by the research group Tropical Agroforestry of Brawijaya University. Some activities of AF Smart Farming have been soil fertility management of AF land after the eruption of Kelud Mountain, and the management of AF and soil fertility by farmers in Ngantang, Malang [205].

Based on the numerous significant contributions of SAF practices to community welfare, soil and water conservation, bioenergy, climate change responses, landscape management, and biodiversity conservation and enhancement, as well as the variety of best practices, scaling up AF implementation should be mainstreamed for both on forest land or private land. It is expected that SAF practices will become the mainstay of forest and landscape management in Indonesia, especially through the SF scheme, in order to achieve several Sustainable Development Goals. Hence, the main key to the successful mainstreaming of SAF for SF implementation lies in suitable policy innovation, the use of appropriate technology, and compromises or trade-offs among the benefits, risks, and available resources.

4. The Way Forward

The academic world only began paying attention to the AF system in the 1970s–1980s, when it was realized there was a need to find alternatives to increase agricultural productivity, reduce degraded land, and increase the benefits for small landowners. Since then, the importance of implementing AF systems has become more urgent, especially in tropical rural areas around the world [206,207]. Although it was an old technique, AF has been introduced as a new science and many publications have presented various results of technical studies related to AF systems.

Developing SAF is not only aimed at increasing profits and resilience for farmers, but also improving environmental parameters, including biodiversity enhancement, soil and water conservation, and sustainable landscape management. Hence, the aim is to obtain collective and comprehensive environmental and economic benefits. This confirms a study revealing that in addition to the multiple benefits derived from current AF practices, SAF also supports sustainability and dynamic land management [208]. We consider that the agriculture and silviculture knowledge employed in SAF are based not only on modern technology, but also on local indigenous knowledge proven to survive over generations, to realize sustainable management of natural resources, including forests. Furthermore, SAF is a win-win land-use option between productive and environmental functions, and it plays an important role as part of a so-called climate-smart landscape approach that simultaneously embraces mitigation and adaptation policies and programs [209].

However, despite the contribution of multiple functions and benefits, the development of SAF faces various challenges, especially related to limited land availability and a lack of alternative sources of income for farmers. The demands of life necessities and lifestyles, global climate change, and government policies are also triggering factors for land conversion. The tendency for land conversion also occurs in the Mamar system [78], although it does not always provide a significant increase in income, as it often requires substantial capital beyond the capacity of farmers. A study in Bungotebo, Jambi, also found that rubber and oil palm prices have induced farmers with limited land ownership and no other on-farm income sources to change their AF land into rubber and oil palm plantations [210]. Another example is in Bantaeng, South Sulawesi, where farmers who previously cultivated candlenut (*Aleurites moluccana*) and kapuk randu (*Ceiba pentandra*) on their land converted it to maize cultivation in order to more quickly obtain a cash income [211].

In addition to the challenges mentioned above, the expansion of oil palm, which in many cases involves the clear-cutting of forested areas in order to open areas for farming, is an activity that might hinder AF development. Many farmers are reluctant to plant oil palm in combination with forestry crops because they believe it will reduce the productivity of their plantations. To overcome this, farmers need assistance with developing commodity diversity by using forestry plants that are multipurpose tree species (MPTS). Social forestry in Indonesia, which focuses on developing business governance for AF farmers, aims to make them no longer depend on oil palm. The completion of monoculture oil palm plantations in forest areas can be addressed by applying the concept of palm oil management (jangka benah).

There are two stages of improvement: The first is to enrich forest plant species by forming mixed oil palm plantations (oil palm AF), aiming to improve the structure and function of forest areas previously disturbed by monoculture plantations. The second is to increase the species and number of mixed plants to form a more complex, multilayer vegetation structure (complex AF) that is close to the structure of natural forest. Meanwhile, coffee in AF practice requires shade, which increases biodiversity and builds multistrata SAF. This is supported by a study from Cambodia [212], which reported that shade-grown coffee delivers the same yields as coffee grown without shading. Therefore, coffee AF is profitable, both ecologically and economically. Incentives for farmers also affect farmers' decisions related to tree species and biodiversity, as shown by an ecological–economic assessment model [213].

Ministry of Environment and Forestry Regulation 9/2021 concerning social forestry stipulates forest criteria and types of social forestry businesses that can be developed. Permitted activities for social forestry in areas with good vegetation cover or conservation areas should be more in the form of environmental services, such as ecotourism activities, whereas AF activities are directed at unproductive forest land or land that has been degraded in order to increase land cover. Since SAF applied to social forestry aims to improve the welfare of the community around the forest and increase community participation in forest rehabilitation activities [22], its implementation is prioritized on degraded forest lands, and it is considered as vegetative conservation for watershed rehabilitation as well [214]. A study found that one of the causes of forest destruction in Indonesia is the conflict between forest communities and forest managers. Social forestry is a form of comanagement that can improve community welfare and sustain the forest while reducing conflicts between communities and forest managers [215].

SAF development is also challenged by farmers' low capacity regarding farming techniques, low awareness, and low motivation to integrate tree or tree crop species into their monoculture farming in forest areas [128]. A study from Rwanda [216,217] also revealed that a lack of skills and technical know-how, limited capital, and the need for quality seeds have decreased the adoption of new AF practices by farmers. Other challenges are related to a lack of trust in the government's commitment to SF implementation [218], limited finances to practice good farming (business capital), limited market options for AF products, and limited post-harvest AF product processing techniques and capacity [219,220]. Similarly, another study [221] confirmed that developing AF with SAF in Indonesia will be more challenging with the complexities of socioeconomic and ecological problems along with broader goals to solve problems at the local to global scale. Climate change, for example, has indirectly affected farming patterns [222]. In addition, inadequate policy support for providing comprehensive and intensive mentoring facilities was also reported to be a cause of the decrease in adopting AF practices [222–224].

The weaknesses of SAF often make it difficult to implement at the farmer level for several reasons: (i) From an economic perspective, AF sometimes harms farmers due to unfavorable market conditions. (ii) From a social perspective, planting trees on agricultural land can sometimes be difficult for farmers, due to unsupportive government policies (e.g., the sandalwood planting policy in NTT). (iii) In terms of soil fertility, farmers do not like to plant trees, because they require a lot of nutrients to grow. (iii) In terms of the growth of other plants, trees that grow tall with spreading branches can harm other plants [225], and need to be altered, such as by pruning and thinning.

In the context of good practices, the customary/traditional AF system developed in various regions in Indonesia is still often neglected. We also consider this system as smart agroforestry (SAF) practice based on local indigenous knowledge. For further development, indigenous SAF would need scientific improvements and innovative treatments to respond to social economic dynamics, including the decrease in farm land due to factors such as increased population and higher standards for product marketing [221].

In spite of these challenges, the sustainability of indigenous SAF can be observed in the social, economic, ecological, technological, and institutional dimensions of dryland farming, with a sustainability level of 54% for mixed gardens, 52% for Mamar agroforests, and 51% for silvopasture, which is classified as moderate [76]. The moderate value represents the impact of the balance of environmentally friendly social–ecological functions that contribute positively to the subsistence needs of farmers [78]. The sustainability dimensions of the traditional AF system can also be seen in its function as an important habitat for key cultural species that are rare in nature. The Mamar AF system is an example of community awareness to independently maintain the sustainability of AF, while non-Mamar agricultural land requires intervention in various ways, such as guidance and counseling on sustainable land use [226].

Considering that the biophysical characteristics and resources as well as the socioeconomic aspects of communities vary, the success of SAF development needs mainstream supporting policies, taking into account sociocultural, economic, and ecological suitability. The best traditional practices for building community economic resilience need to be encouraged as a power node for community-based development. On the other hand, various negative impacts related to the development of market-oriented agricultural and plantation technology that can lead to the simplification of indigenous AF values should also be anticipated, especially a reduction in the values that make up SAF as a node for sociocultural, economic, and ecological adaptation. Technological development is a necessity, but it needs to be harmonized to strengthen the sociocultural values of the community, including the cultural and ecological identity in AF management. Market-oriented simplifications can lead to humans becoming mere economic beings, even though they are socioecological creatures who interact with nature.

As of May 2022, the development of the SF program through five schemes (community forestry, village forest, people plantation forest, customary forest, and forestry partnership) achieved 5.004 million ha for 9223 forest farmer groups (data obtained from Directorate General of Social Forestry and Environmental Partnership, MOEF) out of a targeted 12.7 million ha by 2030 [227]. The large areas allocated for the SF program represent a big opportunity to apply SAF technologies. In addition, the relatively long period of SF agreement (35 years) can be extended for one more period to provide a guarantee for communities that implement AF systems with long-term plants or perennials to get more secure access rights to manage the land and harvest the plant yields [42,228].

Even though in Act 11/2020, Government Regulation 23/2021, and MoEF Regulations 8 and 9/2021, the legal framework for "smart agroforestry" is clearly stipulated as a land-use management option to increase land productivity, there is still a need to anticipate and resolve various challenges in SAF development. Mainstreaming SAF technology on a large scale in SF areas should be followed by enhancing community capacity in farming technologies coupled with providing other supporting facilities for sustainable land-use management [40,229,230]. Kiyani et al. [217] confirmed that corrective actions needed for the success of SAF implementation include providing subsidies and regular training and informal education to help farmers build their skills and capacities, while ensuring their involvement in decision-making related to SAF policies and programs. In addition, it is necessary to establish tree nurseries to increase the production of quality seeds and ensure the successful cultivation of tree crops in SAF development. The concept and advantages of SAF technologies and its success story need to be intensively socialized not only for the community, but also for related stakeholders. Furthermore, technology and innovation should be provided for postharvest treatments and diversified product processing, which would allow some AF products that are perishable to be preserved and marketed more widely.

Subsequently, the mainstreaming and scaling up of SAF as part of SF program development need support not only from the government, but also from all related stakeholders in order to overcome structural barriers and limited community capacity [231]. Multi-stakeholder collaboration is needed, involving NGOs, research institutions, the private sector, international donor agencies, and good government, to encourage innovation in designing, pioneering, and implementing further farming initiatives in Indonesia. In many cases, collaboration can form networks and forums to strengthen partnerships and achieve goals [222]. Community participation can be facilitated by providing more and easier access to technology, capital, markets, information, and supporting infrastructure to increase the economic added value of products generated from AF. In the 4.0 era, market and marketing websites for AF products should be created for farmers with government facilitation, mainly local governments and related stakeholders. With this technology, AF products can be recognized and made available to a wider market [33].

The increase in AF products of various types and functions will in turn encourage the plant-based forest commodity industry to develop postharvest and product processing businesses, on both a small and large scale. This will create many jobs and business opportunities and strengthen the economy of communities. Increased coordination among stakeholders at different levels of bureaucracy, an increased understanding of the challenges faced by smallholder farmers, and the adoption of innovative approaches to managing resources are critical to facilitating farmers' capacity and organizational improvement. Mainstreaming SAF within the implementation of SF programs would provide both a backward and forward linkage in the development of forest-based agribusiness that would significantly contribute to achieving the Sustainable Development Goals in Indonesia especially related to alleviating poverty and hunger, promoting good health and well-being,

27. Direktorat Jenderal Perhutanan Sosial dan Kemitraan Lingkungan. *Rencana Strategis 2015–2019*; Direktorat Jenderal Perhutanan Sosial dan Kemitraan Lingkungan Kementerian Lingkungan Hidup dan Kehutanan: Jakarta, Indonesia, 2015.
28. Perhutanan Sosial dan Kemitraan Lingkungan (PSKL). Gokups. Available online: http://sinav.usahahutan.id/index.php/frontend/beranda (accessed on 23 May 2022).
29. Qurniati, R.; Febryano, I.G.; Zulfiani, D. How Trust Influence Social Capital to Support Collective Action in Agroforestry Development? *Biodiversitas* **2017**, *18*, 1201–1206. [CrossRef]
30. Hasannudin, D.A.L.; Nurrochmat, D.R.; Ekayani, M. Agroforestry Management Systems through Landscape-Life Scape Integration: A Case Study in Gowa, Indonesia. *Biodiversitas J. Biol. Divers.* **2022**, *23*, 4. [CrossRef]
31. Njurumana, G.N.; Ginoga, K.; Octavia, D. Sustaining Farmers Livelihoods through Community Forestry in Sikka, East Nusa Tenggara, Indonesia. *Biodivers. J. Biol. Divers.* **2020**, *21*, 8. [CrossRef]
32. Octavia, D.; Nugroho, N.P. Potency of Diversity and Utilization of Rattan in Paru Village Forest, Sijunjung Regency, West Sumatra. *IOP Conf. Ser. Earth Environ. Sci.* **2020**, *415*, 012002. [CrossRef]
33. Yeny, I.; Octavia, D.; Ginoga, K.; Suharti, S. The Impact of Community Capacity Building on Forest Management at Paru Village Forest, Sijunjung Regency, West Sumatera. *IOP Conf. Ser. Earth Environ. Sci.* **2021**, *914*, 012010. [CrossRef]
34. Fagariba, C.J.; Song, S.; Soule, S.K.G. Livelihood Economic Activities Causing Deforestation in Northern Ghana: Evidence of Sissala West District. *Open J. Ecol.* **2018**, *8*, 57–74. [CrossRef]
35. Wisudayati, T.A.; Danu; Octavia, D.; Hendarto, K.A.; Sianturi, R.U.D.; Megawati; Desmiwati. The Capacity Building of an Agroforestry Demonstration Plot: Participatory Market Chain Approach. *IOP Conf. Ser. Earth Environ. Sci.* **2021**, *917*, 012036. [CrossRef]
36. Republik Indonesia. *Undang Undang Nomor 41/1999 Tentang Kehutanan*; Republik Indonesia: Jakarta, Indonesia, 1999.
37. Republik Indonesia. *Undang-Undang No.32/2009 Tentang Perlindungan dan Pengelolaan Lingkungan Hidup*; Republik Indonesia: Jakarta, Indonesia, 2009.
38. Republik Indonesia. *Undang-Undang No. 37/2014 Tentang Konservasi Tanah dan Air*; Republik Indonesia: Jakarta, Indonesia, 2014.
39. Republik Indonesia. *Peraturan Pemerintah No. 26/2020 Tentang Rehabiltasi dan Reklamasi Hutan*; Republik Indonesia: Jakarta, Indonesia, 2020.
40. Pujo, P.; Sofhani, T.F.; Gunawan, B.; Syamsudin, T.S. Community Capacity Building in Social Forestry Development: A Review. *J. Reg. City Plan.* **2018**, *29*, 113–126. [CrossRef]
41. Direktorat Jenderal Pengendalian Daerah Aliran Sungai dan Hutan Lindung. *Petunjuk Teknis Pelaksanaan Agroforestri No. P.7/2017*; Direktorat Jenderal Pengendalian Daerah Aliran Sungai dan Hutan Lindung: Jakarta, Indonesia, 2017.
42. Murniati; Suharti, S.; Yeny, I.; Minarningsih. Cacao-Based Agroforestry in Conservation Forest Area: Farmer Participation, Main Commodities and Its Contribution to the Local Production and Economy. *For. Soc.* **2022**, *6*, 243–274. [CrossRef]
43. Roshetko, J.M. *Smallholder Tree Farming Systems for Livelihood Enhancement and Carbon Storage*; University of Copenhagen: Frederiksberg, Denmark, 2013.
44. Duffy, C.; Gregory, G.; Toth; Robert, P.O.; Hagan; McKeown, P.C.; Rahman, S.A.; Widyaningsih, Y.; Sunderland, T.C.H.; Spillane, C. Agroforestry Contributions to Smallholder Farmer Food Security in Indonesia. *Agrofor. Syst.* **2021**, *95*, 1109–1124. [CrossRef]
45. Warren-Thomas, E.; Nelson, L.; Juthong, W.; Bumrungsri, S.; Brattström, O.; Stroesser, L.; Chambon, B.; Penot, E.; Tongkaemkaew, U.; Edwards, D.P.; et al. Rubber Agroforestry in Thailand Provides Some Biodiversity Benefits without Reducing Yields. *J. Appl. Ecol.* **2020**, *57*, 17–30. [CrossRef]
46. Mahata, A.; Samal, K.T.; Palita, S.K. Butterfly Diversity in Agroforestry Plantations of Eastern Ghats of Southern Odisha, India. *Agrofor. Syst.* **2019**, *93*, 1423–1438. [CrossRef]
47. Udawatta, R.P.; Rankoth, L.R.; Jose, S. Agroforestry and Biodiversity. *Sustainability* **2019**, *11*, 2879. [CrossRef]
48. Santos, M.; Cajaiba, R.L.; Bastos, R.; Gonzalez, D.; Bakış, A.L.P.; Ferreira, D.; Leote, P.; da Silva, W.B.; Cabral, J.A.; Gonçalves, B.; et al. Why do Agroforestry Systems Enhance Biodiversity? Evidence from Habitat Amount Hypothesis Predictions. *Front. Ecol. Evol.* **2022**, *9*, 630151. [CrossRef]
49. Muschler, R.G. Agroforestry: Essential for Sustainable and Climate-Smart Land Use? In *Tropical Forestry Handbook*; Pancel, L., Köhl, M., Eds.; Springer: Berlin/Heidelberg, Germany, 2016. [CrossRef]
50. Kumar, A.; Hasanain, M.; Singh, R.; Verma, G.; Meena, D.K.; Mishra, F.R.; Role of Agroforestry Measures for Soil and Water Conservation. Food and Scientific Reports. 2020. Available online: https://foodandscientificreports.com/assets/uploads/issues/1584756739role_of_agroforestry_for_soil_and_water_conservation.pdf (accessed on 14 March 2022).
51. Kaushal, R.; Mandal, D.; Panwar, P.; Rajkumar; Kumar, P.; Tomar, J.M.S.; Mehta, H. Chapter 20—Soil and Water Conservation Benefits of Agroforestry. In *Forest Resources Resilience and Conflicts*; Shit, P.K., Pourghasemi, H.R., Adhikary, P.P., Bhunia, G.S., Sati, V.P., Eds.; Elsevier: Amsterdam, The Netherlands, 2021; pp. 259–275.
52. Dollinger, J.; Jose, S. Agroforestry for Soil Health. *Agrofor. Syst.* **2018**, *92*, 213–219. [CrossRef]
53. Hillbrand, A.; Borelli, S.; Conigliaro, M.; Olivier, A. *Agroforestry for Landscape Restoration. Exploring the Potential of Agroforestry to Enhance the Sustainability and Resilience of Degraded Landscapes*; Food and Agriculture Organization of the United Nations: Rome, Italy, 2017.
54. Murniati; Suharti, S.; Minarningsih; Nuroniah, H.S.; Rahayu, S.; Dewi, S. What Makes Agroforestry a Potential Restoration Measure in a Degraded Conservation Forest? *Forests* **2022**, *13*, 267. [CrossRef]

55. Sahoo, G.; Wani, A.; Sharma, A.; Rout, S. Agroforestry for Forest and Landscape Restoration. *Int. J. Adv. Study Res. Work.* **2020**, *9*, 536–542. [CrossRef]
56. Lopez-Ridaura, S.; Sanders, A.; Barba-Escoto, L.; Wiegel, J.; .Mayorga-Cortes, M.; Gonzales-Esquilevel, C.; Lopez-Remirez, M.A.; Escoto-Masis, M.; Morales-Galindo, E.; Garcia-Barcena, T.S. Immediate impact of COVID-19 pandemic on farming systems in Central America and Mexico. *Agric. Syst.* **2021**, *192*, 103178. [CrossRef]
57. Fisher, R. Tropical forest monitoring, combining satellite and social data, to inform management and livelihood implications: Case studies from Indonesian West Timor. *Int. J. Appl. Earth Obs. Geoinf.* **2012**, *16*, 77–84. [CrossRef]
58. Minang, P.A.; Duguma, L.A.; Piabui, S.M.; Foundjem, D.; Tchoundjeu, S. Community forestry as a green economy pathway in Cameroon. In *Sustainable Development through Trees on Farms: Agroforestry in Its Fifth Decade*; Van Noordwijk, M., Ed.; Centre for Research in Agroforestry: Bogor, Indonesia, 2019; p. 269.
59. Palma, J.; Graves, A.R.; Burgess, P.J.; Van der Werf, W.; Herzog, F. Integrating environmental and economic performance to assess modern silvoarable agroforestry in Europe. *Ecol. Econ.* **2007**, *63*, 759–767. [CrossRef]
60. Toral, J.N.; Valdivieso, A.; Jimenez, J.R.A.; Cordova, J.C.; Grande, D. Silvopastoral systems with traditional management in southeastern Mexico: A prototype of livestock agroforestry for cleaner production. *J. Clean. Prod.* **2013**, *57*, 266–279. [CrossRef]
61. Pokorny, B.; Robiglio, V.; Reyes, M.; Vargas, R.; Carrera, C.F.P. The potential of agroforestry concessionsto stabilize Amazonia forest frontiers: A case study on the economic and environmental robustness of informally settled small-scale cocoa farmers in Peru. *Land Use Policy* **2021**, *102*, 105242. [CrossRef]
62. Smith, M.S.; Mbow, C. Editorial overview: Sustainability challenges: Agroforestry from the past into the future. *Curr. Opin. Environ. Sustain.* **2014**, *6*, 134–137. [CrossRef]
63. Villa, P.; Martins, S.V.; Neto, S.N.O.; Rodrigues, A.C.; Hernandez, E.P.; Kim, D.G. Policy forum: Shifting cultivation and agroforestry in the Amazon: Premises for REDD+. *For. Policy Econ.* **2021**, *118*, 102217. [CrossRef]
64. Barlagne, C.; Bezard, M.; Drillet, E.; Larade, A.; Alexandre, J.L.D.; Vinglassalon, A.; Nijnik, M. Stakholders engagement platform to identify sustainable pathways for development of multifunctional agroforestry in Guadeloupe, French West Indies. *Agrofor. Syst.* **2021**, *95*, 1–17. [CrossRef]
65. Van Noordwijk, M.; Khasanah, N.; Garrity, D.P.; Njenga, M.; Tjeuw, J.; Widayati, A.; Liyana, M.; Minang, P.A.; Oborn, I. Agroforestry's role in an energy transformation for human and planetary health: Bioenergy and climate change. In *Sustainable Development through Trees on Farms: Agroforestry in Its Fifth Decade*; Van Noordwijk, M., Ed.; Centre for Research in Agroforestry: Bogor, Indonesia, 2019; p. 283.
66. Mutiani, C.; Febriansyah, R.; Mahdi. Agroforestry management practices in relation to tenure security in Koto Tangah Subdistrict, West Sumatra, Indonesia. In *Natural Resource Governance in Asia*; Lander, G., Janco, C.G., Eds.; Elsevier: Amsterdam, The Netherlands, 2021; pp. 27–37. [CrossRef]
67. Ngongo, Y.; Basuki, T.; de Rosari, B.; Hosang, E.; Nulik, J.; de Silva, H.; Hau, H.D.; Sitorus, A.; Kotta, N.R.E.; Njurumana, G.N.; et al. Local Wisdom of West Timorese Farmers in Land Management. *Sustainability* **2022**, *14*, 6023. [CrossRef]
68. Chengjun, S.; Renhua, S.; Zuliang, S.; Yinghao, X.; Jiuchen, W.; Zhiyu, X.; Shangbin, G. Construction process and development trend of ecological agriculture in China. *Acta Ocologica Sin.* **2021**, in press. [CrossRef]
69. Mawazin; Octavia, D. *Restorasi Ekosistem Lahan Gambut Terdagradasi di KPH Tasik Besar Serkap, Riau*; Pros Sem Nas MasyBiodiv Indon; Smujo International: Solo, Indonesia, 2019; Volume 5, pp. 330–335.
70. Carolina; Wijayanti, F. Traditional Agroforestry Ecosystem for Rural Prosperity. *SHS Web Conf.* **2020**, *86*, 01021. [CrossRef]
71. Nair, P.K.R. State-of-the-Art of Agroforestry Systems. *For. Ecol. Manag.* **1991**, *45*, 5–29. [CrossRef]
72. Hanif, M.D.; Bari, M.D.; Mamun, M.D.; Roy, P.; Roy, R. Potentiality of Organic Rice Production Fertilized with Different Agroforestry Tree Leaf Litter in Northern Bangladesh. *Int. J. Plant Soil Sci.* **2015**, *8*, 1–6. [CrossRef]
73. Chakraborty, M.; Haider, Z.M.; Rahaman, M.M. Socio-Economic Impact of Cropland Agroforestry: Evidence from Jessore District of Bangladesh. *Int. J. Res. Agric. For.* **2015**, *2*, 11–20.
74. Sharmin, A.; Rabbi, S. Assessment of Farmers Perception of Agroforestry Practices in Jhenaidah District of Bangladesh. *J. Agric. Ecol. Res. Int.* **2016**, *6*, 1–10. [CrossRef]
75. Hanif, M.A.; Roy, R.M.; Bari, M.S.; Ray, P.C.; Rahman, M.S.; Hasan, M.F. Livelihood Improvements through Agroforestry: Evidence from Northern Bangladesh. *Small-Scale For.* **2018**, *17*, 505–522. [CrossRef]
76. Pujiono, E.; Raharjo, S.A.S.; Njurumana, G.N.; Prasetyo, B.D.; Rianawati, H. Sustainability Status of Agroforestry Systems in Timor Island, Indonesia. *E3S Web Conf.* **2021**, *305*, 04003. [CrossRef]
77. Ngaji; Kuala, A.U.; Baiquni, M.; Suryatmojo, H.; Haryono, E. Assessing the Sustainability of Traditional Agroforestry Practices: A Case of Mamar Agroforestry in Kupang-Indonesia. *For. Soc.* **2021**, *5*, 438–457. [CrossRef]
78. Ngaji; Kuala, A.U.; Baiquni, M.; Suryatmojo, H.; Haryono, E. Sustaining Subsistence Culture in Mamar Agroforestry Management in West Timor, Is It Possible? *E3S Web Conf.* **2020**, *200*, 02023. [CrossRef]
79. Njurumana, G.N.; Sadono, R.; Marsono, D.; Irham. Ecosystem Services of Indigenous Kaliwu Agroforestry System in Sumba, Indonesia. *E3S Web Conf.* **2021**, *305*, 04002. [CrossRef]
80. Simelton, E.; Dam, B.V.; Catacutan, D. Trees and Agroforestry for Coping with Extreme Weather Events: Experiences from Northern and Central Viet Nam. *Agrofor. Syst.* **2015**, *89*, 1065–1082. [CrossRef]

81. Agroforestry Network. *Agroforestry and Water for Resilient Landscapes*; Agroforestry Network: Stockholm, Sweden, 2020. Available online: https://agroforestrynetwork.org/database_post/agroforestry-and-water-for-resilient-landscapes/ (accessed on 27 February 2022).
82. Mugure, A.; Oino, P.; Phil, P. Benefits of Agroforestry Farming Practices among Rural Households in Kenya: Experiences among Residents of Busia County. *Int. J. Sci. Res.* **2013**, *2*, 442–449.
83. Jamnadass, R.; Place, F.; Torquebiau, E.; Malézieux, E.; Iiyama, M.; Sileshi, G.W.; Kehlenbeck, K.; Masters, E.; McMullin, S.; Weber, J.C.; et al. *Agroforestry, Food and Nutritional Security*; Icraf Working Paper No. 170; World Agroforesty Centre: Nairobi, Kenya, 2013. [CrossRef]
84. Wibawa, G.; Hendratno, S.; Van Noordwijk, M. Permanent Smallholder Rubber Agroforestry Systems in Sumatra, Indonesia. In *Slash-and-Burn Agriculture: The Search for Alternative*; Palm, C.A., Vosti, S.A., Sanchez, P.A., Ericksen, P.J., Eds.; Columbia University Press: New York, NY, USA, 2005. Available online: http://www.asb.cgiar.org/PDFwebdocs/Slash_and_Burn.pdf (accessed on 28 February 2022).
85. Rahman, S.A.; Jacobsen, J.B.; Healey, J.R.; Roshetko, J.M.; Sunderland, T. Finding Alternatives to Swidden Agriculture: Does Agroforestry Improve Livelihood Options and Reduce Pressure on Existing Forest? *Agrofor. Syst.* **2017**, *91*, 185–199. [CrossRef]
86. Toth, G.; Nair, P.K.R.; Duffy, C.P.; Franzel, S.C. Constraints to the Adoption of Fodder Tree Technology in Malawi. *Sustain. Sci.* **2017**, *12*, 641–656. [CrossRef]
87. Leimona, B.; Amaruzaman, S.; Arifin, B.; Yasmin, F.; Hasan, F.; Agusta, H.; Sprang, P.; Jaffee, S.; Frias, J. *Indonesia's 'Green Agriculture Strategies' and Policies: Closing the Gap between Aspirations and Application*; World Agroforestry Centre: Nairobi, Kenya, 2015. Available online: http://apps.worldagroforestry.org/sea/Publications/files/occasionalpaper/OP0003-15.pdf (accessed on 3 March 2022).
88. Nair, P.K.R. Managed Multi-Strata Tree + Crop Systems: An Agroecological Marvel. *Front. Environ. Sci.* **2017**, *5*, 88. [CrossRef]
89. Nyberg, G.; Tobella, A.B.; Kinyangi, J.; Ilstedt, U. Soil Property Changes over a 120-Yr Chronosequence from Forest to Agriculture in Western Kenya. *Hydrol. Earth Syst. Sci.* **2012**, *16*, 2085–2094. [CrossRef]
90. Sun, H.; Tang, Y.; Xie, J. Contour Hedgerow Intercropping in the Mountains of China: A Review. *Agrofor. Syst.* **2008**, *73*, 65–76. [CrossRef]
91. Kabir, A.; Hematzade, Y.; Barani, H. Agroforestry and Its Role in Soil Conservation and Erosion Protec. In Proceedings of the XXI International Grassland Congress/VIII International Rangeland Congress, Hohhot, China, 29 June–5 July 2008.
92. Bargues, T.; Reese, A.H.; Almaw, A.; Bayala, J.; Malmer, A.; Laudon, H.; Ilstedt, U. The Effect of Trees on Preferential Flow and Soil Infiltrability in an Agroforestry Parkland in Semiarid Burkina Faso. *Water Resour. Res.* **2014**, *50*, 3342–3354. [CrossRef]
93. Bayala, J.; Prieto, I. Water Acquisition, Sharing and Redistribution by Roots: Applications to Agroforestry Systems. *Plant Soil* **2020**, *453*, 17–28. [CrossRef]
94. Idris, M.H.; Mahrup, M. Changes in Hydrological Response of Forest Conversion to Agroforestry and Rainfed Agriculture in Renggung Watershed, Lombok, Eastern Indonesia. *J. Manaj. Hutan Trop.* **2017**, *23*, 102–110. [CrossRef]
95. Wu, J.; Liu, W.; Chen, C. Below-Ground Interspecific Competition for Water in a Rubber Agroforestry System May Enhance Water Utilization in Plants. *Sci. Rep.* **2016**, *6*, 19502. [CrossRef]
96. Yang, B.; Meng, X.; Singh, A.K.; Wang, P.; Song, L.; Zakari, S.; Liu, W. Intercrops Improve Surface Water Availability in Rubber-Based Agroforestry Systems. *Agric. Ecosyst. Environ.* **2020**, *298*, 106937. [CrossRef]
97. Van Noordwijk, M.; Verbist, B. *Overstory #104—Soil and Water Conservation*; Agroforestry Net, Inc.: Holualoa, HI, USA, 2002. Available online: https://agroforestry.org/the-overstory/165-overstory-104-soil-and-water-conservation (accessed on 31 March 2022).
98. Gusli, S.; Sumeni, S.; Sabodin, R.; Muqfi, I.H.; Nur, M.; Hairiah, K.; Useng, D.; van Noordwijk, M. Soil Organic Matter, Mitigation of and Adaptation to Climate Change in Cocoa–Based Agroforestry Systems. *Land* **2020**, *9*, 323. [CrossRef]
99. Sharma, N.; Bohra, B.; Pragya, N.; Ciannella, R.; Dobie, P.; Lehmann, S. Bioenergy from Agroforestry Can Lead to Improved Food Security, Climate Change, Soil Quality, and Rural Development. *Food Energy Secur.* **2016**, *5*, 165–183. [CrossRef]
100. Thevathasan, N.; Gordon, A.; Simpson, J.; Peng, X.; Silim, S.; Soolanayakanahally, R.; de Gooijer, H. Sustainability Indicators of Biomass Production in Agroforestry Systems. *Open Agric. J.* **2014**, *8*, 1–11. [CrossRef]
101. Faße, A.; Winter, E.; Grote, U. Bioenergy and Rural Development: The Role of Agroforestry in a Tanzanian Village Economy. *Ecol. Econ.* **2014**, *106*, 155–166. [CrossRef]
102. Kralik, T.; Vavrova, K.; Knapek, J.; Weger, J. Agroforestry Systems as New Strategy for Bioenergy—Case Example of Czech Republic. *Energy Rep.* **2022**, *8*, 519–525. [CrossRef]
103. Ripple, W.J.; Wolf, C.; Newsome, T.M.; Barnard, P.; Moomaw, W.R. World Scientists' Warning of a Climate Emergency. *BioScience* **2019**, *70*, 8–12. [CrossRef]
104. IPCC. *IPCC Special Report on Climate Change, Desertification, Land Degradation, Sustainable Land Manage—Ment, Food Security, and Greenhouse Gas Fluxes in Terrestrial Ecosystems*; Intergovernmental Panel on Climate Change: Geneva, Switzerland, 2019.
105. Van Noordwijk, M.; Bayala, J.; Hairiah, K.; Lusiana, B.; Muthuri, C.W.; Khasanah, N.; Mulia, R. Agroforestry Solutions for Buffering Climate Variability and Adapting to Change. In *Climate Change Impact and Adaptation in Agricultural System*; CAB-International: Wallingford, UK, 2014. Available online: http://blog.worldagroforestry.org/wp-content/uploads/2014/09/Agroforestry-solutions-for-buffering-climate-variability-and-adapating-to-change.van-Noordwijk-et-al.pdf (accessed on 25 April 2022).

106. IPCC. IPCC Special Report on Land Use, Land Use Change and Forestry. *Paper Presented at the Summary for Policy Makers, Geneva, Switzerland.* 2000. Available online: https://www.ipcc.ch/report/land-use-land-use-change-and-forestry/ (accessed on 23 March 2022).
107. Agroforestry Network. *Scaling up Agroforestry: Potential, Challenges and Barriers: A Review of Environmental, Social and Economic Aspects on the Farmer, Community and Landscape Level*; Agroforestry Network: Stockholm, Sweden, 2018.
108. Murthy, K.I.; Gupta, M.; Tomar, S.; Munsi, M.; Tiwari, R. Carbon Sequestration Potential of Agroforestry Systems in India. *J. Earth Sci. Clim. Chang.* **2013**, *4*, 131. [CrossRef]
109. Singh, V.S.; Pandey, D.N. *Multifunctional Agroforestry Systems in India: Science-Based Policy Options*; Climate Change and CDM Cell, Rajasthan State Pollution Control Board: Jaipur, India, 2011.
110. Nair, P.K.R.; Nair, V.D.; Kumar, B.; Showalter, J.M. Chapter Five—Carbon Sequestration in Agroforestry Systems. In *Advances in Agronomy*; Donald, L.S., Ed.; Academic Press: Cambridge, MA, USA, 2010; pp. 237–307. [CrossRef]
111. Matocha, J.; Schroth, G.; Hills, T.; Hole, D. Integrating Climate Change Adaptation and Mitigation through Agroforestry and Ecosystem Conservation. In *Agroforestry—The Future of Global Land Use*; Nair, P.K.R., Garrity, D., Eds.; Springer: Dordrecht, The Netherlands, 2012; pp. 105–126. [CrossRef]
112. Kaczan, D.; Asrlan, A.; Leslie, L. Climate-Smart Agriculture? A Review of Current Practice of Agroforestry and Conservation Agriculture in Malawi and Zambia. In *Working Paper No 13-07*; FAO: Rome, Italy, 2013. [CrossRef]
113. Ellison, D.; Morris, C.E.; Locatelli, B.; Sheil, D.; Cohen, J.; Murdiyarso, D.; Gutierrez, V.; van Noordwijk, M.; Creed, I.F.; Pokorny, J.; et al. Trees, Forests and Water: Cool Insights for a Hot World. *Glob. Environ. Change* **2017**, *43*, 51–61. [CrossRef]
114. Republic Indonesia. *Indonesia's Technology Needs Assessment for Climate Change Mitigation in 2012*; Kardono, Winanti, W.S., Riyadi, A., Purwanta, W., Eds.; Ministry of Environment and Forestry: Jakarta, Indonesia, 2012.
115. Republic of Indonesia. *Indonesia's Technology Needs Assessment for Climate Change Adaptation in 2012*; Kardono, Winanti, W.S., Riyadi, A., Purwanta, W., Eds.; Ministry of Environment and Forestry: Jakarta, Indonesia, 2012.
116. Plieninger, T.; Muñoz-Rojas, J.; Buck, L.E.; Scherr, S.J. Agroforestry for Sustainable Landscape Management. *Sustain. Sci.* **2020**, *15*, 1255–1266. [CrossRef]
117. Arifin, H.S.; Wulandari, C.; Pramkoto, Q.; Kaswanto, R.L. *Analisiis Lanskap Agroforestri: Konsep, Metode dan Pengelolaan Agroforestry Skala Lanskap dengan Studi Kasus Indonesia, Filipina, Laos, Thailand dan Vietnam*; IPB Press: Bogor, Indonesia, 2019.
118. Laumonier, Y.; Simamora, T.; Manurung, A.; Narulita, S.; Pribadi, U.; Simorangkir, A.; Kharisma, S.; Shantiko, B. *Sentinel Landscapes Initiative*; FTA Work, Paper 1–74; FTA: Paris, France, 2020.
119. Butarbutar, T.; Hakim, I.; Sakuntaladewi, N.; Dwiprabowo, H.H.; Rumboko, L.; Irawant, S. Land Suitability Assessment of Nine Species for Agroforestry in Nambo, West Java. *J. Penelit. Hutan Tanam.* **2018**, *15*, 1–66.
120. Mindawati, N.; Darwo, P.; Darmawan, I.D. *Teknik Silvikultur untuk Lahan Terdegradasi di Indonesia*; IPB Press: Bogor, Indonesia, 2019.
121. Antoh, A.A.; Arifin, N.H.; Chozin, M.A.; Arifin, H.S. Short Communication: Agricultural Biodiversity and Economic Productivity of the Yards in Arguni Bawah, Kaimana District, West Papua Province, Indonesia. *Biodivers. J. Biol. Divers.* **2019**, *20*, 1020–1026. [CrossRef]
122. Anandi, A.M.; Yuliani, E.L.; Moeliono, M.; Laumonier, Y.; Narulita, S. Kapuas Hulu: A Background Analysis to Implementing Integrated Landscape Approaches in Indonesia. In *Operationalizing Integrated Landscape Approaches in the Tropics*; CIFOR: Bogor, Indonesia, 2020.
123. Putra, A.N.; Nita, I.; Jauhary, M.R.A.; Nurhutami, S.R.; Ismail, M.H. Landslide Risk Analysis on Agriculture Area in Pacitan Regency in East Java Indonesia Using Geospatial Techniques. *Environ. Nat. Resour. J.* **2021**, *19*, 141–152. [CrossRef]
124. Cannell, M.G.R. Plant Management in Agroforestry: Manipulation of Trees, Population Densities and Mix-Tures of Trees and Herbaceous Crops. In *Plant Research and Agroforestry*; ICRAF: Nairobi, Kenya, 1983.
125. Nair, P.K.R. Classification of Agroforestry Systems. *Agrofor. Syst.* **1985**, *3*, 97–128. [CrossRef]
126. Michon, G.; Mary, F.; Bompard, J. Multistoried Agroforestry Garden System in West Sumatra, Indonesia. *Agrofor. Syst.* **1986**, *4*, 315–338. [CrossRef]
127. Michon, G.; de Foresta, H.; Widjayanto, N. Complex Agroforerstry System in Sumatra. In Proceedings of the Sumatra, Lingkungan dan Pembangunan yang Lalu, Sekarang dan yang akan datang, Bogor, Indonesia, 16–18 September 1992.
128. Murniati. Penguatan Teknologi Agroforestri selama Daur dalam Pengelolaan Hutan Berbasis Masyarakat. In *Orasi Pengukuhan Profesor Riset Bidang Teknologi Agroforestri*; Kementerian Lingkungan Hidup dan Kehutanan: Jakarta, Indonesia, 2020.
129. Smith, J.; Pearce, B.D.; Wolfe, M.S. Reconciling Productivity with Protection of the Environment: Is Temperate Agroforestry the Answer? *Renew. Agric. Food Syst.* **2013**, *28*, 80–92. [CrossRef]
130. Rodriguez, C.; Carlsson, G.; Englund, J.; Flöhr, A.; Pelzer, E.; Jeuffroy, M.H.; Makowski, D.; Jensen, E.S. Grain Legume-Cereal Intercropping Enhances the Use of Soil-Derived and Biologically Fixed Nitrogen in Temperate Agroecosystems. A Meta-Analysis. *Eur. J. Agron.* **2020**, *118*, 126077. [CrossRef]
131. Campera, M.; Balestri, M.; Manson, S.; Hedger, K.; Ahmad, N.; Adinda, E.; Nijman, V.; Budiadi; Imron, M.A.; Nekaris, K.A.I. Shade Trees and Agrochemical Use Affect Butterfly Assemblages in Coffee Home Gardens. *Agric. Ecosyst. Environ.* **2021**, *319*, 107547. [CrossRef]
132. Verret, V.; Gardarin, A.; Pelzer, E.; Médiène, S.; Makowski, D.; Valantin-Morison, M. Can Legume Companion Plants Control Weeds without Decreasing Crop Yield? A Meta-Analysis. *Field Crops Res.* **2017**, *204*, 158–168. [CrossRef]

133. Liu, C.; Wang, Q.; Jin, Y.; Tang, J.; Lin, F.; Olatunji, O. Perennial Cover Crop Biomass Contributes to Regulating Soil P Availability More Than Rhizosphere P-Mobilizing Capacity in Rubber-Based Agroforestry Systems. *Geoderma* **2021**, *401*, 115218. [CrossRef]
134. Alegre, J.; Olivares, C.L.; Silva, C.; Schrevens, E. Recovering Degraded Lands in the Peruvian Amazon by Cover Crops and Sustainable Agroforestry Systems. *Peruv. J. Agron.* **2017**, *1*, 1. [CrossRef]
135. Pande, M.; Gupta, G.N. Scope of Biofertilizers in Agroforestry. *Indian J. Agrofor.* **2020**, *2*, 23–32.
136. Singh, B.; Upadhyay, A.K.; Al-Tawaha, T.W.; Al-Tawaha, A.R.; Sirajuddin, S.N. Biofertilizer as a Tool for Soil Fertility Management in Changing Climate. *IOP Conf. Ser. Earth Environ. Sci.* **2020**, *492*, 012158. [CrossRef]
137. Diene, O.; Wang, W.; Narisawa, K. Pseudosigmoidea Ibarakiensis Sp. Nov., a Dark Septate Endophytic Fungus from a Cedar Forest in Ibaraki, Japan. *Microbes Environ.* **2013**, *28*, 381–387. [CrossRef] [PubMed]
138. Bhattacharjee, R.; Dey, U. Biofertilizer, a Way Towards Organic Agriculture: A Review. *Afr. J. Microbiol. Res.* **2014**, *8*, 2332–2343. [CrossRef]
139. Wako, F.; Shiferaw, D.; Mebratie, M.A. The Need of Integrated Nutrient Management for Coriander (*Coriandrum sativum* L.) Production. *Int. J. Zamrud* **2019**, *4*, 1–13.
140. Verma, A.; Kumar, P.; Saresh, N.V. Secondary Metabolites: Harvesting Short Term Benefits from Arid Zone Agroforestry Systems in India. *Agrofor. Syst.* **2021**, *95*, 515–532. [CrossRef]
141. Saleh, A.M.; Selim, S.; Al Jaouni, S.; AbdElgawad, H. CO_2 Enrichment Can Enhance the Nutritional and Health Benefits of Parsley (*Petroselinum crispum* L.) and Dill (*Anethum graveolens* L.). *Food Chem.* **2018**, *269*, 519–526. [CrossRef]
142. Mariska, I. Metabolit Sekunder: Jalur Pembentukan Dan Kegunaannya. Available online: http://biogen.litbang.pertanian.go.id/?p=56130 (accessed on 24 March 2022).
143. Alwiyah, A.; Manuhara, Y.S.W.; Utami, E.S.W. Pengaruh Intensitas Cahaya terhadap Biomassa dan Kadar Saponin Kalus Ginseng Jawa (Talinum Paniculatum Gaertn.) pada Berbagai Waktu Kultur. *J. Ilm. Biol.* **2015**, *3*, 56–66.
144. Ariandi, E.A.; Duryat, D.; Santoso, T. Analisis Rendemen Atsiri Biji Pala (Myristica Fragrans) pada Berbagai Kelas Intensitas Cahaya Matahari di Desa Batu Keramat Kecamatan Kota Agung Kabupaten Tanggamus. *J. Sylva Lestari* **2018**, *6*, 24–31. [CrossRef]
145. Ratnadass, A.; Fernandes, P.; Avelino, J.; Habib, R. Plant Species Diversity for Sustainable Management of Crop Pests and Diseases in Agroecosystems: A Review. *Agron. Sustain. Dev.* **2012**, *32*, 273–303. [CrossRef]
146. Perfecto, I.; Vandermeer, J.; Philpott, S.M. Complex Ecological Interactions in the Coffee Agroecosystem. *Annu. Rev. Ecol. Evol. Syst.* **2014**, *45*, 137–158. [CrossRef]
147. Boudreau, M.A. Diseases in Intercropping Systems. *Annu. Rev. Phytopathol.* **2013**, *51*, 499–519. [CrossRef]
148. Ememwa, I.K.; Were, H.; Ogemah, V.; Obiero, H.; Wekesa, M. Effect of Modified Spacing Arrangements, Fertilizer Use and Legume Intercrop on Prevalence of Cassava Brown Streak Disease in Western Kenya. *Afr. J. Agric. Res.* **2017**, *12*, 3181–3188. [CrossRef]
149. Sudiono, S.H.S.; Wijayanto, N.; Hidayat, P.; Kurniawan, P. Vegetation Diversity and Intensity of Plant Pest and Diseases in Two Polyculture System in Tanggamus District. *J. HPT Trop.* **2017**, *17*, 137–146.
150. Tomlinson, I.; Potter, C.; Bayliss, H. Managing Tree Pests and Diseases in Urban Settings: The Case of Oak Processionary Moth in London, 2006–2012. *Urban For. Urban Green.* **2015**, *14*, 286–292. [CrossRef]
151. Pumariño, L.; Sileshi, G.W.; Gripenberg, S.; Kaartinen, R.; Barrios, E.; Muchane, M.N.; Midega, C.; Jonsson, M. Effects of Agroforestry on Pest, Disease and Weed Control: A Meta-Analysis. *Basic Appl. Ecol.* **2015**, *16*, 573–582. [CrossRef]
152. Legesse, A.; Negash, M. Species Diversity, Composition, Structure and Management in Agroforestry Systems: The Case of Kachabira District, Southern Ethiopia. *Heliyon* **2021**, *7*, e06477. [CrossRef]
153. Afentina; Yanarita; Indrayanti, L.; Rontinsulu, J.A.; Hidayat, N.; Sianipar, J. The Potential of Agroforestry in Supporting Food Security for Peatland Community—A Case Study in the Kalampangan Village, Central Kalimantan. *J. Ecol. Eng.* **2021**, *22*, 123–130. [CrossRef]
154. Hernández, T.M.; Tscharntke, T.; Tjoa, A.; Anshary, A.; Cyio, B.; Wanger, T.C. Landscape and Farm-Level Management for Conservation of Potential Pollinators in Indonesian Cocoa Agroforests. *Biol. Conserv.* **2021**, *257*, 109106. [CrossRef]
155. Eyasu, G.; Tolera, M.; Negash, M. Woody Species Composition, Structure, and Diversity of Homegarden Agroforestry Systems in Southern Tigray, Northern Ethiopia. *Heliyon* **2020**, *6*, e05500. [CrossRef]
156. Juliarti, A.; Wijayanto, N.; Mansur, I.; Trikoesoemaningtyas. The Growth of Lemongrass (*Cymbopogon nardus* (L.) Rendle) in Agroforestry and Monoculture System on Post-Coal Mining Revegetation Land. *J. Manaj. Hutan Trop.* **2021**, *27*, 15–23. [CrossRef]
157. Ashraf, M.; Zulkifli, R.; Sanusi, R.; Tohiran, K.A.; Terhem, R.; Moslim, R.; Norhisham, A.R.; Butt, A.A.; Azhar, B. Alley-Cropping System Can Boost Arthropod Biodiversity and Ecosystem Functions in Oil Palm Plantations. *Agric. Ecosyst. Environ.* **2018**, *260*, 19–26. [CrossRef]
158. Sari, D.; Budiadi, F.; Imron, M.A. The Utilization of Trees by Endangered Primate Species Javan Slow Loris (*Nycticebus javanicus*) in Shade-Grown Coffee Agroforestry of Central Java. *IOP Conf. Ser. Earth Environ. Sci.* **2020**, *449*, 012044. [CrossRef]
159. Smith, C.; Barton, D.; Johnson, M.; Wendt, C.; Milligan, M. Bird Communities in Sun and Shade Coffee Farms in Kenya. *Glob. Ecol. Conserv.* **2015**, *4*, 479–490. [CrossRef]
160. Torralba, M.; Fagerholm, N.; Burgess, P.J.; Moreno, G.; Plieninger, T. Do European Agroforestry Systems Enhance Biodiversity and Ecosystem Services? A Meta-Analysis. *Agric. Ecosyst. Environ.* **2016**, *230*, 50–61. [CrossRef]
161. Bardhan, S.; Jose, S.; Biswas, B.; Kabir, K.A.; Rogers, W. Homegarden Agroforestry Systems: An Intermediary for Biodiversity Conservation in Bangladesh. *Agrofor. Syst.* **2012**, *85*, 29–34. [CrossRef]

162. Lopez, R.; Sotomayor, D.; Amador, J.; Schroder, E. Contribution of Nitrogen from Litter and Soil Mineralization to Shade and Sun Coffee (*Coffea arabica* L.) Agroecosystems. *J. Trop. Ecol.* **2014**, *56*, 21–33.
163. Behera, L.K.; Nayak, M.R.; Gunaga, R.P.; Dobriyal, M.J.; Patel, D.P. Land Use System Agroforestry for Biodiversity Conservation of Native Species. *Multilogic Sci.* **2015**, *4*, 125–130.
164. De Beenhouwer, M.; Aerts, R.; Honnay, O. A Global Meta-Analysis of the Biodiversity and Ecosystem Service Benefits of Coffee and Cacao Agroforestry. *Agric. Ecosyst. Environ.* **2013**, *175*, 1–7. [CrossRef]
165. Mamani-Pati, F.; Clay, D.E.; Clay, S.A.; Smeltekop, H.; Yujra-Callata, M.A. The Influence of Strata on the Nutrient Recycling within a Tropical Certified Organic Coffee Production System. *ISRN Agron.* **2012**, *2012*, 389290. [CrossRef]
166. Lisnawati, A.; Lahjie, A.M.; Simarangkir, B.D.A.S.; Yusuf, S.; Ruslim, Y. Agroforestry System Biodiversity of Arabica Coffee Cultivation in North Toraja District, South Sulawesi, Indonesia. *Biodiversitas* **2017**, *18*, 741–751. [CrossRef]
167. Jose, S. Agroforestry for Conservin.; Enhancing Biodiversity. *Agrofor. Syst.* **2012**, *85*, 1–8. [CrossRef]
168. Sodik, M.; Pudyatmoko, S.; Semedi, P.; Imron, M. Resource Selection by Javan Slow Loris Nycticebus Javanicus E. Geoffroy, 1812 (Mammalia: Primates: Lorisidae) in a Lowland Fragmented Forest in Central Java, Indonesia. *J. Threat. Taxa* **2019**, *11*, 13667–13679. [CrossRef]
169. Haggar, J.; Pons, D.; Saenz, L.; Vides, M. Contribution of Agroforestry Systems to Sustaining Biodiversity in Fragmented Forest Landscapes. *Agric. Ecosyst. Environ.* **2019**, *283*, 106567. [CrossRef]
170. Sistla, S.; Roddy, A.; Williams, N.; Kramer, D.; Stevens, K.; Allison, S. Agroforestry Practices Promote Biodiversity and Natural Resource Diversity in Atlantic Nicaragua. *PLoS ONE* **2016**, *11*, e0162529. [CrossRef] [PubMed]
171. Bucheli, P.; Julián, V.; Bokelmann, W. Agroforestry Systems for Biodiversity and Ecosystem Services: The Case of the Sibundoy Valley in the Colombian Province of Putumayo. *Int. J. Biodivers. Sci. Ecosyst. Serv. Manag.* **2017**, *13*, 380–397. [CrossRef]
172. Ortolani, G. Agroforestry: An Increasingly Popular Solution for a Hot, Hungry World. Available online: https://news.mongabay.com/2017/10/agroforestry-an-increasingly-popular-solution-for-a-hot-hungry-world/ (accessed on 4 May 2022).
173. Tomar, J.M.S.; Ahmed, A.; Bhat, J.A.; Kaushal, R.; Shukla, G.; Kumar, R. Potential and Opportunities of Agroforestry Practices in Combating Land Degradation. In *Agroforestry*; Shukla, G., Chakravarty, S., Panwar, P., Bhat, J.A., Eds.; IntechOpen: London, UK, 2021. [CrossRef]
174. Nair, P.R.K. *An Itroduction to Agroforestry*; Kluwer Academic Publisher: Alphen aan den Rijn, The Netherlands, 1993; 439p.
175. Lelamo, L.L. A Review on the Indigenous Multipurpose Agroforestry Tree Species in Ethiopia: Management, Their Productive and Service Roles and Constraints. *Heliyon* **2021**, *7*, e07874. [CrossRef] [PubMed]
176. Reed, M.S.; Dougill, A.J.; Taylor, M.J. Integrating Local and Scientific Knowledge for Adaptation to Land Degradation: Kalahari Rangeland Management Options. *Land Degrad. Dev.* **2007**, *18*, 249–268. [CrossRef]
177. Zerihun, M.F. Institutional Analysis of Adoption of Agroforestry Practices in the Eastern Cape Province of South Africa. *S. Afr. J. Environ. Educ.* **2020**, *36*. [CrossRef]
178. FAO. Strengthening Sector Policies for Better Food Security and Nutrition Results-Forestry. In *Policy Guidance Note 13*; FAO: Rome, Italy, 2017.
179. Janra, M.N. Socio-Ecological Aspects of Mandailing Natal People in Buffer Zone of Batang Gadis National Park, North Sumatra. *Redefin. Divers. Dyn. Nat. Resour. Manag. Asia* **2017**, *4*, 141–155.
180. Steffen, W.; Crutzen, P.J.; McNeill, J.R. The Anthropocene: Are Humans Now Overwhelming the Great Forces of Nature? *Ambio* **2007**, *36*, 8. [CrossRef]
181. Partasasmita, R.; Cahyani, N.T.; Iskandar, J. Local Knowledge of the Community in Cintaratu Village, Pangandaran, Indonesia on Traditional Landscapes for Sustainable Land Management. *Biodivers. J. Biol. Divers.* **2020**, *21*, 8. [CrossRef]
182. Kurniasih, H.; Ford, R.M.; Keenan, R.J.; King, B. The Evolution of Community Forestry through the Growth of Interlinked Community Institutions in Java, Indonesia. *World Dev.* **2021**, *139*, 105319. [CrossRef]
183. Njurumana, G.N.; Pujiono, E.; da Silva, M.M.; Oematan, O.K. Ecological Performance of Local Initiatives on Water Resources Management in Timorese Communities, Indonesia. *IOP Conf. Ser. Earth Environ. Sci.* **2021**, *914*, 012031. [CrossRef]
184. Njurumana, G.N. Village Community and Flora Biodiversity Management in Home Garden System at Central of Sumba Regency). *J. Penelit. Kehutan. Wallacea* **2016**, *5*, 25–36. Available online: https://jurnal.balithutmakassar.org/index.php/wallacea/article/viewFile/89/pdf_16 (accessed on 8 June 2022). [CrossRef]
185. Njurumana, G.N.; Marsono, D.; Irham; Sadono, R. Konservasi Cendana (Santalum Album Linn) Berbasis Masyarakat pada Sistem Kaliwu di Pulau Sumba. *J. Ilmu Lingkung.* **2013**, *11*, 51–61. [CrossRef]
186. Iryadi, R.; Sutom; Darma, I.D.P. Multistoried Agroforestry System of Gaharu (Gyrinops Verstegii (Gilg.) Domke) in Flores Island East Nusa Tenggara. *IOP Conf. Ser. Earth Environ. Sci.* **2021**, *648*, 012024. [CrossRef]
187. Njurumana, G.N.; Octavia, D. Conservation Species of Ntfps through Agroforestry for Community Livelihoods in Sikka, East Nusa Tenggara. *J. Sylva Indones.* **2020**, *3*, 1–16. [CrossRef]
188. Turua, U.; Hadi, S.; Juanda, B.; Murniningtyas, E. Ekologi dan Budaya Petani Asli Papua dalam Usaha Tani di Kabupaten Keerom. *Sosiohumaniora* **2014**, *16*, 234–241. [CrossRef]
189. Cairns, M. *Shifting Cultivation Policies: Balancing Environmental and Social Sustainability*; CABI: Boston, MA, USA, 2017. [CrossRef]
190. Limba, R.S.; Lio, A.; Husain, Y.S. Shifting Cultivation System of Indigenous Moronene as Forest Conservation on Local Wisdom Principles in Indonesia. *J. Sustain. Dev.* **2017**, *10*, 121–129. [CrossRef]

191. Siburian, R.H. Sistim Agroforestri pada Lahan Bekas Hutan Sagu di Kampung Baraway Kabupaten Kepulauan Yapen Papua. In Proceedings of the Seminar Hasil Penelitian Kanoppi (Optimalisasi pengelolaan Hutan Berbasis Agroforestry Untuk Mendukung Peningkatan Produktifitas Kayu Dan HHBK, Serta Pendapatan Petani), Bogor, Indonesia, 1 December 2016; Available online: http://repository.pertanian.go.id/bitstream/handle/123456789/15847/Sinergi-Inovasi-Pertanian-pangan-holtikultura.pdf?sequence=1&isAllowed=y (accessed on 20 April 2022).
192. Hegemur; Hanafiah, S.; Leti, S. *Analisis Pola Agroforestri Tradisional di Distrik Kokas Kabupaten Fakfak Propinsi Papua Barat*; Institut Pertanian Bogor: Bogor, Indonesia, 2013.
193. Ihalauw, M.I.; Rahmadaniarti, A.; Panambe, N. Kontribusi Agroforestri Herbal terhadap Penerimaan Tunai Masyarakat Lokal di Sekitar Manokwari Utara (Studi Kasus di Kampung Bremni, Nyoom, dan Lebau). *J. Kehutan. Papuasia* **2020**, *6*, 133–140. [CrossRef]
194. Soekotjo. *Teknik Silvikultur Intensif (Silin)*; Gadjah Mada University Press: Yogyakarta, Indonesia, 2009.
195. Katz, E.; Goloubinoff, M.; Perez, M.R.; Michon, G. Experiences in Benzoin Resin Production in Sumatra. In Proceedings of the Conservation Management and Utilization of Plant Gums, Resin and Oils, Nairobi, Kenya, 6–10 October 1997.
196. Harada, K.; Wiyono; Munthe, L. Production and Commercialization of Benzoin Resin: Exploring the Value of Benzoin Resin for Local Livelihoods in North Sumatra, Indonesia. *Trees For. People* **2022**, *7*, 100174. [CrossRef]
197. Gockowski, J.; Asten, P.V. Agricultural Intensification as a Climate Change and Food Security Strategy for Sub-Saharan Africa. In *Climate Change Mitigation and Agriculture*; ICRAF: London, UK; CIAT: New York, NY, USA, 2012.
198. Place, F.; Ajayi, O.C.; Torquebiau, E.; Detlefsen, G.; Gauthier, M.; Buttoud, G. Improved Policies for Facili-Tating the Adoption of Agroforestry. In *Agroforestry for Biodiversity and Ecosystem Services—Science and Practice*; Kaonga, M., Ed.; IntechOpen: London, UK, 2012. [CrossRef]
199. Yulianti, T.; Abdoellah, S.; Suprayogo, D.; Sari, N.P. Cocoa Production as Affected by Shade Tree Species and Soil Quality. *Pelita Perkeb.* **2018**, *34*, 2. [CrossRef]
200. Suprayogo, D.; Van Noordwijk, M.; Hairiah, K.; Cadisch, G. The Inherent 'Safety-Net' of an Acrisol: Measuring and Modelling Retarded Leaching of Mineral Nitrogen. *Eur. J. Soil Sci.* **2002**, *52*, 185–194. [CrossRef]
201. Craswell, E.T.; Sajjapongse, A.; Howlett, D.; Dowling, A. Agroforestry in the Management of Sloping Lands in Asia and the Pacific. *Agrofor. Syst.* **1998**, *38*, 121–137. [CrossRef]
202. Kugedera, A.T.; George, N.; Mandumbu, R.; Nyamangara, J. Potential of Field Edge Rainwater Harvesting, Biomass Transfer and Integrated Nutrient Management in Improving Sorghum Productivity in Semi-Arid Regions: A Review. *Agrofor. Syst.* **2022**, *96*, 909–924. [CrossRef]
203. USAID & World Agroforestry Centre. Scaling-Up Climate-Smart Agroforestry Technologies in Mali (Smat-Scaling). Available online: https://www.usaid.gov/sites/default/files/documents/1860/USAID%20AEG%20-%20Agroforesty%20Fact%20Sheet%20-%20FTF%20-%20Nov%20%2716_FINAL.pdf (accessed on 28 May 2022).
204. de Foresta, H.; Kusworo, A.; Michon, G.; Djatmiko, W.A. *Ketika Kebun Berupa Hutan: Agroforest Khas Indonesia, Sebuah Sumbangan Masyarakat*; International Centre for Research in Agroforestry: Bogor, Indonesia, 2020.
205. Research Group of Tropical Agroforestry (RGTA). Agroforestry Smart Farming. Brawijaya University. Available online: https://agroforestri.ub.ac.id/activities/agroforestry-smart-farming (accessed on 28 May 2022).
206. CGIAR. Cgiar Research Program on Forests, Trees and Agroforestry. In *Annual Report 2011*; Center for International Forestry Research: Bogor, Indonesia, 2012.
207. Montagnini, F. Introduction: Challenges for Agroforestry in the New Millennium. In *Integrating Landscapes: Agroforestry for Biodiversity Conservation and Food Sovereignty*; Springer: Berlin/Heidelberg, Germany, 2017; pp. 3–10. [CrossRef]
208. Patel-Weynand, T.; Bentrup, G.; Schoeneberger, M.; Karel, T.H.; Nair, P.K.R. Challenges and Opportunities. In *Agroforestry Enhancing Resiliency in U.S. Agricultural Landscapes under Changing Conditions*; USDA: Washington, DC, USA, 2017. [CrossRef]
209. Van Noordwijk, M.; Minang, P.A.; Dewi, S.; Duguma, L.; Bernard, F. *Transforming Redd+ and Achieving the Sdgs through Support for Adaptation-Mitigation Synergy*; ASB Policy Brief 46; ASB Partnership for the Tropical Forest Margins: Nairobi, Kenya, 2015.
210. Kubitza, C.; Krishna, V.V.; Alamsyah, Z.; Qaim, M. The Economics Behind an Ecological Crisis: Livelihood Effects of Oil Palm Expansion in Sumatra, Indonesia. *Hum. Ecol.* **2018**, *46*, 107–116. [CrossRef]
211. Rahmanulloh, A.; Sofiyuddin, M.; Suyanto. *Agroforestry and Forestry in Sulawesi Series: Profitability of Land-Use Systems in South Sulawesi and Southeast Sulawesi*; Working Paper No. 157; World Agroforestry Centre: Bogor, Indonesia, 2012.
212. Ehrenbergerová, L.; Klimková, M.; Cano, Y.G.; Habrová, H.; Lvončík, S.; Volařík, D.; Khum, W.; Němec, P.; Kim, S.; Jelínek, P.; et al. Does Shade Impact Coffee Yield, Tree Trunk, and Soil Moisture on Coffea Canephora Plantations in Mondulkiri, Cambodia? *Sustainability* **2021**, *13*, 13823. [CrossRef]
213. Nishizawa, T.; Kay, S.; Schuler, J.; Klein, N.; Herzog, F.; Aurbacher, J.; Zander, P. Ecological–Economic Modelling of Traditional Agroforestry to Promote Farmland Biodiversity with Cost-Effective Payments. *Sustainability* **2022**, *14*, 5615. [CrossRef]
214. Narendra, B.H.; Siregar, C.A.; Dharmawan, I.W.S.; Sukmana, A.; Pratiwi; Pramono, I.B.; Basuki, T.M.; Nugroho, H.Y.S.H.; Supangat, A.B.; Purwanto; et al. A Review on Sustainability of Watershed Management in Indonesia. *Sustainability* **2021**, *13*, 11125. [CrossRef]
215. Erbaugh, J.T. Responsibilization and Social Forestry in Indonesia. *For. Policy Econ.* **2019**, *109*, 102019. [CrossRef]
216. Cioffo, G.D.; Ansoms, A.; Murison, J. Modernising Agriculture through a 'New' Green Revolution: The Limits of the Crop Intensification Programme in Rwanda. *Rev. Afr. Political Econ.* **2016**, *43*, 277–293. [CrossRef]

217. Kiyani, P.; Andoh, J.; Lee, Y.; Lee, D. Benefits and Challenges of Agroforestry Adoption: A Case of Musebeya Sector, Nyamagabe District in Southern Province of Rwanda. *For. Sci. Technol.* **2017**, *13*, 174–180. [CrossRef]
218. Yeny, I.; Murniati, M.; Suharti, S. Partisipasi Masyarakat dalam Pengembangan Agroforestri di Kesatuan Pengelolaan Hutan (Kph) Gedong Wani. *J. Penelit. Sos. Dan Ekon. Kehutan.* **2020**, *17*, 49–66. [CrossRef]
219. Winara, A. Prospek Jalawure (*Tacca leontopetaloides* (L.) Kuntze) sebagai Sumber Pangan Alternatif dari Hutan Pantai. In *Bunga Rampai Optimalisasi Pemanfaatan Sumber Pangan Dari Hutan: Sumbangsih untuk Ketahanan Pangan Nasional*; IPB Press: Bogor, Indonesia, 2009; pp. 89–109.
220. Agustarini, R.; Kartikaningtyas, D. Iwi, Pangan Lokal Andalan Masyarakat Sumba Timur. In *Bunga Rampai Optimalisasi Pemanfaatan Sumber Pangan Dari Hutan: Sumbangsih untuk Ketahanan Pangan Nasional*; IPB Press: Bogor, Indonesia, 2019.
221. Budiadi; Aqmal, N.J.; Lina, D.L. An Overview and Future Outlook of Indonesian Agroforestry: A Bibliographic and Literature Review. *E3S Web Conf.* **2021**, *305*, 07002. [CrossRef]
222. Wheeler, T.; von Braun, J. Climate Change Impacts on Global Food Security (Review). *Science* **2013**, *341*, 508–513. [CrossRef]
223. Dawson, N.; Martin, A.; Sikor, T. Green Revolution in Sub-Saharan Africa: Implications of Imposed Innovation for the Wellbeing of Rural Smallholders. *World Dev.* **2016**, *78*, 204–218. [CrossRef]
224. De Royer, S.; van Noordwijk, M.; Roshetko, J.M. Does Community-Based Forest Management in Indonesia Devolve Social Justice or Social Costs? *Int. For. Rev.* **2018**, *20*, 167–180. [CrossRef]
225. Wulandari, C.; Sugeng, P.H.; Destia, N. *Pengembangan Agroforestri Yang Berkelanjutan Dalam Menghadapi Perubahan Iklim*; Pustaka Media: Lampung, Indonesia, 2020.
226. Suek, J.; Irham; Hartono, S.; Waluyati, L.R. Production Risks of Mamar, a Traditional Agroforestry System in Timor Island, East Nusa Tenggara Province, Indonesia. *Asian J. Agric. Biol.* **2018**, *6*, 434–441.
227. Kementerian Lingkungan Hidup dan Kehutanan (KLHK). *Peraturan Menteri Lingkungan Hidup dan Kehutanan No P.41/2019 Tahun 2019 Tentang Rencana Kehutanan Tingkat Nasional Tahun 2011–2030*; Kementerian Lingkungan Hidup dan Kehutanan: Jakarta, Indonesia, 2019.
228. He, S.; Yang, L.; Min, Q. Community Participation in Nature Conservation: The Chinese Experience and Its Implication to National Park Management. *Sustainability* **2020**, *12*, 4760. [CrossRef]
229. Parhusip, S.; Suharti, S.; Sukandi, T.; Amano, M. Economic Analysis of Local People's Involvement in Community-based Forest Management (CBFM) in Desa Ciomas, Indonesia. *J. For. Plan.* **2019**, *25*, 1–14. [CrossRef]
230. Brandt, R.; Staiss, C. *Enhancing the Sustainability of Agroforestry Systems Lessons Learned from Agroforestry Activities in the Districts of Malinau and Kapuas Hulu (North and West Kalimantan, Indonesia)*; Deutsche Gesellschaft für Internationale Zusammenarbeit (GIZ) GmbH, Forests and Climate Change Programme (FORCLIME): Jakarta, Indonesia, 2019.
231. Rahman, S.A.; Sunderland, T.; Kshatriya, M.; Roshetko, J.M.; Pagella, T.; Healey, J.R. Towards Productive Landscapes: Trade-Offs in Tree-Cover and Income across a Matrix of Smallholder Agricultural Land-Use Systems. *Land Use Policy* **2016**, *58*, 152–164. [CrossRef]

MDPI

Article

Analysis of Agro Alternatives to Boost Cameroon's Socio-Environmental Resilience, Sustainable Development, and Conservation of Native Forests

Tomas Gabriel Bas [1,*], Jacques Gagnon [2], Philippe Gagnon [2] and Angela Contreras [3]

1 Escuela de Ciencias Empresariales, Catholic University of the North, Antofagasta 1780000, Chile
2 École de Gestion, Université de Sherbrooke, Sherbrooke, QC J1K 2R1, Canada; jacques.gagnon@usherbrooke.ca (J.G.); philippe.gagnon.10@gmail.com (P.G.)
3 Facultad de Ciencias Agrarias y Forestales, Catholic Maule University, Talca 3480094, Chile; acontreras@ucm.cl
* Correspondence: tomas.bas@ucn.cl

Abstract: Located in Central Africa, Cameroon is a country with strong social inequalities and fragile governance and institutions. This has a direct impact on the sustainable development of its territory, communities, and native forest, which are subject to constant socio-environmental and economic pressures due to overexploitation. This research has three purposes: (1) to conduct a comparative theoretical/empirical diagnosis on the quality of Cameroon's institutional framework, governance, and public policies related to territorial sustainability; (2) to assess the impact of the three clusters identified among the 44 stakeholders interviewed (forestry companies/certifiers; NGOs/communities; and banks/public institutions) on each other; and (3) to analyze the contribution of the use of cassava (*Manihot esculenta*) as an agro alternative to Cameroon's socio-ecological resilience, sustainable development, and conservation of native forests. The research found: (1) the need for mixed governance with joint accountability to find equitable and lasting sustainable solutions for the parties involved, making communities/ethnic groups visible in the decision-making process; and (2) the agro use of cassava has a positive impact on socio-ecological resilience by contributing to employment, the protection of devastated soils, and the provision of quality food, and by reducing pollution from the cement industry through using cassava waste as an input.

Keywords: sustainable development; theoretical; governance/institutions; socio-ecological resilience; Cameroon forests

Citation: Bas, T.G.; Gagnon, J.; Gagnon, P.; Contreras, A. Analysis of Agro Alternatives to Boost Cameroon's Socio-Environmental Resilience, Sustainable Development, and Conservation of Native Forests. *Sustainability* **2022**, *14*, 8507. https://doi.org/10.3390/su14148507

Academic Editor: Victor Rolo

Received: 15 June 2022
Accepted: 5 July 2022
Published: 12 July 2022

1. Introduction

Cameroon's forests cover approximately 23 million hectares, representing almost 50% of the country's total land area [1]. Furthermore, the same authors estimate that Cameroon is home to some 8260 plant species (of which 156 are endemic) and approximately 2000 wildlife species, making it the fifth largest country in Africa in terms of biodiversity.

The country's economy is based almost exclusively on the exploitation of its natural resources, which include oil and gas, high-value timber, and non-timber forest products, as well as minerals and agricultural products such as coffee, cotton, cocoa, maize, and cassava [2,3].

Cameroon's native forests are highly threatened by human-induced deforestation and degradation with a significant annual loss of vegetation cover, which has been gradually increasing from 0.34% during the 2000–2005 period to 0.90% in the 2010–2013 period and there are no concrete indications that this will be reversed in the short to medium term [4].

Logging in Cameroon's forests is often selective and at the same time extensive [5,6]. Along these lines, it is estimated that the largest area of timber exploitation is found in the eastern and southern regions of the country [3,7,8]. According to Auzel et al. [9], this exploitation is historically concentrated on around twenty species, which account for 75% of the country's timber production. The species most exploited historically for their commercial value are: *Triplochiton scleroxylon*, *Entandrophragma cylindricum*, *Terminalia superba*, *Lophira alata*, and *Milicia excelsa* [9]. Forests, as generators of essential biodiversity, make significant contributions to the local economy, providing multiple inputs (biodiversity, timber, charcoal, medicinal plants, various foods, tourism, etc.), which in some way should contribute to the fight against extreme poverty of local communities [1].

Table 1 shows a description of the relationship between the forest area and other economic zones in Cameroon between 1999 and 2015. The forest area has decreased from 22 to 19 million hectares, reaching a deforestation rate of 1% per year, which is one of the highest deforestation rates in Central Africa. Only 10% of the total forest is protected by law [10]. On the other hand, Table 1 also shows that the area destined for agricultural exploitation has increased from 19 to 21% of the total Cameroonian productive territory.

Table 1. Description of forest area and its relationship with other economic areas in Cameroon, by observing specific years.

Year	Total Forest Area (MH) ¥	Forest as Percentage of Land Area (%)	Agricultural Land as Percentage of Land Area (%)	Forest Land Decreasing per Year (%)	Log Industry Production (m^3)	Log Industry Permanent Employees	Agricultural Permanent Employees	Total Population Permanent Employed
1999	22	47	19	1	3.0 mill	UN	UN	UN
2009	20	43	20	1	2.4 mill	49,993	26,530	859,000
2015	19	40	21	1	3.2 mill	59,067 $^{\alpha}$	57,522 $^{\alpha}$	1,183,752 $^{\alpha}$

¥: approximate values, MH: million hectares, UN: uninformed. α: value calculated by adding the new accumulated employees with the existing ones in the sector, reported by [11]. Source: [11–13].

Regarding the logging industry, the percentage of territory that practices this activity is not exactly known. Unsustainable and illegal logging has been identified as one of the main drivers of forest degradation in Cameroon. The misuse of certain logging permits in the country is frequent, due to the lack of effective regulation and law enforcement [14]. A slight increase in the production of logs has been observed, but this growth has varied over the years. However, between 2001 and 2019, Cameroon lost the equivalent of 4.2% of its tree cover, contributing 519 Mt of CO_2 emissions. During the same period, this country lost 3.2% of its primary humid forest, representing 47% of the total tree cover [15].

As shown in Figure 1, according to the African Development Bank Database [13], the production of wood for fuel has increased over the years, which would pose not only a threat to the health of native forests (Figure 1A), but also a threat to global warming through the amount of greenhouse emissions released into the atmosphere [16]. At the same time, cassava cultivation has also gradually increased, which would represent an excellent sustainable development opportunity for the country, by making use of already deforested lands with highly degraded soils to be reforested with native species (Figure 1B).

On the other hand, the amount of agricultural land exploited per hectare per inhabitant in Cameroon increased from one to four hectares per inhabitant in the 1990s, with a sharp increase between 2000 and 2005, and then stabilized at levels of 12 hectares per inhabitant on average, a volume that has been maintained to date (Figure 1C).

Forest land in Cameroon has been declining slightly over the years (Figure 1D), leaving little hope for cultural awareness and adaptation by communities and institutions to protect the country's native forest heritage in the medium to long term. However, there could be a bias in these data, due to the constant migration of peasant populations to the cities in search of job opportunities. Along the same lines, and perhaps due to the same phenomenon, employment represented by the agricultural sector has suffered a slight decrease between 1990 and 2020 (Figure 1E). The labor force working in agriculture, which had grown significantly in the 1980s and 1990s, stabilized from 2000 to the present at approximately 3.6 million inhabitants (Figure 1F).

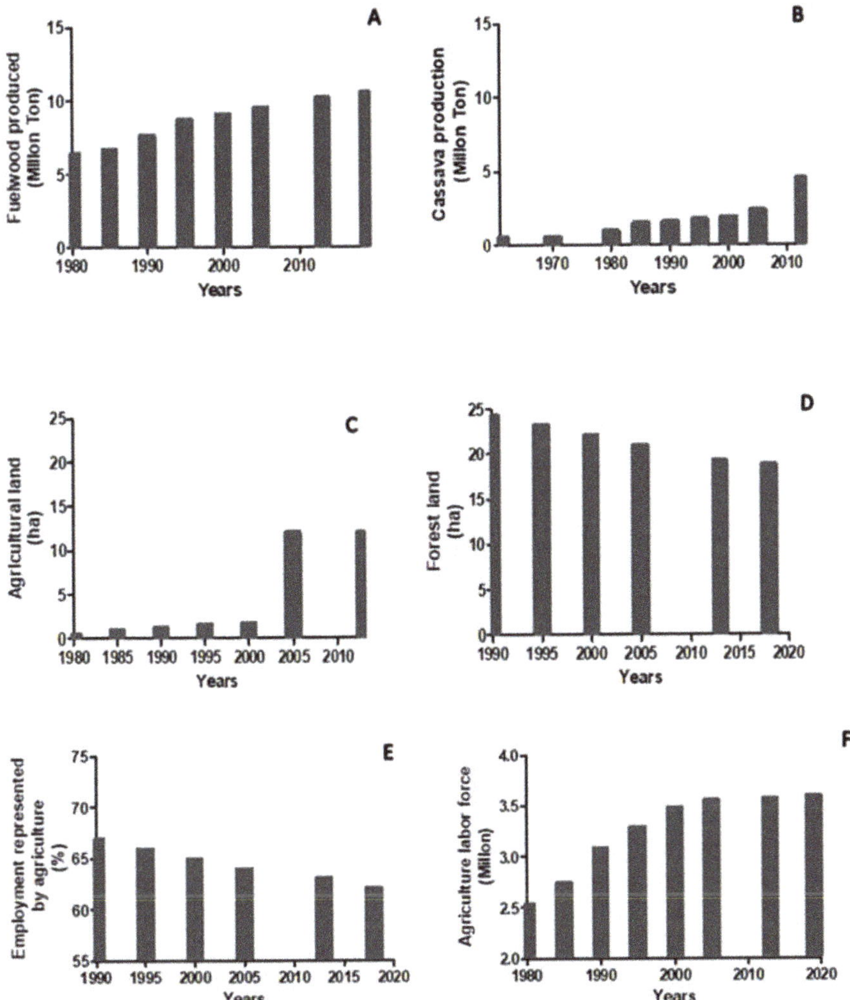

Figure 1. Socio-economic information for Cameroon from 1980 to 2020. (**A**) Fuelwood produced per year. (**B**) Cassava production. (**C**) Agricultural land used permanently. (**D**) Forest land existing in the country. (**E**) Percentage of total employment given by agriculture. (**F**) Agricultural labor force [13].

The concrete industry is among the most highly polluting in the world, using toxic and carcinogenic non-renewable oils, as well as water-soluble polymers that can be easily replaced by organic amylopectin, which is also highly soluble and can be obtained from cassava. The cement industry presents a significant challenge in terms of sustainability, due to the large amount of concrete used on the planet every day and the number of toxic agents released into the atmosphere, greatly contributing to climate change.

Cameroon, like the rest of the African countries, is experiencing significant population growth, with the consequent demand for this type of cement for the manufacture of concrete and mortars, used in the construction of houses, roads, and bridges [17,18]. Cement is currently irreplaceable in large construction sites, opening a large window of opportunity for starch from cassava waste, while enabling a large reduction in the use of its most toxic components, allowing cassava waste to be given a useful life while also generating a highly innovative and friendlier added value for the cement industry [19,20].

Cameroon needs to be analyzed because of its relevance in curbing climate change. The importance of Cameroon's forests for the biosphere is widely unknown, but it is only comparable to the forests of the Amazon and the boreal forests, which is why preserving them is an obligation and a duty for humanity [21].

This article addresses Cameroon's challenge in relation to the sustainable development of its territory and the conservation of its native forests and its rich ecosystem, through alternatives and strategies that help to take pressure off these forests, favoring socio-ecological resilience and mitigating negative impacts based on the opportunity cost. This can be achieved by implementing sustainable growth alternatives that bring together all the parties involved and where cassava (*Manihot esculenta*) cultivation can play a major role, with the multiple uses of this tuber as food, fertilizer, soil protection, and a supply—through its waste—for a more sustainable cement industry. In fact, it would go a long way towards showing the benefits of implementing a circular economy. A study conducted by Spreafico [22] analyzed some common design strategies to set up circular economy options (i.e., design for reducing wastes, design for using renewable energies, design for reuse, design for remanufacturing, design for recycling, design for energy recovery, design for disposal, and design for recovering energy from waste) and their corresponding impact reductions, concluding that it is useful to research the relationships between the CE options and the types of environmental impacts.

2. Materials and Methods

The research intended to answer two questions: (1) What would be the best strategy capable of generating sustainable development of Cameroon's native forests, considering the interaction of very critical aspects such as socio-cultural, economic, agricultural, and institutional and governance aspects? and (2) How could cassava (*Manihot esculenta*) cultivation boost socio-ecological resilience in Cameroon through a specific opportunity cost, thus enabling the conservation of native forests and sustainable local development, and, at the same time, giving life to an industry as polluting as cement?

Two different methodological techniques were used: one for the primary source in the field and the other for the secondary sources of referential information. The aim was to contrast and strengthen the study and, thus, give greater validation to the empirical results.

The primary source data collection consisted of a series of formal on-site interviews with 44 of the most significant stakeholders willing to participate (from 82 potential stakeholders contacted). They make up the Cameroonian decision-making ecosystem in terms of their hierarchical, social, political, business, and ethnic positions related to the object of study. The qualitative information obtained from the interviews was processed as follows: full transcription of the interviewees' statements, coding, and manual categorization.

The study model used was that of "grounded theory" described by Penalva et al. [23]. This is a method that allows for the discovery of theories, concepts, hypotheses, and propositions based directly on the data, and not on a priori assumptions, other research, or existing theoretical frameworks. The interviews were unstructured (open-ended questions) and a latent content analysis was conducted [24]. The data coding was performed manually and divided by themes to avoid leaving out any of the opinions raised during the interviews. The interviews revealed a saturation of themes by equivalence, i.e., it was observed that the different stakeholders had similar opinions on the vast majority of the topics discussed, regardless of the sector to which they belonged.

It is important to note the degree of difficulty in carrying out this activity, due to distances, in many cases in unsafe situations, and the complexity of establishing contact with the different groups interviewed. The profiles of the 44 interviewees were distributed among representatives of the eight sectors identified as the most important involved in the sustainability of Cameroon's native forests (Table 2).

Table 2. Profile of the 44 stakeholders interviewed.

NGO	Banks Sponsors	Forestry Companies	Collective Communities	Political Stakeholders	Political Employees	Others	Direction
6 NGO Directors	1 Fund manager and funder	1 Forest operator	1 Chief Bantou	1 Official of the Ministry of Forestry	1 Departmental delegate	1 Delegate of a community forest	1 Regional Director of company
3 NGO Representatives	2 Funder representative	1 Company president	2 Village chiefs (Baka community)	1 Departmental delegate	1 Sub-prefect	1 Forestry Consultant—Former Forester	2 Site Director, Group
1 NGO Forest-Environment advisor	1 Funder	3 Company	1 Community Forest in the central region			1 Head of certification office	
	1 Project coordinator and funder	1 Managing Director of the company	1 Community Forest in the southern region (village chiefs)			1 Certification office coordinator	
		1 Internal certification director of the company	3 Community Forest in the south-east region (Chief, president of the GIC, advisor)			1 Model forests	
		1 Managing Director of the company	1 Community Forest of the southern region (Member of the GIC)				
10	5	8	9	2	2	5	3
44 Stakeholders							

N = 44 stakeholders (82 potential stakeholders were contacted, of which 44 finally agreed to be interviewed in person).

These instruments made it possible to define components of Cameroon's organizational culture, as well as the organizational structure of companies involved in logging and timber certification, their management style, and the vision of the role of government and institutions in relation to environmental protection, in terms of the intervention of companies responsible for the exploitation of the country's forest resources. The data coding and analysis were carried out through interview summary sheets and collected notes that helped to enhance the understanding, perception, and views of each representative interviewed. These tools were coded, as mentioned above, in a sequential and iterative manner. Finally, a cross-sectoral analysis was carried out, taking into account stakeholders and elements of institutional culture. To that end, meetings were held with stakeholders involved in decision-making throughout the research process. Finally, a database was created to describe the different organizations and their objectives for the sustainable development of the territory, the communities involved, and the Cameroonian forests, from the point of view of the stakeholders directly linked to the local ecosystem.

The second methodological technique was based on the exploration of secondary data known as qualitative documentary, which involved the selection, compilation, and review of scientific literature through reading and critiquing the theoretical and bibliographical instruments found according to the "Boolean" operators. These seek to interpret an experience related to predetermined keywords from different sources of information and are characterized by an interpretative approach to the findings and gaps of the research itself [25,26].

To make the search effective, documentary sources or bibliographic databases were chosen by discipline or subject: Science Direct, Agricola, Compendex, Derwent, Statistics Canada, Research Gate, Scopus, Web of Science, Innovation Index, and GeoIndex. These were combined to counteract the degree of data overlap that might exist between them.

The search was carried out based on the selection profile by means of relations between descriptors, using logical or "Boolean" operators (AND, OR). The bibliographic search was structured considering a degree of sensitivity (recovery rate) and specificity (precision rate) based on the definition of the construction of the research questions and objectives, as well as the interviews with the 44 stakeholders to delimit the search for information [26]. Sources were selected and organized through the identification of the following keywords: Theoretical Framework; Native Forest; Conservation; Sustainable; Local Development; Governance and Institutionality; Cassava and Agriculture; and finally, Cement Engineering. Complementary words of the same root as the mother words were added to them, to help understand and explain the dynamics related to the sustainable local development of the territory and the theoretical tools and alternatives of empirical analysis for socio-ecological resilience (use of cassava and its waste in the cement industry) and the conservation of the native forest in Cameroon. Based on this logic, we tried to cover as broad a spectrum as possible to bridge the gap between what is available as secondary information and what was discovered through field interviews.

The empirical analysis was contrasted with the theoretical documentary information, in order to validate the cultural, environmental, institutional, and governance aspects, through the broadest participation of key stakeholders available. To do this, a causal model was used, which is a method that consists of establishing causal links or interactions between different phenomena through a formal process that uses evidence to infer causal relationships or link effects and causes, in this case, the relationship between organizational culture and all the aspects analyzed in the theoretical discussion, in terms of sustainable development [27].

This model is based on contingency theory, which is an approach to the study of organizational behavior that explains how contingent stakeholders such as culture, environment, institutional framework, and governance, among others, influence the design and function of organizations as a social structure [28]. The underlying assumption of contingency theory is that no one type of organizational structure, or another, is equally replicable to all others. Organizational and institutional effectiveness depends on a fit or match between the volatility of the environment and the size and function of the organization in which it is located. The aim is to look for patterns in order to try to understand, rather than predict, the specific political, institutional, and community realities in Cameroon, i.e., how one variable affects another and how they are all under the same institutional umbrella. Eisenhardt's [29] recommendations to triangulate data collection were followed to increase the robustness of the research.

3. Results and Discussion

3.1. Theoretical-Bibliographic Review

The information need chosen for the literature search strategy is expressed in a word or short phrase (e.g., sustainable development). Some of the concepts are split in half to have more research scope, "Forests and climate change" into "Forests" and "Change". We used the words that came up most in the interviews to find keywords for each of the required concepts, as well as using various types of documents (dictionaries, encyclopedias, and well-known articles) to further enrich the search. All words or expressions of a concept are synonyms or quasi-synonyms and, as necessary, there have been cross-references of related words. We have used Boolean operators and parentheses, e.g., (word1 * OR word2 * OR ...) AND (word7 * OR ...). In addition, exact expressions ("xx") and appropriate field codes were employed (e.g., in Compendex > WN KY). The "{English} WN LA" was also used, but also literature in Spanish and French, which finally yielded more than 9800 results. Subsequently, to optimize the search strategy, different options were tested, eliminating some keywords that were too generic or not directly related to the search. It was found that the number of results was too high, so, in order to narrow down the search more specifically, the abstracts and titles that matched the object of research were used. Based on the analysis of the results obtained (reading the titles and detailed references, if necessary), we were able to analyze and classify the different articles, considering those that came closest to our research questions, and then we dynamically modified the selection strategy until we reached 217 results, which were considered sufficient to classify and consult. The time window selected was from 2001 to 2021, mainly due to the need to have current data related to the object of the search. In addition, a time window was opened between 1989 and 2000 to include articles that could be classics in the subject analyzed or of great timeless contribution, and even exceptional articles from previous years if required.

3.1.1. Sustainable Development

The concept of "sustainable development" is relatively complex, due to its breadth and lack of consensus among researchers on both the form and substance of what should or should not be considered sustainable. The roots of this more contemporary concept date back to the 1950s [30] with the awareness of socio-environmental problems, in relation to the increasingly abrupt changes caused in the environment, which are manifested in different ecosystems damaged by human intervention and even by nature itself or the combination of both [31–35]. Later, Carson [36] and Meadows et al. [37] emphasized the footprint of human intervention on the biosphere and its various impacts on people's quality of life and the accelerated deterioration of nature in general, pointing out the limits imposed by natural resources as finite in regards to socio-ecological intervention.

On the other hand, the concept of "sustainable territorial development", as explained by Ashraf et al. [38] seeks a convergence towards concepts such as ecological capacity/carrying capacity, natural resource/environment, biosphere, environmental technology, and zero/slow growth, which were already used before the term sustainability or sustainable development became more popular. Thus, the concept of sustainable territorial development seems to be a bold approach, basically because it would only remain a grand theoretical principle that adheres little to the actual practices of the organizational and social environment and, therefore, there is no clear definition that brings together all streams of thought on the subject [39–42].

Goslinga et al. [43] established what they call "critical habitat" to designate the massive deterioration of different ecosystems that represent areas of high biodiversity value and that have significant importance for different species and strongly threatened or unique spaces. Depending on how well natural resources are managed, this will have a transformative impact, which may be positive and/or negative, on the development trajectory of a nation and its society [44]. Figure 2 shows the close interaction between natural resources and the environment.

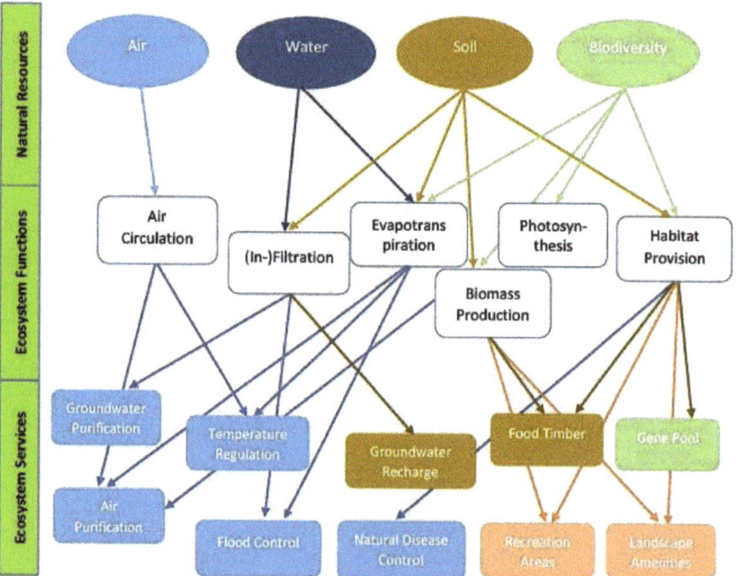

Figure 2. The relationships between natural resources, ecosystem functions, and ecosystem services. Ecosystem services are color-coded by category: blue for regulating services; brown for provisioning services; rose for cultural services; green for supporting services. Source: [45].

However, several authors agree that the components of sustainability are economy, society, and environment, which together work towards the long-term prosperity of territories and their communities [33,46,47]. From this trilogy, terms such as circular economy, social innovation, green innovation, eco-innovation, green consumption, among others, have been incorporated, which gives an idea of the plethora of interpretations that accompanies sustainable development [33,47–59].

Crucially, many authors are inclined to give superlative weight to economic development, to the detriment of the environment and social and environmental development [38,60]. In this regard, Kjellén's proposal of a "diamond of sustainability" [61,62] provides more dynamic links. Kjellén removes the environment from the classical paternalistic and hegemonic views of all natural and non-natural, physical, and biological stakeholders, including society and its culture, emancipating natural resources from it. All these dimensions are directly related to each other, and the most sustainable and equitable development, both of different societies and of their territories, often depends closely on their positive interaction, within the same plane of transcendence.

Figure 3 shows how natural resources occupy an elementary place in the diamond, already separated from the environment, although on the same plane as a value structure [63]. Then, there are energy, public policy, economics, technology, and public health, with their respective dependencies such as lifestyle, land, food, water, development, employability, and poverty. While they appear to be independent stakeholders at the level, they are all interconnected and ancillary to each other, in the manner of a complex ecosystem.

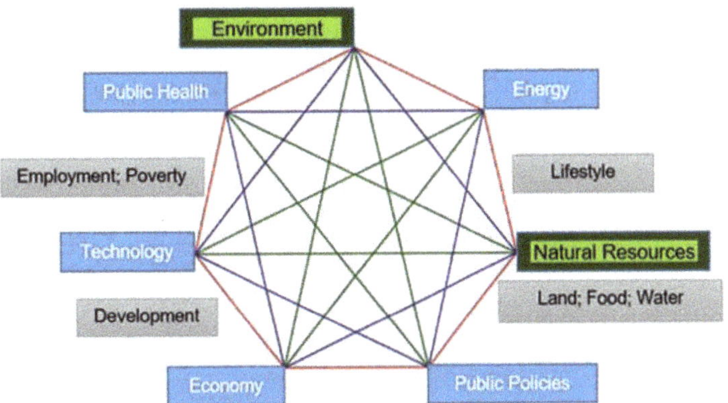

Figure 3. The Kjellén diamond of sustainability. Source: [61].

Jovovic et al. [33] explore models related to sustainability from another perspective, based on the three classic dimensions involved in sustainable development: economic, environmental, and social, which form the vertices of a triangle as can be seen in Figure 4. Each vertex involves finding answers that balance the three dimensions mentioned, bringing to the surface issues such as social and political impacts, which are often overlooked in the most common sustainable development design processes. In this model, the interaction between the three components can be integrated with an intermediate "sustainability zone" that recognizes the interdependence of biological, economic, and social systems. Through these three zones or vertices, the aim is for all parties involved to gain benefits, in the style of a virtuous winner/winner circle, where optimized returns at all three vertices are achieved through eco-resilience, allowing for greater sustainability while improving the lives of communities.

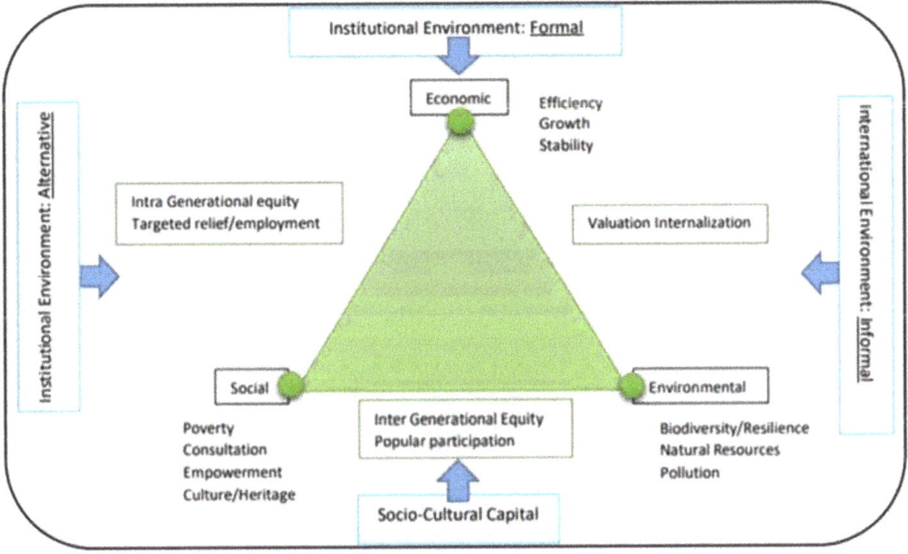

Figure 4. Munasinghe's approach to sustainable development. Source: [33].

3.1.2. Governance

The term "governance" is often used in reference to the formal and informal interactions between different levels of government (whether local, regional, or national), individuals, associations, and organizations, with the aim of reducing potentially existing gaps [64]. However, for some years now, there has been an intellectual debate, in reference to the integral development of society, to understand governance from a globalized, interdisciplinary, and multifaceted world, supported by a multitude of criteria, and no longer focused on previous traditional hierarchical governance models.

Today, the aim is to structure governance in a more participatory and interactive way between the various stakeholders that make up an environmental ecosystem and its community, as these not only have an impact at the local level, but also at the global level and, therefore, require simultaneous solutions at different scales [65–68]. The basis of good governance is provided by a platform of strong institutions, enabling countries that enjoy them to narrow the gaps between different stakeholders and develop harmoniously over time [68].

According to Kardos [69] and Ndeh [70], good governance promotes, among other things, transparency, timely accountability, quality education that makes society prosper, more effective management of resources (human, natural, economic, and financial), and the participation of civil society in decision-making, so that governance and sustainable development should be closely linked, enabling better development of society as a whole and the sustainability of the local territory for present and future generations. On the contrary, poor governance is often associated with corruption, cronyism, clientelism, and various forms of favoritism, lack of quality education, accompanied by a poor or even absent institutional framework, which leads to a lack of effective control mechanisms that in turn regulate investments and the distribution of the benefits or profits generated in this territory that have been left by the state, and can even jeopardize the freedom of the communities involved and democracy itself [70–74].

Governance, seen from a more traditional hierarchical perspective, regardless of whether it is a developed or developing country, has similar historical characteristics in terms of the verticality of its mandate, but not in terms of its institutional capacity. This consists of top-down decision-making, headed by a high-ranking public authority, without the participation of private or mixed organizations in the adoption of decisions, which clearly reveals serious shortcomings, especially in developing countries, with weaker and in many cases almost non-existent institutions.

The incremental incorporation of governance concepts according to the need for progress is more inclusive and takes into account the different ambivalences that can arise, the uncertainty and power manifested through the evolution of society, aiming for change, based on the analysis of the socio-ecological and socio-economic systems present in any community and which are particular to each society [75]. Because of this, since the end of the 20th century, mixed systems of governance have been emerging, which include the environment in decision-making, represented in governmental (public), non-governmental (private), and mixed entities, as one of the solutions to the growing complexity of institutional variables manifested at the beginning of this century [76].

However, despite this evolution, Ndeh [70] believes that Cameroon still suffers from a very marked unidirectional public governance that has been carried over from the colonial era, added to the complications caused by its heterogeneous Anglo-French culture with its historical differences and more than 250 ethnic groups, which further fragment equitable decision-making that allows for the prosperity and sustainability of Cameroonian society as a whole. In addition to this, and the weakness of its institutions, there is a kind of clientelist government that further complicates the process of governance in this country. According to Turner [77] and MacCormick [78], sometimes governance can directly influence culture, while at other times the reverse is true, which can lead to poor organizational culture, leading to poor governance, leading to poor decision-making. Culture, as well as organizational culture, is often presented as intangible elements until they are confronted [79].

In this regard, Cameroon is a country immersed in complex social, economic, and environmental interests and, at the same time, with a very pronounced deficit at the educational level, with large lines of corruption, which could be corrected with "mixed" governance systems that include governmental and non-governmental stakeholders, through the local civil community, trade union members, NGOs, and self-monitoring and auditing companies [67]. This could also strengthen the country's institutions in general and give them greater credibility in the eyes of the public. On the other hand, there is evidence that from this innovative concept of mixed governance, understanding has been gained about different formal and informal components, which often directly affect environmental and forest governance in these countries [80,81]. Many organizations involved in environmental governance often have to cope with and manage multiple variables, related to social constraints and expectations of the various stakeholders involved, usually due to a lack of competence or capacities of the ones holding responsibility [64,82].

It is essential to highlight the importance of forest resource governance in a country like Cameroon, whose GDP is based on the exploitation of natural resources and particularly forest resources, so it is important to consider mixed participation in its governance. In this regard, for Adeyeye et al. [83], it is key to ensure the participation of indigenous communities or ethnic minorities and greater participation of women in environmental governance, not only for ethical reasons, but also because of the knowledge and interests of the region's inhabitants, who should be able to express themselves on the present and future of the forest ecosystem.

A closer look at the state of the tropical forests of Cameroon reveals that, for several decades now, numerous legal reflections have been performed, mainly involving the decentralization and reform of the forestry regime in the tropics. The objective is to create favorable conditions for the sustainable management of forests through the stakeholders directly involved, including of a social, private, public, and mixed nature, from the various local communities [68,84,85]; at present, some theoretical and empirical research suggests that collaboration and participation of communities, especially regarding local residents, are essential for the design and implementation of a successful and sustainable long-term development plan, where organizational culture is part of the equation. In that regard, Cameron and Quinn [79] developed six key elements related to the development of organizational culture, such as: most dominant characteristics of culture; organizational leadership; human capital management; organizational "glue"; strategic emphases; and success criteria. Moreover, a key element is the enhancement of local capacities to implement and manage initiatives through a supportive network of regional economic development organizations, as well as environmental protection, with a focus on greater social equity objectives [86–88].

Public policies play a crucial role in the decisions taken to protect the different communities involved, especially the most vulnerable ones, without forgetting the impact that climate change will have on them [1]. Cohen and Levinthal [89] argue that public science and technology policies should not only generate capacities within a nation, territory, or region at different levels of resource management, but should also make it possible to put these capacities at the service of organizations and society as a whole. These economies, much more convergent at the institutional level, are the ones that place a greater emphasis on value and knowledge creation and, therefore, on innovation and technological development, which is reflected in the high economic and financial performance of developed nations, also called knowledge-based economies [90,91].

3.1.3. The Natural Resource "Curse" and Socio-Ecological Resilience

Local sustainable development is specific to each region and society, its ecosystem, its culture, and its needs, and is often dependent on existing local knowledge [92]. Countries that base their economies exclusively on natural resources are, paradoxically, often at a disadvantage compared to those that lack them or do not base their economies on their exploitation.

This phenomenon is reflected in the case of the countries known as the "Asian tigers" (Singapore, Hong Kong, Taiwan, and South Korea), which lack significant natural resources [93], but which developed their industries and technology rapidly between 1945 and 1990, leading them to thrive in the knowledge-based economy [94]. Similarly, countries such as the United States, Canada, Australia, and Finland have an abundance of natural resources, but their economies were not based exclusively on the exploitation of these resources, but rather diversified, underpinning them with a more knowledge-based economy, managing to industrialize and develop to become world technological leaders. The key to the development of these countries lies in the quality of the institutions responsible for the governance of these nations [95].

In contrast, numerous authors elaborate on the negative spillover effects that are often caused by abundant natural resources, precisely because of the poor quality of the institutions responsible for their management [96–98]. At the same time, several authors have observed the existence of a higher degree of corruption in these developing countries. Moreover, revenues from natural resources can fluctuate considerably from year to year, depending on many uncontrollable variables, such as the volatility of international commodity prices, the abundance or scarcity of a given resource, climatic variations, natural or man-made disasters, and even pandemics such as COVID-19, among others [94,96,98].

In this regard, it is essential to integrate socio-ecological resilience, which will make it possible to recover the loss of environmental balance caused by the permanent disturbance to which the fragile biodiversity of Cameroon's forests and lands has been subjected for decades [99,100].

The theory of resilience was developed in the early 1950s by physicists, engineers, and mathematicians to describe a dynamic event that started with an object in equilibrium, which was subjected to a perturbation that removed it from that equilibrium phase, and then the time elapsed was measured until it returned to its original equilibrium [101]. Later on, this concept underwent some changes in its interpretation and management, and was used by ecologists to explain adaptive capacity, in which an environmental ecosystem is capable of absorbing a change, maintaining a similar structure, or even returning to its original carrying capacity, after having suffered several major disturbances [102,103]. Along the same lines, González-Quintero and Avila-Foucatla [104] define socio-ecological resilience as the capacity of a system to absorb a given disturbance and then reorganize itself again by undergoing a change that allows it to continue conserving essentially the same function, structure, and identity (Figure 5).

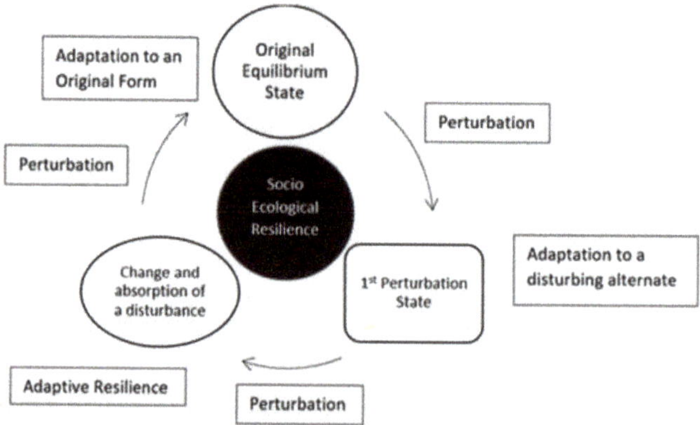

Figure 5. Socio-ecological resilience. Source: [104].

This theory of resilience has evolved in its interpretation. At the beginning, it was considered to be a transforming entity, then over time it moved away from this characteristic and towards a description of two cycles; one of them was known as basic and the other as adaptive, the latter being the most widely used due to its flexibility in the continuous change of socio-ecological systems [102]. According to Heslinga et al. [105], they believe that the concept of resilience is particularly appropriate for socio-ecological systems because both concepts are at the interface of human and natural processes in time and space, and share an interest in the protection, management, and planning of a given potentially disturbed area. On the other hand, it determines the capacity to adapt to and benefit from change. Resilience, from a community perspective, reflects the ability of the community to adapt to an uncertain environment, being able to respond to adversity through the cooperation of governors and the entire community [106]. However, it is important to value social, political, environmental, and biodiversity agreements, which give legitimacy to the success of the initiative, in order to better plan for more sustainable and long-lasting development [107].

3.1.4. The Opportunity Cost of Cassava

A viable proposal to the overexploitation of Cameroon's native forests is to take advantage of the opportunity cost presented through the socio-ecological resilience of cross-cutting sustainability, managing the cultivation of cassava (*Manihot esculenta*), which is very rich in starch as food and whose waste, equally rich in the glucidcarbohydrate binder, can be used as a component for the cement industry, giving concrete greater strength and structural stability.

Cassava (*Manihot esculenta crantz*, Euphorbiaceae) is a crop that was apparently domesticated in South America's central Amazon approximately 8000–10,000 years ago, being then brought to Africa by Portuguese explorers about five centuries ago [108,109]. This vegetable, a woody tuber, is widely used in the diets of local African communities, to the extent that it provides 30% of the calories consumed by them, making it the most important of all root and tuber crops as a source of calories, but it also ranks fourth in importance in terms of the number of hectares planted, after rice, sugar cane, and maize [110]. More than 800 million people (approximately 11.5% of the world's population) depend on it as an important source of carbohydrates, and it is also one of the products with the highest energy yield per hectare cultivated, in which 100% of the plant is edible, with up to 80% of the total available starch being concentrated in its roots [111–114]. Numerous studies refer to the advantages that the consumption of a starch-rich diet confers to health in general and to the heart in particular, making it a direct benefit for communities to have this important resource [115,116]. In addition, cassava roots are consumed boiled or processed into other products and byproducts, such as flour, modified cassava flour (mocaf), cassava rice, in chips, gari and fufu water, in cassava stick, and in "mitumba", while the leaves of the plant are consumed as vegetables.

While it is not the largest producer of cassava in Africa, Cameroon currently cultivates approximately 341,000 hectares of this fast-growing tuber, which leads the country to produce about 5.4 million tons per year (Table 3), with an estimated revenue of about USD 700 million [117]. Each Cameroonian household consumes approximately 75 kg of cassava per year, representing more than 2.5 million tons of this vegetable annually [118–121]. In other words, Cameroon currently produces twice as much cassava as its population consumes annually, where its use is almost entirely for food and feed only.

Table 3. Production in tonnes of cassava in the World, in Africa, and in Cameroon 2015–2018 and their respective shares [117].

Year	World	Africa	% World	Cameroon	% World	% of Africa
2015	277,072,000	152,822,000	3.27	5,000,000	1.8	3.27
2016	276,510,000	155,607,000	3.32	5,170,000	1.87	3.32
2017	275,655,000	157,453,000	3.39	5,799,000	1.94	3.42
2018 (estimated)	277,070,000	160,730,000	3.36	5,400,000	1.95	3.36

Figure 6 shows how the global cassava market has grown steadily since 1970, despite a virus disease that has been stalking this crop since 2013 and is known as "cassava brown streak disease" [122]. This disease is characteristically found in sub-Saharan Africa and not in Latin America, so germplasm from Colombia is being used to try to generate greater genetic diversity in plants to combat this virus, which causes chlorosis of the veins, streaks on leaves and stems, and necrosis of the roots, making them inedible, with the consequent loss of harvests and socio-economic damage [123]. Establishing an accurate and early sanitary diagnosis of the fields affected by the virus is very costly, so through technological innovation, stochastic epidemiological modelling and simulations are carried out to reduce these costs and be more accurate in the diagnosis and final intervention, in order to reduce the incidence of the virus and potential losses [124,125].

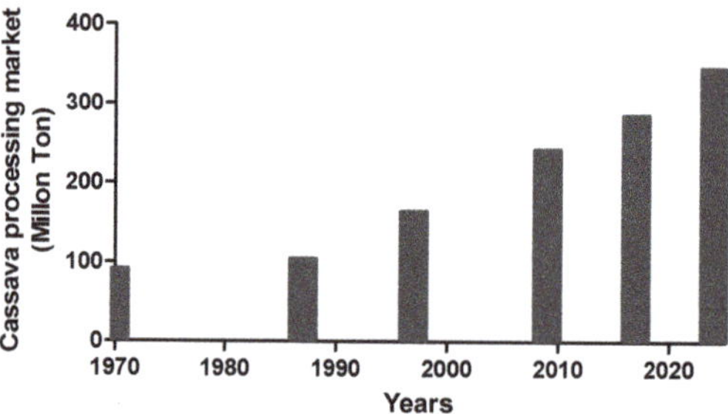

Figure 6. Global cassava market recorded over 54 years, taking into account the demand for processed cassava product. Information obtained from [126–128].

Despite this setback, the global cassava market grew from 91 million tons in 1970 to 277 million tons in 2018 and the market volume is expected to exceed 345 million tons by 2024 in line with the progressive population increase in Cameroon and Africa in general.

Moreover, Cameroon's climate is highly conducive to the cultivation of this tuber, which favors its production throughout the year in the country's five forest zones [129]. It is also important to note that cassava is a hardy plant, capable of withstanding all kinds of inclemency, to the extent that it has a high tolerance to extreme soil conditions and drought, all of them important attributes for it to be chosen as a circular alternative for the socio-ecological resilience of Cameroon's native forests and to cope with the climate change being experienced [111,130,131]. Thus, cassava opens the door to a more sustainable use of the local territory through lands that have already been deforested and their soils devastated or impoverished by highly nutrient-demanding crops and by the action of water and wind erosion that occurs when these soils are stripped of their original natural vegetation cover and are practically irrecoverable to their original rich condition, due to the damaging effects on both the structure and quality of the soil [132]. This means that they are often no longer suitable for reforestation with native trees, which require specific nutrients and particular biodiversity [133].

According to Tieguhong et al. [3], production forests in Cameroon are exploited and managed under four models of forest management and partnership arrangements (Table 4). These four production management models cover 47.04% of Cameroon's forests, representing 23% of the total territory [134]. There are 111 forest management units in Cameroon, covering a total of 7,058,958 ha, where 66% (5,071,000 ha) are managed concessions and 34% (2,393,061 ha) are certified [3,135].

In terms of forest production, efforts are focused on stabilizing the volumes of timber harvested and developing non-timber forest products. In addition, the state promotes projects that consider: (1) the transformation of forestry waste into energy; (2) the application of methods and technologies for the appropriate use of firewood; and (3) the implementation of a development plan for the production of starchy maize and cassava.

It is hoped that through these actions, growing nutritional needs can be met, as well as the development of new productive and industrial niches that will benefit the economy of the country and its regions, while at the same time taking pressure off native forests at all levels.

Table 4. Production management models in Cameroon and their respective shares [3].

Production Management Model	Share of Total Forests	Type of Partnership
Forest concessions/forest management units	30.64%	Government in partnership with the private timber sector
Community forests	8.04%	Government in partnership with local communities
Council forests	6.71%	Government in partnership with local councils
Sale of standing volumes	1.65%	Government in partnership with the private sector

Source: [3].

Cameroon's strategy for sustainable development and poverty reduction in the future includes increasing agricultural production and the area of land used for this purpose by about 30% compared to 2005, in order to ensure food production, strengthen growth and employment in the agricultural sector, and thus reduce pressure on native forests and their communities [94] (GESP, 2009). This plan is implemented through two main actions: (1) Promoting medium- and large-scale farms by facilitating access to farmland; and (2) Encouraging the creation of groups and institute synergies of family businesses in the form of cooperative societies. The latter with the support of the state, including:

- Access to new production techniques through mechanization and agricultural consultancies.
- Access to agricultural credit with the opening of micro-finance establishments and banks interested in this specialized line of credit.
- Access to markets through improved marketing channels and support for the promotion of Cameroonian agricultural products (such as yucca and maize) on the international market.

The overall plan considers special support for youth development in rural areas of Cameroon and the implementation of development programs related to production in forested areas [94] (GESP, 2009), which would greatly open the doors to sustainable agroforestry, using rather "rustic" crops in terms of nutritional requirements at the soil level, as is the case of the cassava (*Manihot esculenta crantz*, Euphorbiaceae).

This activity is closely related to the "opportunity cost" [17,136], which means that while cultivating cassava on land that has already been ravaged and almost deserted by erosion, it also prevents further degradation, while at the same time facilitating the conservation of forests that have not yet been invaded by human activity, ensuring that they remain without or with very little human intervention. This is presented with a view to protecting the country's native primary forests and helping to combat climate change, preserving the atmosphere from large quantities of CO_2 emissions, which is fundamental to achieving the socio-ecological resilience of a forest that is emblematic for the planet. The analysis of all these variables leads us to reflect on cassava as an excellent resource to be considered as a key factor in cross-sustainable development, while raising awareness of the importance of these forests in the ecosystem, not only locally, but globally.

On the other hand, cassava is a very versatile crop, in terms of the multiple planting systems it can support at small, medium, and large scales, making it an important source of employment for local villages that are involved in the cultivation, harvesting, and processing of this plant, which is used for different value-added sub-products in diverse industries [120,137]. Indeed, in addition to feeding the population and its animal production, cassava can also provide economic income through the generation and sale of starch for industrial use from its waste. This industrial starch has an enormous binding capacity, and this is where one of the interests in this research lies, particularly as a specific component in the cement industry, in the relentless pursuit of generating greener and more sustainable concrete production in the medium and long terms [109,138,139]. Thus, the incorporation of starch obtained from cassava residues as a more sustainable input in the manufacture of a more environmentally friendly Portland cement is proposed.

This type of starch contains two categories of macromolecules, called linear amylose and branched amylopectin, which consist of glucose chains [140]. Branched amylopectin accounts for 83% of the carbohydrate content present in cassava roots, while linear amylose occurs in a much smaller percentage, approximately 17% [112]. It is important to bear in mind that the thickening action of starches and their derivatives, as well as their hardness upon solidification, must be analyzed taking into account the relationship between these two macromolecules. The properties of linear amylose are as a thickener, water binder, emulsion stabilizer and gelling agent, while branched amylopectin is the one that constitutes a higher percentage of starch and is highly soluble [112,141].

The ratio of linear amylose to branched amylopectin in cassava yields a starch with higher hardness and strength, which is often used as a biomaterial adjuvant in many industries such as adhesives and in the construction of wood laminates, to provide greater strength in the processing of such laminates, as well as giving them considerable fire resistance. However, the main interest in this research lies in its use in the cement industry, due to its physicochemical characteristics as a binder in concrete when water is added to the mortar, as well as allowing the mixture to considerably lower its temperature during setting [141–145].

This glucose chain (amylose and amylopectin) that makes up cassava starch has been shown to give the cement mix greater strength and flexibility compared to mortars lacking this substance [146,147]. It also improves the durability properties of concrete when exposed to chloride-rich, sulphate-rich, or alkaline environments [140].

The resistance of cement to chemical attack, moisture, abrasion, and other deterioration effects is very important for the life span of the structure in a building. Chemical admixtures are often used to increase the durability of concrete. Examples of such chemicals are non-renewable oil-based products such as polynaphthalene sulphonate (PNS), polycarboxylate (PC) and polyacrylate (PA), and particulate matter (PM), which are not only toxic to health and the environment in general but are also extremely expensive. They all contain formaldehyde, which can have carcinogenic effects on humans [19,141,148–150]. Water is another resource that is generally affected with different concentrations of sulphates and toxic metals such as zinc, lead, and chromium, which can even reach groundwater, as well as rivers and lakes [151]. The level of pollution generated by the cement industry is such that, to obtain one ton of cement, the equivalent of 0.8–1 tons of CO_2 is emitted into the atmosphere, contributing greatly to greenhouse gas emissions and climate change [152].

Because of this, the use of organic additives such as starch from cassava has strongly emerged as a great eco-sustainable green engineering option. This material is renewable, thus contributing to environmental sustainability, has a good amalgamation capacity due to its high solubility, and confers higher hardness and compressive, tensile, and flexural strength than samples lacking it [19,148,149,153–158].

3.2. Empirical Study

The different postulates and interdisciplinary foundations related to sustainable territorial development, addressed in this article from a theoretical perspective, such as governance, institutionalism, critical habitat, the four pillars of sustainability, the curse of natural resources, opportunity costs, as well as the diamond of sustainability, including natural capital and socio-ecological resilience, show a directional convergence with the empirical results obtained in this research [33,61,99,100,104,159,160]. An interesting contribution of this article is that the different variables are addressed for the first time all together to explicitly explain the lack of concrete actions towards local sustainability in the Cameroonian territory.

In this regard, and in order to enrich the discussion based on interviews with 44 stakeholders from sectors directly or indirectly involved in the management, coexistence, and exploitation of the native forest in Cameroon, at least three distinct groups or clusters were identified with different commitments to the sustainability of forest territories. On the one hand, there are the forestry and related companies, together with their certifiers, which make up what we call cluster 1, which have an evidently more economic vision and whose organizational culture seeks to maximize profits and minimize losses, without considering aspects unrelated to this [79,161,162], on the other hand, a second cluster (cluster 2) was formed by local communities, NGOs, and mixed (public/private) partner institutions, which have a greater sensitivity and awareness of the dangers of forest overexploitation and the degradation of biodiversity and soils [14,32,67,163–165]. Finally, a third group (cluster 3) is identified, although it is not as decisive due to its intermediate characteristics, made up of banks and the government itself, with views and positions that are both divergent and convergent in relation to the two preceding clusters [87,88].

The interviews revealed that the main objective of 80% of Cameroonian forestry companies is to exploit the forests to the maximum of their capacity, seeking to extract all the resources available to them, in terms of timber, for as long as possible and at the lowest possible cost. This occurs without considering the consequences of their activities, which impact the sustainability of the local territory and its communities in the medium and long term, as they are not concerned with a balance between their development and the extraction capacity of the available natural resources [35], as well as the socio-ecological resilience of the exploited territories [102], given that natural resources are limited despite their abundance, which would justify the natural resource curse theory [99–101,160]. This way of acting on the part of forestry companies in their exploitation of allocated forests, even if legal, shows a flagrant lack of ethics, but also and most seriously, it exposes once again the lack of comprehensive public policies to warn of potential abuses by companies and thus put a limit to what is happening in Cameroon with the annual loss of almost 1% of native forests. However, this is very difficult to materialize in the short and medium term because of a very top-down, inefficient, and incompetent governance and institutionality, with bad practices rooted from colonial times [1,34,64,69,70,75,84,166]. However, there are some signs of awareness of the importance of migrating from the current top-down governance system towards a mixed system, including reflexive and arguably even environmental governance [37,64,80–82,167,168].

Along the same vein, forestry companies show a worrying lack of commitment to the land they exploit and the communities that live there, hiding behind existing government regulations, many of which are flawed by a certain degree of corruption and social and skill or capacity limitations of those with responsibilities over the various affected communities [64,82,169]. Clearly, the actions of forestry companies are not illegal, as existing regulations for concessions and exploitation of forest territories support them [8]. However, these same authors believe that some international private sector forestry policies may not be the best option for the protection of the Cameroonian forest ecosystem, because they are concentrated in the hands of large forestry corporations to the detriment of small forestry corporations and the communities living there, where community participation in decisions that concern them is essential [68,83]. In doing so, these corporations show a lack of sensitivity and empathy towards the ecosystem and local communities, who are often harmed by their abusive behavior, which, curiously, on many occasions, in turn threatens the future of their own logging business model. The empirical study also shows that local communities understand that forest use is not infinite [100], but at the same time they are aware that they lack self-management and adequate management skills, but are very open to learning. This is an important element, as it is these local communities who mostly know the behavior of the forest and its ecosystem and yet are willing to learn what is necessary about the management and sustainable use of the forest [170–175]. On the other hand, 100% of the consulted NGOs recognize the need to increase communication between communities and their abilities to diversify the production of resources, without forgetting the complexity of managing 250 different ethnic groups fragmented by their history, culture, and even languages and dialects [169]. It is important to highlight NGOs, as they are the ones in charge of seeking and delivering a certain balance of dissemination and protection among stakeholders, but particularly among the most vulnerable [164].

These elements lead us to believe that government authorities and the public policies sanctioned by them [165] have failed to require changes in terms of reference for the allocation of exploitable forests to recipient and certifying companies, including the banks in charge of financing such activities. The key here would be to impose stricter conditions that go beyond the rules in force in 2021. It should not be forgotten, however, that Cameroon has the oldest system of forest concessions in Africa and there is what is known as inertia in the work and concession models, making it very difficult to change deeply rooted habits in the social collective [176]. Cameroon exploits its forests under four management models and partnership agreements, where 66% of the territory is managed by managed concessions and 34% by certified concessions. However, it should be noted that there has also been timid progress in the intention to bring about change through the implementation of the Voluntary Partnership Agreement (VPA), signed in 2010 and ratified by Cameroon in 2011 [177], with which, the aim is to reduce, to a certain extent, those who illegally exploit the timber extracted there, considering the sharp decline in the country's forest cover from 22 to 19 million hectares, mainly due to illegal logging [14]. However, there is a systematic relapse into the vicious cycle of poor institutions and poor-quality governance in the country and human capital is poorly educated to solve this problem, [74,81,178,179].

The results of the empirical study also reveal that certified forestry companies consider that compliance with the law is the best way to protect the environment and thus, they transfer all responsibilities linked to the environment where they operate directly to the State, which must guarantee the formation of communities and the protection of concessioned lands [180,181]. In this regard, it is important to note that the vast majority of forestry companies operating in Cameroon are only interested in maximizing the exploitation of forests they are supported by banks from their economic/financial capital perspective and also have the support of certification companies, but they forget, in this important equation, the natural (or ecological) capital, which is the generator of both natural resource flows and natural services, as suggested by [182]. Likewise, the government becomes practically a mere spectator, relying on the laws that they themselves dictate, which are rarely for the benefit of local communities, but rather in support of large corporations, which intensively exploit, almost to the point of extinction, the concessioned lands, leaving lands razed to the ground [70]. The same author points out that this reality suggests a certain degree of complacency on the part of government authorities, which has been carried over from colonial times and which does not seem to be overly concerned about the consequences that these practices may have for biodiversity, climate change, and the more than 250 local ethnic groups that live there, leading to an imbalance plagued with inequalities between the parties involved [169]. This point is the key, as it leads to the question: What is supposed to be balanced and constant in the sustainable local development of a territory? The answer is clearly geared towards establishing a benefit for future generations that is not resented over time [31,183], which is clearly not met by current logging regulations, where although there is a management regulation for Cameroon's production forests in partnership, almost 31% is given in concessions and barely 8% are considered community forests [3].

The results also show that communities and NGOs, unlike forestry companies, expect to contribute to the sustainable development of forests and have a more long-term vision of forest exploitation. Regarding the responsibilities for managing sustainable exploitation of Cameroon's forests, opinions are divided, with 80% of the forestry companies and NGOs interviewed considering that the state (government) has the greatest responsibility for enforcing the logging law, training communities, and developing a program for the sustainable use of natural resources.

However, on the contrary, 100% of the forest-dwelling communities consider that the forestry companies should be responsible for this action, which shows, at a glance, a power struggle between the different conglomerates involved. Meanwhile, 100% of the banks and certifying companies have a much more balanced opinion on the matter and consider that all parties should have a degree of responsibility, according to the degree of exploitation that corresponds to each one. Therefore, a mixed control would be the most appropriate and thus counteract the lack of credibility of the different governments and institutions and seek a balance between the parties [81,178].

On the other hand, 100% of the sample is related to the banking institutions responsible for financing the work of forestry companies, whose main objective is the exploitation of the forest as a purely economic component, while financially supporting the certification of timber companies, so that they can take advantage of the different available forests granted by the Cameroonian government. In this regard, it should be noted that Cameroon took the decision in 2010 to promote sustainable growth in the country, proposing projects with a long-term development vision until 2035. In this context, the "Growth and Employment Strategy Paper"—GESP (2010–2020) was created to promote development, while seeking to reduce poverty levels by taking advantage of the exploitation of the abundant natural resources available [180]. As can be seen from the interviews with these stakeholders (certifiers, forestry companies and banks, governments), the reality shows that there have been no major changes since 2010 despite attempts to change course. There are intentions, but there is a lack of action.

More than 95% of the companies and institutions consulted indicate that certification and law enforcement is the best strategy to care for the environment in the long term, and so far, they have not speculated on any other type of strategy. In general, cluster 1 and 3 stakeholders consulted think that the forest regenerates itself and that this is sufficient for their aspirations to obtain the resources necessary for their livelihoods. When asked about the sustainability protocols applied in their home institution, they indicated that they apply the law and that this seems to them to be sufficient. Therefore, in the institutional framework, governance and transparency are key to decision-making and compliance with existing protection regulations [184]. The biggest problem lies in the fact that those who make up these two conglomerates believe that the forest regenerates itself infinitely and furthermore without taking into consideration the degradation of biodiversity and the soils that support the forest canopy by indiscriminate logging. Unfortunately, we know that the forest does not regenerate by itself, since when it is cut down, the fertile layers of the soil are rapidly degraded, which leads to desertification or loss of yields, which is why forests must be protected, to safeguard the biodiversity that accompanies them, in addition to the fact that they contribute to the local economy by providing multiple inputs and are an important reservoir even for the pharmaceutical industry [1].

At the same time, 60% of the forestry companies consulted consider that any activity carried out by a private company in relation to the protection or improvement of the environment and biodiversity should not only be financed, but should also be the responsibility of the government of the day, with the former avoiding this equation of responsibilities, claiming that they are certified and that they comply with all current regulations and that they are not obliged to make improvements of any kind to the land they have been awarded to exploit. However, the most surprising aspect of this reflection is that 100% of the certifying companies consider that those forestry companies that have been certified contribute to the care of the environment. This is somewhat contradictory to the current exploitation of Cameroon's forests and the incessant loss of their vegetation cover year after year [99], with an average loss of 4.2% of tree cover annually (adding native and non-native forests), which represents 519 Mt of CO_2 released into the atmosphere and contributes greatly to climate change [13,15]. It follows that the companies responsible for certification claim that certified forestry companies, by the mere fact of being certified, already comply with all the regulations, which from a technical point of view is true, but then the question arises: Does their claim respond to a structural failure of the bidding specifications for concessions based on an inefficient public policy? The answer is clearly yes, due to the deficiency and lightness of the obligations subscribed to in the bidding specifications, which we have addressed in this discussion. Therefore, changing the policies and conditions of concession and use of forest territories is fundamental for a radical change in the way these territories are conserved, and in this regard, the theories of Kjellén [61,62] and Hawkes [159] are powerful tools to consider in this paradigm shift. Another reality may be related to the fact that corporations that exploit forests lack local culture and ancestral attachment to their ecosystem [92,185,186]. Probably for this reason, they do not have an ethical and more ecological view of the impact of their actions. Therefore, they only rely on the weakness of existing policies, without measuring the harmful consequences of their actions on the ecosystem, biodiversity, and the communities that inhabit it, as well as the impact on climate change that such actions could cause in the long term. What is interesting, however, is that the communities and NGOs participating in the study consider that care for the environment is essential to keep logging sustainable in the long term. They appear to have a much greater socio-ecological awareness of the key importance of forest care in maintaining long-term sustainable logging, perhaps underpinned by local knowledge that is critical to the conservation not only of forests, but of biodiversity as a whole [171,175].

Continuing with the empirical results, 100% of timber companies and certifiers responded that they consider the forest to be a renewable resource in its own right. Of all the parties consulted, only the communities (100%) indicated that they are concerned about reforestation. There is a total lack of knowledge among stakeholders about the real regeneration capacity of forests as a non-renewable resource [99–101,160].

At the same time, 100% of the forestry companies and certifiers specify that their organizations are top-down and that decisions are taken by a chief or general manager, which brings them much closer to the system of governance prevailing in Cameroon, where it is precisely a top-down system that allows acts of corruption or at least lack of transparency to permeate [70,81]. In contrast, 100% of the communities, partner organizations, and banks consulted state that their decisions are made by an advisory board, which makes them more mixed in nature, and therefore much more transparent in their actions [67].

When stakeholders were consulted on what actions should be taken most urgently in relation to forest care and sustainability, the general response from NGOs, communities, and banks was that a "Mixed Intermediate Organization" should be created, with the participation of the government of the day and the different actors directly and indirectly involved in forest exploitation. This takes us in the direction of mixed and even reflexive governance [76]. For their part, forestry, certification, and related companies did not respond, because they firmly believe that applying the law as it stands, in addition to being certified, is sufficient for the care of the environment and the biodiversity of the natural resources found there. Furthermore, 100% of the consulted NGOs indicated that there is a need to increase communication and develop skills and competencies in communities, eliminate poverty, and decentralize power, which brings us towards what Dries et al. [76] suggest with mixed governance systems that include the environment in decision-making, represented by public and private as well as mixed entities.

In addition, there is a need to improve the quality of education to increase the generation of knowledge suitable for the country's development. Taking into account this dynamic correlation, the empirical study showed that social, ecological, and territorial vulnerability are intrinsically linked and should be considered as integrated socio-ecological systems, where anthropogenic resilience is considered together with its object (nature) and subject (human activities). These elements are directly related to the so-called "forest use" and "forest dependence" described by Bele et al. [1] in the search for a balance that reduces the poverty of the communities involved and contributes to the country's development through the good management of available natural resources.

Interestingly, 100% of forestry companies and certifiers consider that the country's development is based on the creation and good functioning of hospitals and schools, along with the construction of civil works. However, the communities and NGOs consulted indicate that not only are these works necessary, but that it is also essential to educate people and train them to become critically aware of the work they do. Furthermore, these stakeholders indicate that the economic matrix must be diversified, generating new businesses and thus not depending only on what the forest and its biodiversity provide as a natural resource and avoiding in some way the postulate of the curse of natural resources. This is why the alternative of cassava as a value-added engine, generating labor, food and a complement to industries such as cement, is so interesting to consider. These results indicate that there is a need for better educational training and an increase in the welfare of the population [33].

4. Conclusions

The study suggests the use of balanced alternatives for all the parties involved in the large conglomerate, which in our study was represented by the 44 stakeholders interviewed (forestry companies/certifiers; NGOs/communities; and banks/public bodies). From this problematic approach, we sought to answer the two main questions of this research: (1) What would be the best strategy capable of generating sustainable development of Cameroon's native forests, considering the interaction of very critical aspects such as socio-cultural, economic, agricultural, and institutional and governance aspects? and (2) How could cassava (*Manihot esculenta*) cultivation boost socio-ecological resilience in Cameroon through a specific opportunity cost, thus enabling the conservation of native forests, sustainable local development, and, at the same time, giving life to an industry as polluting as cement?

All participating stakeholders, regardless of their sector, indicated that they expected the government, through public policies, laws, and mixed institutions, to generate actions for the protection of both native forests and the community dependent on them. However, citizen participation at the governance and decision-making levels is extremely limited. The latter is mainly due to the fact that many people are unaware of their rights and duties as citizens. The ability to organize and build group trust is also limited, which generates verticality in the way people deal with each other, leading to nodes of corruption. Moreover, the lack of education at all levels and the development of basic health and welfare services limits the development of critical thinking among the population. All of this favors governmental verticality in decision-making, and therefore favors some sectors such as large forestry corporations in these decisions.

Interestingly, the empirical study revealed that 100% of the participants indicated that there should be a "Mixed Intermediate Organization", involving all stakeholders related to the Cameroon forest, including the government of the day. However, without quality basic education, it is impossible to think about the care of the forest and its ecosystems, and even more so to include the thinking of the communities, who in the study proved to be the only ones who have a view of sustainability underpinned by their culture. It is worrying to know that none of the stakeholders interviewed who were responsible for forest exploitation and who were also important actors in making crucial decisions concerning the forest and its communities were aware that the forest is not renewable on its own. The exception was NGOs and indigenous communities, who were concerned about their future and that of the forests and were willing to educate themselves more and better in order to improve their competencies in sustainable management of native forests. However, mistrust and discrimination among the more than 250 ethnic communities and the displacement to which they are often subjected greatly limits their participation in discussions and decision-making. Changing this balance of power and coexistence between local communities is essential to reduce the social gaps between them.

It is essential to highlight the importance of the "opportunity cost" as a management strategy to promote the cassava (*Manihot esculenta*) as an alternative agro forestry source and as an important agent of change and socio-ecological resilience in Cameroon. This crop not only serves as a staple food for the population and for animal production, but also allows, through its vegetal cover, to protect the soil from erosion and desertification caused by deforestation. At the same time, it offers the opportunity to preserve large extensions of virgin native forests by preventing them from being exploited. Likewise, its waste is also converted into a fundamental input for new green cement engineering, giving it greater robustness and resistance, while eliminating the use of highly polluting products in its manufacture, thus providing sustainable support and a means to foster the development of a most circular economy.

One of the major weaknesses of the study was the relatively small number of stakeholders interviewed. A greater number and diversity would obviously have made the study more robust, not only statistically, but also in terms of a greater diversity of opinions. However, it is not an easy task in Cameroon to bring together people involved in power and willing to participate in critical research because of the costs, distances, and different dialects in this African country.

In future research, it would be very important to obtain a larger and more diversified number of stakeholders (despite the difficulties mentioned above) and to consider three crucial elements. The first would be to analyze the effectiveness of cassava cultivation in the longer term on deforested land and evaluate how this agricultural alternative behaves in the regeneration of eroded soils. The second would be to measure, in the long term, the potential impacts on the opportunity cost explained above, in terms of the positive or negative effects of the use of cassava waste in the cement industry to promote at the same time a permanent circular economy. The third would be to analyze in the medium and long term how the native forests behave by removing the pressure of overexploitation that exists on them today by dissuading the communities towards the agricultural alternative of cassava.

Author Contributions: Conceptualization, J.G.; Data curation, P.G.; Methodology, A.C.; Writing—review & editing, T.G.B. All authors have read and agreed to the published version of the manuscript.

Funding: This research was funded by Social Sciences and Humanities Research Council of Canada: 435-2014-1890.

Institutional Review Board Statement: Not applicable.

Informed Consent Statement: Not applicable.

Data Availability Statement: Data is contained within the article.

Conflicts of Interest: The authors declare no conflict of interest.

References

1. Bele, M.; Somorin, O.; Sonwa, D.; Nkem, J.; Locatelli, B. Forests and climate change adaptation policies in Cameroon. *Mitig. Adapt. Strateg. Glob. Change* **2011**, *16*, 369–385. [CrossRef]
2. Aquilas, N.; Mukong, A.; Kimengsi, J.; Ngangnchi, F. Economic activities and deforestation in the Congo basin: An environmental kuznets curve framework analysis. *Environ. Chall.* **2022**, *8*, 100553. [CrossRef]
3. Tieguhong, J.C.; Ketchatang, P.T.; Chia, E.; Assembe-Mvondo, S.; Oeba, V.O. The role of the private forestry sector in response to climate change in central Africa: The case of Cameroon. *Int. For. Rev.* **2019**, *21* (Suppl. S1), 112–125. [CrossRef]
4. Hansen, M.C.; Potapov, P.V.; Moore, R.; Hancher, M.; Turubanova, S.; Tyukavina, A.; Thau, D.; Stehman, S.; Goetz, S.; Loveland, T. High-resolution global maps of 21st-century forest cover change. *Science* **2013**, *342*, 850–853. [CrossRef]
5. Chia, E.; Tiani, A.; Sonwa, D. The State of Research on Effectiveness and Equity (2Es) in Forests Management Regimes in Cameroon and Its Relevance for REDD+. *Environ. Nat. Resour. Res.* **2013**, *3*, 92–105. [CrossRef]
6. Duguma, L.A.; Minang, P.A.; Foundjem-Tita, D.; Makui, P.; Mandiefe Piabuo, S. Prioritizing enablers for effective community forestry in Cameroon. *Ecol. Soc.* **2018**, *23*, 1. [CrossRef]
7. Alemagi, D. Sustainable development in Cameroon's forestry sector: Progress. Challenges and strategies for improvement. *Afr. J. Environ. Sci. Technol.* **2011**, *5*, 65–72. [CrossRef]
8. Carodenuto, S.; Cerutti, P. Forest Law Enforcement, Governance and Trade (FLEGT) in Cameroon: Perceived private sector benefits from VPA implementation. *For. Policy Econ.* **2014**, *48*, 55–62. [CrossRef]
9. Auzel, G.; Nguenang, M.; Feteké, R.; Delvingt, W. L'exploitation forestière artisanale des forêts communautaires au Cameroun: Vers des compromis écologiquement plus durables et socialement plus acceptable. *Réseau For. Pour Développement Rural.* **2001**, *25*, 1–12.
10. World Resources Institute (WRI). *Cameroon's Forest Estate*; Ministry of Forestry and Wildlife; World Resources Institute: Washington, DC, USA, 2014; Available online: http://wriorg.s3.amazonaws.com/s3fs-public/uploads/CMR_Poster_2014_english.pdf (accessed on 12 July 2021).
11. Institut National de la Statistique (INS). Répertoire et Démographie des Entreprises Modernes en 2015. 3 ème Edition. 2015. Available online: www.statistics-cameroon.org (accessed on 3 September 2021).
12. FAO (Food and Agriculture Organization). *Global Forest Resources Assessment 2020: Main Report*; Food and Agriculture Organization: Rome, Italy, 2020. [CrossRef]

13. African Development Bank (AFDB) Socio Economic Database: Cameroon, from Soci. Available online: https://dataportal.opendataforafrica.org/nbyenxf/afdb-socio-economic-database-1960-2020 (accessed on 20 June 2021).
14. Hoare, A. *Illegal Logging and Related Trade: The Response in Cameroon. Energy, Environment and Resources*; Chatham House, the Royal Institute of International Affairs: London, UK, 2015; ISBN 978-1-78413-0312.
15. Global Forest Review (GFR). 2020. Available online: https://www.globalforestwatch.org/dashboards/country/CMR. (accessed on 17 November 2020).
16. Ickowitz, A.; Sills, E.; de Sassi, C. Estimating smallholder opportunity costs of REDD+: A pantropical analysis from households to carbon and back. *World Dev.* **2017**, *95*, 15–26. [CrossRef]
17. Bayiha, B.; Yamb, E.; Valdès, J.; Sontia, J.; Metekong Ndigui Billong Nzengwa, R. Optimizing the Choice of Limestone Deposits for the Production of Portland Cement in Cameroon. *Int. J. Mater. Sci. Appl.* **2018**, *7*, 174–185. [CrossRef]
18. D'Aquin Biyindi, T.; Kwefeu Mbakop, F.; Toumi, E.; Woumbeng Etonleu, B.; Pondi, J. Comparative Physico Mechanical Study of Cements CEM II 42.5R in Cameroon: Case of DANGOTE and CIMENCAM. *Open J. Civ. Eng.* **2019**, *9*, 18–34. [CrossRef]
19. Crépy, L.; Petit, J.; Joly, N.; Wirquin, E.; Martin, P. First step towards bio-superplasticizers. In Proceedings of the 13th International Congress on the Chemistry of Cement, Madrid, Spain, 3–8 July 2011.
20. Patel, G.; Deo, S. Parametric study of natural organic materials as admixture in concrete. *Int. J. Appl. Eng. Res.* **2016**, *11*, 6271–6277.
21. FAO & UNEP. *The State of the World's Forests 2020. Forests, Biodiversity and People*; FAO & UNEP: Rome, Italy, 2020; Available online: https://doi.org/10.4060/ca8642en (accessed on 8 September 2020).
22. Spreafico, C. An analysis of design strategies for circular economy through life cycle assessment. *Environ. Monit. Assess.* **2022**, *194*, 180. [CrossRef] [PubMed]
23. Penalva, C.; Alaminos, A.; Francés, F.; Santacreu, O. *La Investigación Cualitativa, Técnicas de Investigación y Análisis con atlas.ti, Pydlos Ediciones*; Publisher: Quito, Ecuador, 2015; ISBN 978-9978-14-303-2.
24. Herrera, C. Investigación cualitativa y análisis de contenido temático. Orientación intelectual de revista Universum. *Rev. Gen. Inf. Doc.* **2017**, *28*, 119–142. [CrossRef]
25. Baena Paz, G. *Metodología de la Investigación*; Grupo Editorial Patria: Mexico City, Mexico, 2008.
26. Troncoso Pantoja, C.; Amaya-Placencia, A. Entrevista: Guía práctica para la recolección de datos cualitativos en investigación de salud Interview: A practical guide for qualitative data collection in health research. *Rev. Fac. Med.* **2017**, *65*, 329–332. [CrossRef]
27. Efroymson, R.; Kline, K.; Angelsen, A.; Verburg, P.; Dale, V.; Langeveld, J.; McBride, A. A causal analysis framework for land-use change and the potential role of bioenergy policy. *Land Use Policy* **2016**, *59*, 516–527. [CrossRef]
28. Leeman, J.; Baquero, B.; Bender, M.; Choy-Brown, M.; Ko, L.; Nilsen, P.; Wangen, M.; Birken, S. Advancing the use of organization theory in implementation science. *Prev. Med.* **2019**, *129*, 105832. [CrossRef]
29. Eisenhardt, K. Building Theory from case study research. *Acad. Manag.* **1989**, *4*, 532–550. [CrossRef]
30. Lihtonen, V. *Metsätalouden Suunnittelu ja Järjestely [Planning and Organizing Forestry]*; WSOY: Porvoo, Finland, 1959.
31. Hall, R.; Ashford, N. *Primer on the Emergence and Evolution of Sustainable Development (1951 to 2012)*; Virginia Tech: Blacksburg, VA, USA, 2012; Available online: https://ralphphall.files.wordpress.com/2012/11/primer_sustdev_2012-11-26.pdf (accessed on 22 January 2022).
32. Broman, G.; Robert, K.-H. A framework for strategic sustainable development. *J. Clean. Prod.* **2017**, *147*, 17–31. [CrossRef]
33. Jovovic, R.; Draskovic, M.; Delibasic, M.; Jovovic, M. The concept of sustainable regional development—Institutional aspects, policies and prospects. *J. Int. Stud.* **2017**, *10*, 255–266. [CrossRef] [PubMed]
34. Kopnina, H. Sustainability: New strategic thinking for business. *Environ. Dev. Sustain.* **2017**, *19*, 27–43. [CrossRef]
35. Cherrington, E.; Griffin, R.; Anderson, E.; Hernandez Sandoval, B.; Flores-Anderson, A.; Muench, R.; Markert, K.; Adams, E.; Limaye, A.; Irwin, D. Use of public Earth observation data for tracking progress in sustainable management of coastal forest ecosystems in Belize, Central America. *Remote Sens. Environ.* **2020**, *245*, 111798. [CrossRef]
36. Carson, R. *Silent Spring*; Houghton Mifflin Company: Boston, MA, USA; Riverside Press: Cambridge, UK, 1962.
37. Meadows, D.; Randers, J.; Meadows, D. *The Limits to Growth: The 30-Year Update*; Earthscan: London, UK, 2006.
38. Ashrafi, M.; Adams, M.; Walker, T.; Magnan, G. How corporate social responsibility can be integrated into corporate sustainability: A theoretical review of their relationships. *Int. J. Sustain. Dev. World Ecol.* **2018**, *25*, 672–682. [CrossRef]
39. Johnston, P.; Everard, M.; Santillo, D.; Robèrt, K.-H. Reclaiming the Definition of Sustainability. *Environ. Sci. Pollut. Res.* **2007**, *14*, 60–66. [CrossRef]
40. Missimer, M.; Robèrt, K.-H.; Broman, G. A strategic approach to social sustainability—part 1: Exploring the social system. *J. Clean. Prod.* **2016**, *140*, 32–41. [CrossRef]
41. Peterson, N. Introduction to the special issue on social sustainability: Integration, context, and governance, Sustainability. *Sci. Pract. Policy* **2016**, *12*, 3–7. [CrossRef]
42. Mensah, J. Sustainable development: Meaning, history, principles, pillars, and implications for human action: Literature review. *Cogent Soc. Sci.* **2019**, *5*, 1653531. [CrossRef]
43. Goslinga, J.; Jonesa, M.; Arnella, A.; Watson, J.; Ventere, O.; Baqueroa, A.; Burgess, N. A global mapping template for natural and modified habitat across terrestrial. *Earth Biol. Conserv.* **2021**, *250*, 108674. [CrossRef]
44. Lacroix, P.; Moser, F.; Benvenuti, A.; Piller, T.; Jensen, D.; Petersen, I.; Planque, M.; Ray, N. MapX: An open geospatial platform to manage, analyze and visualize data on natural resources and the environment. *SoftwareX* **2019**, *9*, 77–84. [CrossRef]

45. Tobias, T. Preserving Ecosystem Services in Urban Regions: Challenges for Planning and Best Practice Examples from Switzerland. *Integr. Environ. Assess. Manag.* **2013**, *9*, 243–251. Available online: https://www.researchgate.net/figure/The-relationships-between-the-natural-resources-ecosystem-functions-and-ecosystem_fig1_234105361 (accessed on 26 June 2022). [CrossRef] [PubMed]
46. Juma, C.; Lee, K.; Mathews, J. Innovation Capabilities for Sustainable Development in Africa. UNU-WIDER, Helsinki, Finland. 2014. Available online: https://www.wider.unu.edu (accessed on 6 January 2022).
47. Fadhilah, Z.; Ramayah, T. Behind the Green Doors: What Management Practices Lead to Sustainable Innovation? *Procedia-Soc. Behav. Sci.* **2012**, *65*, 247–252. [CrossRef]
48. Tseng, M.-L.; Chiu, A.; Tan, R.; Siriban-Manalang, A. Sustainable consumption and production for Asia: Sustainability through green design and practice. *J. Clean. Prod.* **2013**, *40*, 1–5. [CrossRef]
49. Rytteri, T.; Peltola, T.; Leskinen, L. Co-production of forestry science and society: Evolving interpretations of economic sustainability in Finnish forestry textbooks. *J. For. Econ.* **2016**, *24*, 21–36. [CrossRef]
50. Stojanović, I.; Ateljević, J.; Stević, R. Good Governance as a Tool of Sustainable Development. *Eur. J. Sustain. Dev.* **2016**, *5*, 558–573. [CrossRef]
51. Roper, S.; Tapinos, E. Taking risks in the face of uncertainty: An exploratory analysis of green innovation. *Technol. Forecast. Soc. Change* **2016**, *112*, 357–363. [CrossRef]
52. Thongplew, N.; Spaargaren, G.; Van Koppen, K. Companies in search of the green consumer: Sustainable consumption and production strategies of companies and intermediary organizations in Thailand. *NJAS—Wagening J. Life Sci.* **2017**, *83*, 12–21. [CrossRef]
53. Zhang, D.; Morse, S.; Kambhampati, U. *Sustainable Development and Corporate Social Responsibility*; Routledge: Abingdon, UK, 2017; p. 286.
54. Shepherd, D.A.; Patzelt, H. *Researching Entrepreneurships' Role in Sustainable Development*; Trailblazing in Entrepreneurship; Palgrave Macmillan: London, UK, 2017. [CrossRef]
55. D'Amatoa, D.; Korhonena, J.; Toppinen, A. Circular, Green, and Bio Economy: How Do Companies in Land-Use Intensive Sectors Align with Sustainability Concepts? *Ecol. Econ.* **2019**, *158*, 116–133. [CrossRef]
56. Dogaru, L. Eco-Innovation and the Contribution of Companies to the Sustainable Development. *Procedia Manuf.* **2020**, *46*, 294–298. [CrossRef]
57. Dogaru, L. Green Economy and Green Growth—Opportunities for Sustainable Development. *Proceedings* **2021**, *63*, 70. [CrossRef]
58. Friant, M.; Vermeulen, W.; Salomone, R. A typology of circular economy discourses: Navigating the diverse visions of a contested paradigm. *Resour. Conserv. Recycl.* **2020**, *161*, 104917. [CrossRef]
59. Patwa, N.; Sivarajah, U.; Seetharaman, A.; Sarkar, S.; Maiti, K.; Patwa, K. Towards a circular economy: An emerging economies context. *J. Bus. Res.* **2021**, *122*, 725–735. [CrossRef]
60. Glinskiy, V.; Serga, L.; Alekseev, L. 17th Global Conference on Sustainable Manufacturing Territorial Differentiation as the Factor of Sustainable Economic Development. *Procedia Manuf.* **2020**, *43*, 263–268. [CrossRef]
61. Kjellén, B. *Multilateral Diplomacy and Science*; the Columbia University Earth Institute: New York, NY, USA, 1999.
62. Kjellén, B. *A New Diplomacy for Sustainable Development: The Challenge of Global Change*; Routledge: Oxford, UK, 2007.
63. Adejumo, A.; Adejumo, O. Prospects for Achieving Sustainable Development through the Millennium Development Goals in Nigeria. *Eur. J. Sustain. Dev.* **2014**, *3*, 33–46. [CrossRef]
64. Gustafsson, K.; Lidskog, R. Boundary organizations and environmental governance: Performance, institutional design, and conceptual development. *Clim. Risk Manag.* **2018**, *19*, 1–11. [CrossRef]
65. Paavola, J. Institutions and environmental governance: A reconceptualization. *Ecol. Econ.* **2007**, *63*, 93–103. [CrossRef]
66. Smedby, N.; Quitzau, M.-B. Municipal Governance and Sustainability: The Role of Local Governments in Promoting Transitions: Municipal Governance and Sustainability. *Environ. Policy Gov.* **2016**, *26*, 323–326. [CrossRef]
67. Hegger, D.; Runhaar, H.; Van Laerhoven, F.; Driessen, P. Towards explanations for stability and change in modes of environmental governance: A systematic approach with illustrations from the Netherlands. *Earth Syst. Gov.* **2020**, *3*, 100048. [CrossRef]
68. Hawkins, C.; Wang, X. Sustainable Development Governance: Citizen Participation and Support Networks in Local Sustainability Initiatives. *Public Work. Manag. Policy* **2012**, *17*, 7–19. [CrossRef]
69. Kardos, M. The reflection of good governance in sustainable development strategies. 8th International Strategic Management Conference. *Procedia—Soc. Behav. Sci.* **2012**, *58*, 1166–1173. [CrossRef]
70. Ndeh, M. Good governance in Cameroon: Perceptions and practice in an emerging Cameroon by 2035. *Glob. J. Political Sci. Adm.* **2015**, *3*, 1–8. Available online: https://www.eajournals.org (accessed on 18 December 2021).
71. Tunyi, A.; Ntim, C. Location Advantages, Governance Quality, Stock Market Development and Firm Characteristics as Antecedents of African M&As. *J. Int. Manag.* **2016**, *22*, 147–167. [CrossRef]
72. Khodary, Y. Good governance: A new perspective for institutional reform—a comparative view of water, education and health institutions in Egypt. *Int. J. Public Policy* **2016**, *12*, 359–377. [CrossRef]
73. Schoneveld, G. Host country governance and the African land rush: 7 reasons why large-scale farmland investments fail to contribute to sustainable development. *Geoforum* **2017**, *83*, 119–132. [CrossRef]
74. Aljarallah, R. Natural resource dependency, institutional quality and human capital development in Gulf Countries. *Heliyon* **2020**, *6*, e04290. [CrossRef] [PubMed]

75. Voß, J.; Bornemann, B. The politics of reflexive governance: Challenges for designing adaptive management and transition management. *Ecol. Soc.* **2011**, *16*, 9. [CrossRef]
76. Kaufmann, M.; Wiering, M. The role of discourses in understanding institutional stability and change – an analysis of Dutch flood risk governance. *J. Environ. Policy Plan.* **2022**, *24*, 1–20. [CrossRef]
77. Turner, R. Investigating how does governmentality and governance influence decision making on projects. *Proj. Leadersh. Soc.* **2020**, *1*, 100003. [CrossRef]
78. MacCormick, J.S. *Governing Organisational Culture*; Paper prepared as part of the Director Tools Series; Australian Institute of Company Directors: Sydney, AU, Australia, 2019.
79. Cameron, K.; Quinn, R. *Diagnosing and Changing Organizational Culture: Based on the Competing Values Framework*; Jossey-Bass: San Francisco, CA, USA, 2011.
80. Johansson, J. Collaborative governance for sustainable forestry in the emerging bio-based economy in Europe. *Curr. Opin. Environ. Sustain.* **2018**, *32*, 9–16. [CrossRef]
81. Mbzibain, A.; Ongolo, S. Complementarity, rivalry and substitution in the governance of forests: Learning from independent forest monitoring system in Cameroon. *For. Policy Econ.* **2019**, *109*, 101981. [CrossRef]
82. González, N.; Kröger, M. The potential of Amazon indigenous agroforestry practices and ontologies for rethinking global forest governance. *For. Policy Econ.* **2020**, *118*, 102257. [CrossRef]
83. Adeyeye, Y.; Hagerman, S.; Pelai, R. Seeking procedural equity in global environmental governance: Indigenous participation and knowledge politics in forest and landscape restoration debates at the 2016 World Conservation Congress. *For. Policy Econ.* **2019**, *109*, 102006. [CrossRef]
84. Brown, H.C.; Nkem, J.N.; Sonwa, D.J.; Bele, Y. Institutional adaptive capacity and climate change response in the Congo Basin forests of Cameroon. *Mitig. Adapt. Strateg. Glob. Change* **2010**, *15*, 263–282. [CrossRef]
85. Cronkleton, P.; Pulhin, J.; Saigal, S. Co-management in Community Forestry: How the Partial Devolution of Management Rights Creates Challenges for Forest Communities. *Conserv. Soc.* **2012**, *10*, 91–102. [CrossRef]
86. Sharp, E.; Daley, D.; Lynch, M. Understanding Local Adoption and Implementation of Climate Change Mitigation Policy. *Urban Aff. Rev.* **2011**, *47*, 433–457. [CrossRef]
87. Trudeau, D. Integrating social equity in sustainable development practice: Institutional commitments and patient capital. *Sustain. Cities Soc.* **2018**, *41*, 601–610. [CrossRef]
88. Norrman, J.; Söderqvist, T.; Volchko, Y.; Back, P.; Bohgard, D.; Ringshagen, E.; Svensson, H.; Englöv, P.; Rosén, R. Enriching social and economic aspects in sustainability assessments of remediation strategies—Methods and implementation. *Sci. Total Environ.* **2020**, *707*, 136021. [CrossRef]
89. Cohen, W.; Levinthal, D. Absorptive Capacity: A New Perspective on Learning and Innovation. *Adm. Sci. Q.* **1990**, *35*, 128–152. [CrossRef]
90. Niosi, J.; Bas, T.G. The Competencies of Regions—Canada's Clusters in Biotechnology. *Small Bus. Econ.* **2001**, *17*, 31–42. [CrossRef]
91. Barkhordari, S.; Fattahi, M.; Ali Azimi, N. The Impact of Knowledge-Based Economy on Growth Performance: Evidence from MENA Countries. *J. Knowl. Econ.* **2018**, *37*, 1–15. [CrossRef]
92. Milán-García, J.; Uribe-Toril, J.; Ruiz-Real, J.; Valenciano, J. Sustainable Local Development: An Overview of the State of Knowledge. *Resources* **2019**, *8*, 31. [CrossRef]
93. Toma, S.G. Learning from the Asian Tigers: Lessons in Economic Growth Annals—Economy Series, Constantin Brancusi University, Faculty of Economics. *Ann.-Econ. Ser.* **2019**, *3*, 63–69.
94. Watson, J.K. Asian Tigers: Singapore, Hong Kong, Taiwan, and South Korea. *Encycl. Postcolonial Stud.* **2016**, *1*, 1–11.
95. Amiri, H.; Samadian, F.; Yahoo, M.; Jamali, S. Natural resource abundance, institutional quality and manufacturing development: Evidence from resource-rich countries. *Resources Policy* **2019**, *62*, 550–560. [CrossRef]
96. Minot, N. Food price volatility in sub-Saharan Africa: Has it really increased? *Food Policy* **2014**, *45*, 45–56. [CrossRef]
97. Fasanya, O.; Odudu, F. Modeling return and volatility spillovers among food prices in Nigeria. *J. Agric. Food Res.* **2020**, *2*, 100029. [CrossRef]
98. Lashitew, A.; Werker, E. Do natural resources help or hinder development? Resource abundance, dependence, and the role of institutions. *Resour. Energy Econ.* **2020**, *61*, 101183. [CrossRef]
99. Carpenter, S.; Walker, B.; Anderies, J.; Abel, N. From metaphor to measurement: Resilience of what to what? *Ecosystems* **2001**, *4*, 765–781. [CrossRef]
100. Westerman, K.; Oleson, K.; Harri, A. Building Socio-ecological Resilience to Climate Change through Community-Based Coastal Conservation and Development: Experiences in Southern Madagascar. *West. Indian Ocean. J. Mar. Sci.* **2012**, *11*, 87–98.
101. Cinner, J.; Barnes, M. Social Dimensions of Resilience in Social-Ecological Systems. *One Earth Primer* **2019**, *1*, 51–56. [CrossRef]
102. Liang, J.; Li, Y. Resilience and sustainable development goals based social-ecological indicators and assessment of coastal urban areas. A case study of Dapeng New District, Shenzhen, China. *Watershed Ecol. Environ.* **2020**, *2*, 6–15. [CrossRef]
103. Plant, R. Landscape as a Scaling Strategy in Territorial Development. *Sustainability* **2022**, *14*, 3089. [CrossRef]
104. González-Quintero, C.; Avila-Foucat, V. Operationalization and Measurement of Social-Ecological Resilience: A Systematic Review. *Sustainability* **2019**, *11*, 6073. [CrossRef]
105. Heslinga, J.; Groote, P.D.; Vanclay, F. Using a social-ecological systems perspective to understand tourism and landscape interactions in coastal areas. *J. Tour. Futures* **2017**, *3*, 23–38. [CrossRef]

106. Kohsaka, R.; Rogel, M. Partnerships for the Goals. Traditional and Local Knowledge for Sustainable Development: Empowering the Indigenous and Local Communities of the World. In *Encyclopedia of the UN Sustainable Development Goals*; Leal Filho, W., Azul, A., Brandli, L., Özuyar, P., Wall, T., Eds.; Springer: Cham, Switzerland, 2019. [CrossRef]
107. Yang, B.; Feldman, M.; Li, S. The status of perceived community resilience in transitional rural society: An empirical study from central China. *J. Rural. Stud.* **2020**, *80*, 427–438. [CrossRef]
108. El-Sharkawy, M. Cassava biology and physiology. *Plant Mol. Biol.* **2004**, *56*, 481–501. [CrossRef] [PubMed]
109. Spencer, D.; Ezedinma, C. Cassava cultivation in sub-Saharan Africa. In *Achieving Sustainable Cultivation of Cassava*; Hershey, C.H., Ed.; Burleigh Dodds Science Publishing Limited: Cambridge, UK, 2017; Volume 1, pp. 123–148. [CrossRef]
110. Akinsanya, A.; Afolami, S.; Kulakow, P.; Parkes, E.; Coyne, D. Popular Biofortified Cassava Cultivars are Heavily Impacted by Plant Parasitic Nematodes, Especially *Meloidogyne* Spp. *Plants* **2020**, *9*, 802. [CrossRef] [PubMed]
111. Ngome, A.F.; Amougou, M.F.; Tata, P.I.; Ndindeng, S.A.; Mfopou, M.Y.; Mapiemfu-Lamare, D.; Njeudeng, T.S. Effects of cassava cultivation on soil quality indicators in the humid forest of Cameroon. *Greener J. Agric. Sci.* **2013**, *3*, 451–457. [CrossRef]
112. Salvador, E.; Steenkamp, V.A.; McCrindle, C. Production, consumption and nutritional value of cassava (Manihot esculenta, Crantz) in Mozambique: An overview. *J. Agric. Biotechnol. Sustain. Dev.* **2014**, *6*, 29–38. [CrossRef]
113. Temegne, N.C.; Ngome, A.F.; Fotso, K.A. Influence de la composition chimique du sol sur la teneur en éléments nutritifs et le rendement du manioc (Manihot esculenta Crantz, uphorbiaceae) dans deux zones agro-écologiques du Cameroun. *Int. J. Biol. Chem. Sci.* **2015**, *9*, 2776–2788. [CrossRef]
114. Carine, T.; Mouafor, B.I.; Ngome, A.F. Agro-morphological Characterization of Cassava (Manihot esculenta Crantz) Collected in the Humid Forest and Guinea Savannah Agro-ecological Zones of Cameroon. *Greener J. Agric. Sci.* **2016**, *6*, 209–225. [CrossRef]
115. McDougall, J.; Thomas, L.; McDougall, C.; Moloney, G.; Saul, B.; Finnell, J.; Richardson, K.; Mae Petersen, K. Effects of 7 days on an ad libitum low-fat vegan diet: The McDougall Program cohort. *Nutr. J.* **2014**, *13*, 1–7. [CrossRef] [PubMed]
116. Jukanti, A.; Pautong, P.; Liu, Q.; Sreenivasulu, N. Low glycemic index rice—A desired trait in starchy staples. *Trends Food Sci. Technol.* **2020**, *106*, 132–149. [CrossRef]
117. FAO. *Food Outlook—Biannual Report on Global Food Markets*; FAO: Rome, Italy, 2018; p. 104.
118. Oyewole, F.; Eforuoku, F. Value Addition on Cassava Wastes among Processors in Oyo State, Nigeria. *J. Agric. Ext.* **2019**, *23*, 135–146. [CrossRef]
119. Ekop, I.; Simonyan, K.; Evwierhoma, E. Utilization of Cassava Wastes for Value Added Products: An Overview. *Int. J. Sci. Eng. Sci.* **2019**, *3*, 31–39.
120. Ndam, Y.; Mounjouenpou, P.; Kansci, G.; Kenfack, M.; Meguia, M.; Natacha, N.; Eyenga, N.; Akhobakoh, M.; Nyegue, A. Influence of cultivars and processing methods on the cyanide contents of cassava (Manihot esculenta Crantz) and its traditional food products. *Sci. Afr.* **2019**, *5*, e00119. [CrossRef]
121. Oteh, O.; Hefferon, K.; Agwu, N. Moving Biofortified Cassava Products Closer to Market in Nigeria. *Front. Sustain. Food Syst.* **2020**, *4*, 1–11. [CrossRef]
122. Tomlinson, K.; Bailey, A.; Alicai, T.; Seal, S.; Foster, G. Cassava brown streak disease: Historical timeline, current knowledge and future prospects. *Mol. Plant Pathol. Br. Soc. Plant Pathol.* **2017**, *19*, 1282–1294. [CrossRef]
123. Sheat, S.; Fuerholzner, B.; Stein, B.; Winter, S. Resistance Against Cassava Brown Streak Viruses from Africa in Cassava Germplasm from South America. *Front. Plant Sci.* **2019**, *10*, 567. [CrossRef]
124. Mohamed, E. Resource Rents, Human Development and Economic Growth in Sudan. *Economies* **2020**, *8*, 99. [CrossRef]
125. Ferris, A.; Stutt, R.O.; Godding, D.; Gilligan, C. Computational models to improve surveillance for cassava brown streak disease and minimize yield loss. *PLoS Comput. Biol.* **2020**, *16*, e1007823. [CrossRef]
126. IMARC Group. *Cassava Starch Market: Global Industry Trends, Share, Size, Growth, Opportunity and Forecast 2019–2024*; Market Report; IMARC Group: Noida, India, 2019.
127. Organización de las Naciones Unidas Para la Alimentación y la Agricultura-FAO. *The Word Cassava Economy-Report*; FAO: Rome, Italy, 1997.
128. Organización de las Naciones Unidas Para la Alimentación y la Agricultura-FAO. *Food Outlook-Global Market Analysis: Cassava*; FAO: Rome, Italy, 2009.
129. Bilong, E.; Abossolo-Angue, M.; Ngome Ajebesone, F.; Anaba, D.; Madong, B.; Nomoa, L.; Bilong, P. Improving soil physical properties and cassava productivity through organic manures management in the southern Cameroon. *Heliyon 8* **2022**, *8*, e09570. [CrossRef] [PubMed]
130. Echebiri, R.; Edaba, M. Production and Utilization of Cassava in Nigeria: Prospect for Food Security and Infant Nutrition. *Prod. Agric. Technol.* **2008**, *4*, 38–52.
131. Hauser, S.; Nolte, C.; Carsky, R.J. What role can planted fallows play in the humid and subhumid zones of West and Central Africa? *Nutr. Cycl. Agroecosyst.* **2006**, *76*, 297–318. [CrossRef]
132. Tellen, V.; Yerima, B. Effects of land use change on soil physicochemical properties in selected areas in the North West region of Cameroon. *Environ. Syst. Resour.* **2018**, *7*, 3. [CrossRef]
133. Cusack, D.; Karpman, J.; Ashdown, D.; Cao, Q.; Ciochina, M.; Halterman, S.; Lydon, S.; Neupane, A. Global change effects on humid tropical forests: Evidence for biogeochemical and biodiversity shifts at an ecosystem scale. *Rev. Geophys.* **2016**, *54*, 1–88. [CrossRef]
134. Tieguhong, J. *Report on Public-Private Partnerships (PPP) in the Forestry Sector in Cameroon*; African Forest Forum: Nairobi, Kenya, 2016.

135. De Wasseige, C.; Tadoum, M.; Eba'aAtyi, R.; Doumenge, C. *The Forests of the Congo Basin—Forests and Climate Change*; Weyrich: Neufchâteau, Belgium, 2015; ISBN 978-2-87489-355-1.
136. Luttrell, C.; Sills, E.; Aryani, R.; Ekaputri, A.; Evinke, M. Beyond opportunity costs: Who bears the implementation costs of reducing emissions from deforestation and degradation? *Mitig. Adapt. Strateg. Glob. Chang.* **2018**, *23*, 291–310. [CrossRef]
137. Versino, F.; López, O.; García, M. Exploitation of by-products from cassava and ahipa starch extraction as filler of thermoplastic corn starch. *Compos. Part B Eng.* **2020**, *182*, 107653. [CrossRef]
138. Suryaningrat, I.; Amilia, W.; Choiron, M. Current Condition of Agroindustrial Supply Chain of Cassava Products: A Case Survey of East Java, Indonesia. *Agric. Agric. Sci. Procedia* **2015**, *3*, 137–142. [CrossRef]
139. Dudu, O.; Ma, Y.; Ajibola, A.; Oyedeji, B.; Oyeyinka, S.; Ogungbemi, J. Bread-making potential of heat-moisture treated cassava flour-additive complexes. *LWT-Food Sci. Technol.* **2020**, *130*, 109477. [CrossRef]
140. Oni, D.; Mwero, J.; Kabubo, C. The Effect of Cassava Starch on the Durability Characteristics of Concrete. *Open Civ. Eng. J.* **2020**, *14*, 289–301. [CrossRef]
141. Akindahunsi, A.A.; Uzoegbo, H.C. Strength and Durability Properties of Concrete with Starch Admixture. *Int. J. Concr. Struct. Mater.* **2015**, *9*, 323–335. [CrossRef]
142. Satin, M. Functional properties of starches. In *Spotlight Tropical Starch Misses Market*; AGSI report, Agriculture 21; FAO-Magazine: Rome Italy, 1998; p. 11.
143. Sindhu, N.; Sivakamasundari, S.; Debojyoti, P. Study on flexural behavior of r.c beam using different starches as an admixture. *Int. J. Pure Appl. Math.* **2018**, *118*, 24.
144. Kariuki, S.; Muthengia, J.; Erastus, M.; Leonard, G.; Marangu, J. Characterization of composite material from the copolymerized polyphenolic matrix with treated cassava peels starch. *Heliyon* **2020**, *6*, e04574. [CrossRef] [PubMed]
145. Yan, Y.; Scrivener, K.; Yu, C.; Ouzia, A.; Liu, J. Effect of a novel starch-based temperature rise inhibitor on cement hydration and microstructure development: The second peak study. *Cem. Concr. Res.* **2021**, *141*, 106325. [CrossRef]
146. Izaguirre, A.; Lanas, J.; Alvarez, J. Behaviour of starch as a viscosity modifier for aerial lime-based mortars. *Carbohydr. Polym.* **2010**, *80*, 222–228. [CrossRef]
147. Akindahunsi, A.; Schmidt, W. Effect of cassava starch on shrinkage characteristics of concrete. *Afr. J. Sci. Technol. Innov. Dev.* **2019**, *11*, 441–447. [CrossRef]
148. Akindahunsi, A.; Schmidt, W.; Uzoegbo, H.; Iyuke, S. The Influence of Starches on some Properties of Concrete. Johannesburg, South Africa. *Adv. Cem. Concr. Technol. Afr.* **2013**, 637–646.
149. Joseph, S.; Xavier, A. Effect of Starch Admixtures on Fresh and Hardened Properties of Concrete. *Int. J. Sci. Eng. Res.* **2016**, *4*, 27–30.
150. Syamala Devi, K.; Vijaya Lakshmi, V.; Alakanandana, A. Impacts of cement industry on environment—An overview. *Asia Pac. J. Res.* **2017**, *1*, 156–161.
151. Arachchige, U.; Alagiyawanna, A.; Balasuriya, B.; Chathumini, K.; Dassanayake, N.; Devasurendra, J. Environmental Pollution by Cement Industry. *Int. J. Res.* **2019**, *6*, 631–635.
152. Baenla, J.; Bike Mbahb, J.B.; Djon Li Ndjock, I.; Elimbi, A. Partial replacement of low reactive volcanic ash by cassava peel ash in the synthesis of volcanic ash based geopolymer. *Constr. Build. Mater.* **2019**, *227*, 116689. [CrossRef]
153. Luke, K.; Luke, G. Effect of sucrose on the retardation of Portland cement. *Adv. Cem. Res.* **2000**, *12*, 9–18. [CrossRef]
154. Peschard, A.; Govin, A.; Grosseau, P.; Guilhot, B.; Guyonnet, R. Effect of polysaccharides on the hydration of cement paste at early ages. *Cem. Concr. Res.* **2004**, *34*, 2153–2158. [CrossRef]
155. Okafor, F. The Performance of Cassava Flour as a Water-Reducing Admixture for Concrete. *Niger. J. Technol.* **2010**, *29*, 106–112.
156. Akindahunsi, A.A.; Schmidt, W.; Uzoegbo, H.C.; Iyuke, E.S. Technological advances in the potential uses of cassava starch in concrete. In Proceedings of the 6th International Conference of Africa Materials Research Society, Victoria Falls, Zimbabwe, 11–16 December 2011.
157. Akindahunsi, A.; Uzoegbo, H.; Iyuke, S. Use of starch modified concrete as a repair material. In Proceedings of the 3rd International Conference on Concrete Repair, Rehabilitation and Retrofitting, Cape Town, South Africa, 3–5 September 2012.
158. Lasheras-Zubiate, M.; Navarro-Blasco, I.; Fernández, J.M.; Iverez, J.I. Effect of the addition of chitosan ethers on the fresh state properties of cement mortars. *Cem. Concr. Compos.* **2012**, *34*, 964–973. [CrossRef]
159. Hawkes, J. *The Fourth Pillar of Sustainability: Culture's Essential Role in Public Planning*; Common Ground P/L: Melbourne, Australia, 2001.
160. Feindt, P.; Weiland, S. Reflexive governance: Exploring the concept and assessing its critical potential for sustainable development. Introduction to the special issue. *J. Environ. Policy Plan.* **2018**, *20*, 661–674. [CrossRef]
161. Dyck, B.; Walker, K.; Caza, A. Antecedents of sustainable organizing: A look at the relationship between organizational culture and the triple bottom line. *J. Clean. Prod.* **2019**, *231*, 235–1247. [CrossRef]
162. Isensee, C.; Teuteberg, F.; Griese, K.-M.; Topi, C. The relationship between organizational culture, sustainability, and digitalization in SMEs: A systematic review. *J. Clean. Prod.* **2020**, *275*, 122944. [CrossRef]
163. Burford, G.; Hoover, E.; Velasco, I.; Janoušková, S.; Jimenez, A.; Piggot, G.; Podger, D.; Harder, M. Bringing the "Missing Pillar" into Sustainable Development Goals: Towards Intersubjective Values-Based Indicators. *Sustainability* **2013**, *5*, 3035–3059. [CrossRef]
164. Taffa Ariti, A.; van Vliet, J.; Verburg, P. What restrains Ethiopian NGOs to participate in the development of policies for natural resource management? *Environ. Sci. Policy* **2018**, *89*, 292–299. [CrossRef]

165. Brehony, P.; Tyrrell, P.; Kamanga, J.; Waruingi, L.; Kaelo, D. Incorporating social-ecological complexities into conservation policy. *Biol. Conserv.* **2020**, *248*, 108697. [CrossRef] [PubMed]
166. Cwikła, M.; Góral, A.; Bogacz-Wojtanowska, E.; Dudkiewicz, M. Project-Based Work and Sustainable Development—A Comparative Case Study of Cultural Animation Projects. *Sustainability* **2020**, *12*, 6519. [CrossRef]
167. Piabuo, S.M.; Foundjem-Tita, D.; Minang, P.A. Community forest governance in Cameroon: A review. *Ecol. Soc.* **2018**, *23*, 34. [CrossRef]
168. Kimengsi, J.; Mukong, A.; Giessen, L.; Pretzsch, J. Institutional dynamics and forest use practices in the Santchou Landscape of Cameroon. *Environ. Sci. Policy* **2022**, *128*, 68–80. [CrossRef]
169. Sikod, F.; Teke, J. *Governance and Economic Growth in Cameroon*; Cameroon AERC Research Paper 250; African Economic Research Consortium: Nairobi, Kenya, 2012.
170. Altieri, M.; Nicholls, C.; Henao, A.; Lana, M. Agroecology and the design of climate change-resilient farming systems. *Agron. Sustain. Dev.* **2015**, *35*, 869–890. [CrossRef]
171. Garnett, S.T.; Burgess, N.D.; Fa, J.E.; Fernandez-Llamazares, A.; Molnar, Z.; Robinson, C.J.; Watson, J.E.; Zander, K.K.; Austin, B.; Brondizio, E.S.; et al. A spatial overview of the global importance of Indigenous lands for conservation. *Nat. Sustain.* **2018**, *1*, 369–374. [CrossRef]
172. Caballero-Serrano, V.; McLaren, B.; Carrasco, J.; Alday, J.; Fiallos, L.; Amigo, J.; Onaindia, M. Traditional ecological knowledge and medicinal plant diversity in Ecuadorian Amazon home gardens. *Glob. Ecol. Conserv.* **2019**, *17*, e00524. [CrossRef]
173. Hill, S.; Arnell, A.; Maney, C.; Butchart, S.; Hilton-Taylor, C.; Ciciarelli, C.; Davis, C.; Dinerstein, E.; Purvis, A.; Burgess, N. Measuring forest biodiversity status and changes globally. *Front. For. Glob. Chang.* **2019**, *2*, 70. [CrossRef]
174. Aquilas, N.A.; Forgha, N.G.; Agbor, M.S. Testing the Environmental Kuznets Curve hypothesis for the depletion of natural resources in Cameroon. *J. Econ. Manag. Sci.* **2019**, *1*, 122–142.
175. Kimengsi, J.N.; Balgah, S.N. Colonial hangover and institutional bricolage processes in forest use practices in Cameroon. *For. Policy Econ.* **2021**, *125*, 102406. [CrossRef]
176. Bruggeman, D.; Meyfroidt, P.; Lambin, E. Production forests as a conservation tool: Effectiveness of Cameroon's land use zoning policy. *Land Use Policy* **2015**, *42*, 151–164. [CrossRef]
177. Official Journal of the European Union (OJEU). Voluntary Partnership Agreement between the European Union and the Republic of Cameroon on Forest Law Enforcement, Governance and Trade in Timber and Derived Products to the European Union (FLEGT), L 92/4. 2011. Available online: https://eur-lex.europa.eu/legal-content/EN (accessed on 17 January 2021).
178. Minang, P.; McCall, M.; Skutsch, M.; Verplanke, J. A data support infrastructure for Clean Development Mechanism forestry implementation: An inventory perspective from Cameroon. *Mitig. Adapt. Strateg. Glob. Chang.* **2007**, *13*, 157–178. [CrossRef]
179. Ahlborg, H.; Boräng, F.; Jagers, S.; Söderholm, P. Provision of electricity to African households: The importance of democracy and institutional quality. *Energy Policy* **2015**, *87*, 125–135. [CrossRef]
180. Growth and Employment Strategy Paper (GESP). *Reference Framework for Government Action for the Period 2001–2020*; Republic of Cameroon; Internat Monetary Fund: Washington, DC, USA, 2010; p. 172.
181. Arslan, M.; Alqatan, A. Role of institutions in shaping corporate governance system: Evidence from emerging economy. *Heliyon* **2020**, *6*, e03520. [CrossRef] [PubMed]
182. Acosta, L.; Maharjan, P.; Peyriere, H.; Jill Mamiit, R. Natural capital protection indicators: Measuring performance in achieving the Sustainable Development Goals for green growth transition. *Environ. Sustain. Indic.* **2020**, *8*, 100069. [CrossRef]
183. Fernandez, L.; Gutierrez, M. Bienestar Social, Económico y Ambiental para las Presentes y Futuras Generaciones. *Inf. Tecnológica* **2013**, *24*, 121–130. [CrossRef]
184. Bas, T.G.; Oliu, C. Innovation strategy management survey of the Chilean biomedical industry. Assessment of windows of opportunities to reduce technological gaps. *Int. J. Health Plan. Manag.* **2018**, *33*, e512–e530. [CrossRef]
185. Joa, B.; Winkel, J.; Primmer, E. The unknown known—A review of local ecological knowledge in relation to forest biodiversity conservation. *Land Use Policy* **2018**, *79*, 520–530. [CrossRef]
186. Joa, B.; Schraml, U. Conservation practiced by private forest owners in Southwest Germany—The role of values, perceptions and local forest knowledge. *For. Policy Econ.* **2020**, *115*, 102141. [CrossRef]

Article

Ecological–Economic Modelling of Traditional Agroforestry to Promote Farmland Biodiversity with Cost-Effective Payments

Takamasa Nishizawa [1,*], Sonja Kay [2], Johannes Schuler [1], Noëlle Klein [2,3], Felix Herzog [2], Joachim Aurbacher [4] and Peter Zander [1]

1 Farm Economics and Ecosystem Services, Leibniz Centre for Agricultural Landscape Research (ZALF) e.V., Eberswalder Strasse 84, 15374 Muecheberg, Germany; schuler@zalf.de (J.S.); peter.zander@zalf.de (P.Z.)
2 Department of Agroecology and Environment, Agroscope, Reckenholzstrasse 191, 8046 Zurich, Switzerland; sonja.kay@agroscope.admin.ch (S.K.); noelle.klein@agroscope.admin.ch (N.K.); felix.herzog@agroscope.admin.ch (F.H.)
3 Planning of Landscape and Urban Systems (PLUS), ETH Zurich, Stefano-Franscini-Platz 5, 8093 Zurich, Switzerland
4 Institute of Farm and Agribusiness Management, Justus-Liebig-University Giessen, Senckenbergstrasse 3, 35390 Giessen, Germany; joachim.aurbacher@agrar.uni-giessen.de
* Correspondence: takamasa.nishizawa@zalf.de; Tel.: +49-(0)33432-82-490

Abstract: Orchard meadows, a traditional agroforestry system in Switzerland combining the dual use fruit and fodder production, are declining, even though the farmland managed under agri-environmental schemes (AES) has been expanding. Despite increasing interest in agroforestry research for developing sustainable agriculture, it is poorly understood how subsidies contribute to the maintenance of trees on agricultural land and the promotion of farmland biodiversity. Therefore, the objective of the present study is to examine the effects of incentive-based AES on both farmers' decisions regarding trees and biodiversity by developing an ecological–economic assessment model. To explore cost-effective AES, we explicitly consider the heterogeneity of farm types. We apply this integrated model to the farms in Schwarzbubenland, a small hilly region in Northern Switzerland. Results show that the adoption of AES and the compliance costs of participating in AES considerably vary among farm types, and the current AES do not provide farmers with sufficient payments to maintain any type of orchard meadows, despite the ecological benefits of orchard meadows. The integrating modeling developed in this study enables us to better understand the relationship between subsidies and biodiversity through farmers' decisions on land use and facilitates the design of cost-effective payments for the maintenance of agroforestry.

Keywords: agroforestry; biodiversity; agri-environmental schemes; integrated ecological–economic modeling; cost effectiveness

Citation: Nishizawa, T.; Kay, S.; Schuler, J.; Klein, N.; Herzog, F.; Aurbacher, J.; Zander, P. Ecological–Economic Modelling of Traditional Agroforestry to Promote Farmland Biodiversity with Cost-Effective Payments. *Sustainability* **2022**, *14*, 5615. https://doi.org/10.3390/su14095615

Academic Editor: Luigi Aldieri

Received: 8 April 2022
Accepted: 1 May 2022
Published: 6 May 2022

1. Introduction

Given the increasing awareness of biodiversity degradation in Switzerland [1,2], Swiss agricultural policies have developed agri-environmental schemes (AES) to promote biodiversity on farmland, including orchards, vineyards, vegetables, etc. [3]. AES are voluntary programs that provide financial incentives for farmers. In the case of Switzerland, AES are a part of direct payments. Considering the high share of direct payments to the total farm income in Switzerland (around 50% on average in 2018–2020 [4]), land-use decisions can be assumed to be highly dependent on public payments. One of the requirements for receiving direct payments is that at least 7% of a farm's production area needs to be managed under AES. These areas are referred to as ecological focus areas (EFA) [5], similar to EFA in the European Union, but related to different conditions, and focus on the provision of farmland biodiversity.

The majority of EFA are implemented on grassland and high-stem orchards, regarded as a traditional agroforestry system in Switzerland [6], combining the dual use for fruit and fodder production. Particularly, orchard meadows are likely to play a key role in biodiversity promotion as agrobiodiversity hotspots under AES [7,8]. Due to their diverse structure, they supply habitats for various species, including small mammals, reptiles and several insect groups [9]. Along with biological diversity, orchard meadows provide socio-cultural features, such as landscape aesthetics, recreation and regional identity [10,11]. However, maintaining orchard meadows has become increasingly challenging, due to higher production costs, mechanized farming, increasing quality requirements and the infestation of invasive fruit flies [12,13]. The decline in orchards may trigger the loss of not only the traditional characteristics of the regional landscape, but also the habitats for various species. Therefore, the recent decline in orchard meadows in Switzerland [14,15] is of great concern. It is vital to investigate to what extent AES have an impact on the maintenance of orchards and farmland biodiversity, given a certain level of AES payments. The cost of adopting AES is a critical factor in farmers' decisions of whether to participate [16]. Additionally, the different effects of AES across specific farm types should be considered, as varied adoption costs are expected due to the differences in farm management.

There is accumulating evidence that farm types influence the effects of AES differently [17,18]. Bamière et al. [19] argued that a detailed representation of farm management can provide us with valuable insights into designing agri-environmental policies and AES. Indeed, Mack et al. [5] revealed that the implementation of action-based AES was strongly influenced by farm types. Therefore, simplified AES can lead to less ecologically beneficial effects, failing their conservation potential [20], although they may be readily implemented by farmers [18]. However, more research is needed about the direct relationship between heterogeneous farm types and their consequences on the cost-effectiveness of AES. Fewer than 15% of studies evaluate cost-effectiveness when assessing AES [21]. Investigating the cost-effectiveness of payment programs can be a key to providing relevant implications for optimizing AES and ensuring the sustainability of agricultural policy [22]. Additionally, unless such programs prove to be cost effective, some legitimacy concerns may arise: governmental bodies, taxpayers and users of ecosystem services may be reluctant to pay [23].

To address this science-policy gap, we developed an ecological–economic assessment model by integrating the results of the expert system for farmland biodiversity assessment, SALCA-BD (Swiss Agricultural Life Cycle Assessment—Biodiversity), into an optimization-based bio-economic farm model (BEFM). The objective of this study is to provide policymakers with insight into the design of cost-effective AES, taking into account different farm types, for maintaining orchard meadows and promoting farmland biodiversity. To that end, we investigated the feedback mechanisms between AES and their subsequent ecological and economic effects per farm type, via farmers' decisions on land use. Our study can be distinguished from studies published to date in that we evaluated the cost-effectiveness of agri-environmental schemes while taking into account the impacts of both agroforestry and heterogeneous farms. We addressed the following research questions:

1. What role do AES play in land use and sustaining orchard meadows?
2. Which types of farms are more likely to implement AES and which measures?
3. How does the cost effectiveness of AES differ between farm types?
4. How would AES change the regional land use and affect the diversity of individual species?

2. Materials and Methods

Figure 1 illustrates our methodological approach for this study, in which we integrated the SALCA-BD and the BEFM, and describes the flow of model inputs and outputs. The following subsections explain each of these methodologies in depth.

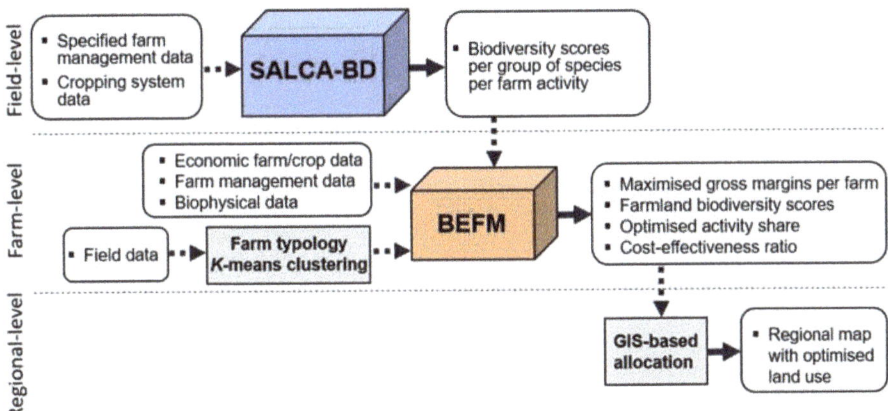

Figure 1. Structure of the ecological–economic model with the flow of inputs and outputs, integrated the results of each activity from the expert system for farmland biodiversity assessment (SALCA-BD) into the optimization-based bio-economic farm model (BEFM).

2.1. Case Study Region and Data

The study region, Schwarzbubenland (Figure 2), is located in Canton Solothurn, characterized by gently rolling hills (elevation 430 to 670 m). The average temperature is between 7.7 °C and 9.1 °C with annual precipitation of 800 to 1000 mm. Forestry (44%) and farmland (43%) are the main land uses. The area size is approximately 50 km^2, of which 1783 hectares are used as farmland, consisting of 32% arable land, 20% grassland and 48% orchard meadows. The study region is characterized by traditional high-stem cherry orchards combined with permanent grasslands. They have been established for subsistence and commercial fruit production, and the permanent grasslands are grazed by cattle and occasionally mown. Orchard meadows are recognized as agro-biodiversity hotspots. However, the decline of orchard trees can also be observed in this region [14].

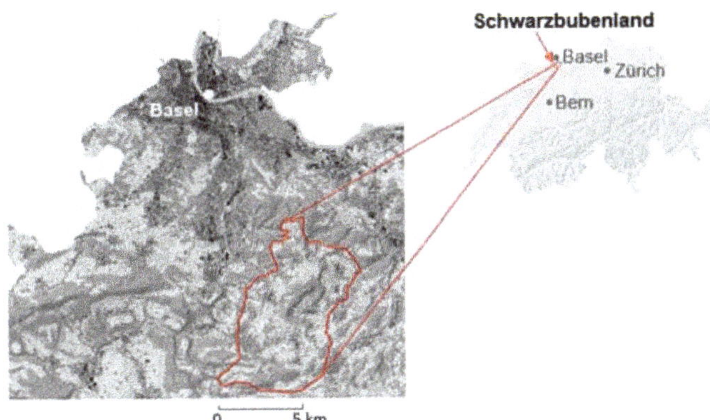

Figure 2. The study region: Schwarzbubenland (Switzerland) (Source: SwissImage ©Swisstopo).

Canton Solothurn provided 4698 spatially explicit field data spots in 2020 on the type of livestock, crops, management, the number of trees, area size, and the average slope degree [24]. Of the recorded 74 farms, over half of the farming enterprises are mixed farms, with combinations of arable crops, animal husbandry (mostly cattle for milk and meat production) and some fruit production. The average farm size is 24.1 ha, slightly larger than the average Swiss farm (21 ha, [25]), with approximately 0.77 livestock units per hectare.

2.2. SALCA-BD (Swiss Agricultural Life Cycle Assessment—Biodiversity)

The expert system SALCA-BD [26] evaluates the habitat suitability and favorable or adverse effects of agricultural activities on terrestrial species diversity at field scale [27]. Farmland biodiversity is represented by a set of indicator species groups (ISGs) that are sensitive to land use and farm management: vascular plants, birds, mammals, amphibians, snails, spiders, carabids, butterflies, wild bees, and grasshoppers. SALCA-BD assessed farm activities on both arable land and grassland as well as EFA. Along with the assessment of habitats' suitability on each ISG, management options, such as fertilizer and plant protection use, soil tillage, sowing, irrigation, the number and timing of mowing, etc., were explored. The SALCA-BD scores are calculated per ISG per farm activity and range between 0 and 50. The evaluation is non-spatially explicit. Results from the model have been validated in Switzerland and neighboring countries [27]. Jeanneret et al. [26] explain the method in more detail. Appendix A presents the biodiversity scores of the farm activities at a field-level evaluated with the model in this study.

For the aggregation of the habitats at a farm-level, we assumed a linear relationship between the biodiversity score of each farm activity and its area. Hence, we calculated the farmland biodiversity (FBD) score as follows:

$$\text{FBD score per farm type} = \sum_{i}^{n}\sum_{j}^{m} \text{BD score}_{ij} * \text{Area}_{ij} / \text{total farm size} \tag{1}$$

where BD score is the biodiversity estimated with SALCA-BD, i is a farm activity, and j is a management option. To obtain the biodiversity score at the regional level, the FBD scores of each farm type are aggregated by applying the weight of aggregation.

2.3. Farm Typology

We used a centroid-based clustering analysis, k-means clustering, to identify typical farm types in the study region. The number of clusters needs to be determined a priori. To determine the optimal number of centroids, we used the elbow method [28], while observing the performance of the cluster method at the same time. Supplementary Material S1 outlines the methods in detail. Based on the expert knowledge and the collected field data, we selected the following six explanatory variables for the identification of representative farms: the number of suckler cows and dairy cows (LSU), area of arable land, grassland, and orchard (ha) and stock intensity (LSU ha^{-1}). We considered the intensity of livestock and the area of extensive farmland habitats in the explanatory variables, as our typology of farms should reflect environmental impacts [29].

2.4. BEFM (the Bio-Economic Farm Model)

2.4.1. General Approach

The BEFM we developed can determine the optimal production pattern and level of land use by maximizing the total gross margin (GM), given the available resources and restrictions [30,31] and the assumption that the farm behaves as a profit maximizer. Therefore, it follows the general form of a linear programming model for n activities and m structural restrictions [32]:

$$\begin{aligned} &\text{maximize } Z = \textstyle\sum_{j=1}^{n} c_j x_j, \\ &\text{subject to}: \textstyle\sum_{j=1}^{n} a_j x_j \leq b_i \\ &\text{and} \quad x_j \geq 0, \end{aligned} \tag{2}$$

where $i = 1, 2, \ldots, m$, $j = 1, 2, \ldots, n$, Z is the total GM at a farm level, x is the farm activities, c is the gross margins or costs per unit of activity, a is the technical coefficients, and b is the resource availability or upper/lower limits of activities. We developed the BEFM for each of the identified typical farm types with the same formula above. The add-in COIN-OR CBC linear solver (OpenSolver 2.9.3) in Excel was used to find the optimal solution in the linear programming model [33].

2.4.2. Farm Activities

The BEFM covers the main activities observed in the farm types that we identified: crop production (cash and fodder production), grassland production (meadows and pasture), livestock production and AES. Some of the activities belong to both fodder production and AES (e.g., less intensive meadows, extensive meadows and orchard meadows). In reality, farmers can choose any measures from the list of AES, but we selected the most relevant measures in our model (Table 1). We also considered different management options for each activity that distinguish the intensity level of inputs. Extensive management must be free of fungicides, plant growth regulators, insecticides, or chemical–synthetic stimulators of natural resistance [34].

Table 1. List of the production activities modeled in the bio-economic farm model.

Grassland (Fodder)		Arable Land (Fodder)		Arable Land (Cash Crops)	
Intensive	Meadow	Intensive	Fodder wheat	Intensive	Spelt wheat
	Pasture		Triticale		Winter Wheat
Less intensive	Meadow [1]		Oats		Spring wheat
Extensive	Meadow [1]		Winter barley		Rye
	Pasture [1]		Ley pasture	Extensive	Spelt wheat
	Orchard-Meadow [1]	Extensive	Fodder wheat		Winter Wheat
			Triticale		Spring wheat
			Oats		Rye
Livestock			Winter barley		Flower strips [1]
	Dairy cow		Ley pasture		
	Suckler cow		White peas		
	Young stock		Silo-green corn		

Less intensive meadow [1], extensive meadow/pasture [1], orchard meadow [1] and flower strips [1] are eligible to receive the payments from AES.

To obtain crop yields across intensity levels, we referred to the yearly, average regional yield data (2003–2020) in Canton Solothurn [35] and the gross margin report of AGRIDEA: "Deckungsbeiträge DBKAT" [36]. For grass yields, we referred to the formula in GRUD [37] and estimated the yields of meadows and pastures at different intensities given the elevation in the study region. Supplementary Materials S2 (Table S2, 1,2) provides all activities modeled in this study, including their yields, variable costs, GMs, etc.

2.4.3. Modules

The BEFM was constructed in a modular way. It is recommended that BEFMs should be modular to enhance the use of evidence in policymaking processes [38,39]. There are four modules exogenously given in the model (Figure 3). Each module is a subset of a larger section of the linear programming and comprises a set of constraints that serve to optimize the farm's gross margin.

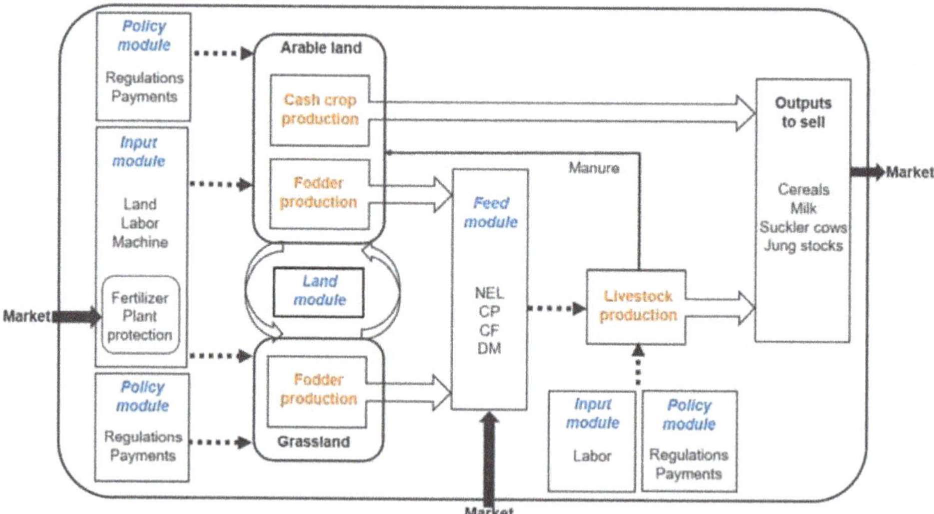

Figure 3. Activity flow inside the BEFM in the case of the dairy farm. NEL is net energy lactation; CP, crude protein; CF, crude fiber; and DM, dry matter.

Land module: We assumed that farmers could convert grassland into arable land depending on the payment level of direct payments as a result of maximizing the GM. Hence, this land module allows the model to convert the initial area of grassland into crop production and determines the optimal ratio of land use (i.e., the share of grassland, orchard and arable land). Under Swiss agri-environmental regulations, this conversion is possible as long as erosion events can be avoided. Nonetheless, we assumed that permanent grassland would remain on steep slopes (>24%), regardless of the payment level.

Input module: This module explicitly specifies the required labor hours and the level of input usage of different fertilizers and plant protection for each activity. For fertilizers, we considered N, P_2O_5, K_2O and Mg, and for plant protection, we included herbicides, fungicides, insecticides, growth regulators and trichogramma treatment. The costs of seeds, machinery and other miscellaneous items were also included [36]. Supplementary Materials S2 (Table S2, 3) details all the categories of variable costs included in this module for each farm activity. Note that we assumed manure to be used only within the farm without any exchanges with other farms.

Feed module: This feed module balances the supply and demand of livestock for feed in a nutritional form. We selected net energy lactation (NEL) in MJ DM-kg^{-1} (DM = dry matter), crude protein in kg DM-kg^{-1} and crude fiber in kg DM-kg^{-1}. We referred to the database of feed nutrition developed in Switzerland [40] to identify the nutritional values of the modeled crops and grasses. Supplement B describes these values per activity. To estimate the demand for livestock nutrition, we assumed the specific weight and performance of an adult milk cow: a 600 kg cow produces 8000 L of milk per year and young stocks (offspring of cows). We referred to the feed requirement tables [41] and determined the minimal requirement of the selected nutritional values as well as the maximum intake of dry matter per day per cow. For young stocks, we aggregated the nutritional requirements over different developmental phases of heifers, calves and bulls. Table 2 shows the results of the calculation for the nutritional constraints in the feed module.

Table 2. Nutritional constraints in the feed module to balance the supply and demand of livestock for nutrition.

	Dairy Cow		Suckler Cow	
	Adult Cow	Youngstock	Adult Cow	Youngstock
Maximum DM intake per day (kg)	16.8	12.5	14.0	7.8
Minimum NEL per day (MJ)	105.0	47.1	80.0	36.1
Minimum crude-protein per day (kg)	2.3	1.2	1.9	0.9
Minimum crude-fiber per day (kg)	3.4	2.4	2.8	1.4

Note that the unit kg is referred to the weight of dry matter (DM). NEL is net energy lactation.

Agricultural policy module: This module captures the role of direct payments, including the AES payments in farmers' land-use decisions by incorporating the obligatory measures and payments. We selected 14 different payment types [3]. Supplementary Materials S2 (Table S2, 4) details the total amount of direct payments that each activity receives and the breakdown of the sum. Table 3 lists all the restrictions that were implemented in the BEFM.

Table 3. List of the modeling restrictions.

Type of Restrictions	Explanation
Restrictions to qualify for direct payments	Crop rotation cereals (without corn and oats < 66% of AL), crop ration wheat, spelt and triticale (<50% of AL), crop rotation oats (<25% of AL), crop rotation corn (<40% of AL), crop rotation white peas (<15% of AL), flower strips (<50% of AL), biodiversity measure (>7% of total farmland), minimal livestock intensity, grassland-based milk and meat program [1]
Restrictions based on expert knowledge	Pasture limitation (less than 50% of grassland), nutritional balance (upper limit of DM intake, minimum NEL, crude protein and crude fiber), permanent GL (slope degree ≥ 24%), crop rotation limit cereals (<80% of AL)
Restrictions based on statistics	Total farm size, area of permanent GL, GL and AL, area of flexible land, labor hour, youngstock balance (share of offspring to adult cows), stable capacity

GL/AL are grassland and arable land. DM/NEL mean dry matter and net energy lactation. [1] The participation of the grassland-based milk and meat program was assumed to be subjected to only the large dairy farm and the suckler farm.

Table 4 shows the current payment level for EFA. While quality measures QI is an action-based measure rewarding farmers for adopting designated EFA, quality measures QII is a result-based measure for fulfilling specific goals [5]. For this study, we only considered the payment level of QI. This is because not all fields are eligible for QII, as they require specific site and biophysical conditions. Yet, for high-stem fruit trees, we considered both payments of QI and QII. This is because the payments for trees are primarily determined by the age of a tree (QI for 0 to 10 years and QII after 11 years) and additional measures (nesting boxes for birds, extensive grasslands, etc.). Thus, the extra costs to qualify for QII can be assumed to be negligible. Therefore, given an assumed tree's life of 60 years [42], we averaged the payments over 60 years per year and calculated the annual payment. In this study, we selected two types of orchard meadows, as explained in the next section.

Table 4. The current payments of AES in Switzerland and the payment level calculated for the model of this study.

Biodiversity Measures (EFA)	Quality Measure I	Quality Measure II	Modeled Payment
Extensive meadow (CHF ha^{-1})	860	1840	860
Less intensive meadow (CHF ha^{-1})	450	1200	450
Extensive pasture (CHF ha^{-1})	450	700	450
High-stem fruit trees (CHF tree^{-1})	13.5	31.5	39.8
Orchard meadow Type A (CHF ha^{-1}) (CHF/ha)	-	-	1642
Orchard meadow Type B (CHF/ha^{-1})	-	-	2052
Flower strips (CHF ha^{-1})	2500	-	2500

Only high-stem fruits trees consider both QI and QII payments. Orchard meadows receive payments for the corresponding meadow production as well as payments for trees (assumed 30 trees ha^{-1}). Orchard meadow Types A and B are explained in the next subsection.

2.4.4. Orchard Meadows

The orchards in the study region are mainly high-stem cherry trees (Kay et al., 2018). To model orchard meadows in the BEFM, we made several assumptions. First, we assumed two types of orchard meadows available in the model: orchard meadows with and without commercial cherry production. For orchard meadows with commercial cherry production (orchard meadow Type A), we assumed that they were managed on less intensive meadows. The gross margin of Type A is based on three price levels (low, medium, and high). For orchard meadows without commercial cherry production (orchard meadow Type B), we assumed that farmers did not harvest cherries, but kept the trees only to receive subsidies. We also assumed that orchard meadow Type B is managed on extensive meadows. Based on our available data, we assumed that 30 trees were planted per hectare for both types of orchard meadows. In the model, both types of orchard meadows are available for all farm types. Table 5 describes the detailed gross margin calculation for the modeled orchard meadows. Supplementary Materials S2 (Table S2, 5) contains comprehensive gross margin calculations of orchard meadows with further disaggregated items.

Table 5. GM calculation for the modeled orchard meadows.

	Orchard Meadow Type A	Orchard Meadow Type B	Source
Description	Commercial cherry production	No cherry production (maintaining trees for AES)	
Trees	30 trees ha^{-1}	30 trees ha^{-1}	Own source
Cherry yield	30 kg tree^{-1}	-	Giannitsopoulos 2020
Cherry price	1.5/1.2/0.7 CHF kg^{-1}	-	Giannitsopoulos 2020
Meadow management	Less intensive (2 cuts year^{-1})	Extensive (1 cut year^{-1}, no fertilizer)	
Forage yield loss	−15% less (yield: 54 dt ha^{-1})	−10% less (yield: 23 dt ha^{-1})	According to a local expert
Annual replanting	0.5 tree ha^{-1}	0.5 tree ha^{-1}	Schönhart 2011a
Establishment cost	140 CHF tree^{-1}	140 CHF tree^{-1}	Giannitsopoulos 2020
Maintenance cost [2]	6585 CHF ha^{-1}	630 CHF ha^{-1}	Giannitsopoulos 2020
Clearing cost [1]	60 CHF tree^{-1}	60 CHF tree^{-1}	Giannitsopoulos 2020
Labor	44 h ha^{-1}	27 h ha^{-1}	Giannitsopoulos 2020
Subsidy for trees	39.8 CHF tree^{-1}	39.8 CHF tree^{-1}	Bundesrat 2016
Total subsidy	2393 CHF ha^{-1}	2803 CHF ha^{-1}	Bundesrat 2016
Total revenues	3743 CHF ha^{-1} (1.5 CHF kg^{-1}) 3473 CHF ha^{-1} (1.2 CHF kg^{-1}) 3050 CHF ha^{-1} (0.7 CHF kg^{-1})	2803 CHF ha^{-1}	
Total costs [1]	6837 CHF ha^{-1}	882 CHF ha^{-1}	
GM with subsidy	−3094 CHF ha^{-1} (1.5 CHF/kg) −3364 CHF ha^{-1} (1.2 CHF/kg) −3787 CHF ha^{-1} (0.7 CHF/kg)	1921 CHF ha^{-1}	

Establishing cost [1], clearing cost [1] and total costs [1] are converted into annual equivalent costs based on a discount rate of 4% [42] and the average inflation rate over the last 30 years in Switzerland (0.9%). Maintenance cost [2] includes input use, harvesting and machinery for orchard meadow Type A and machinery and miscellaneous costs for orchard meadow Type B.

2.5. Policy Scenarios

Figure 4 illustrates how we ran the model with the baseline scenario and the policy scenarios. We first ran the model to retrieve the optimal baseline solution given the current payment level and the fixed land assumption. Next, we ran the model, while increasing the payment level from the current level to 200% by 10% increments. For this policy scenario, we applied the flexible land assumption. Therefore, the model can determine the optimal share of grassland and arable land at each level of payments of AES and corresponding cropping patterns.

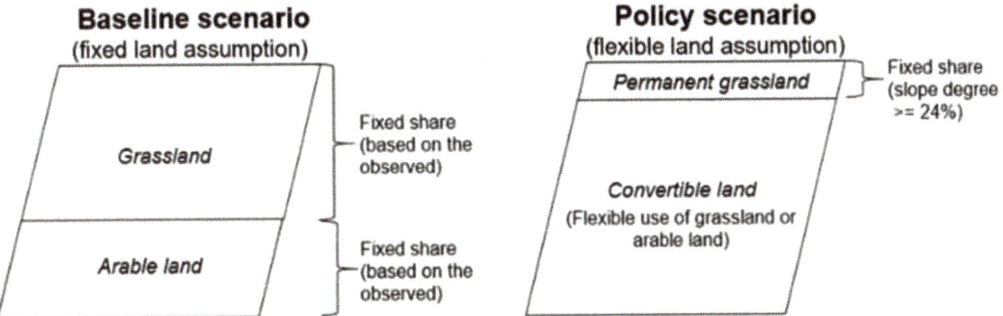

Figure 4. Land use determined under the land module with the baseline and policy scenarios. Note that both grassland and permanent grassland can be also used as orchard meadows in the model.

2.6. Evaluation of Cost Effectiveness

We measured the cost effectiveness of AES using the following two indicators: the cost-effectiveness ratio (CER) [43] and the producer rent. The CER represents the maximum FBD (farmland biodiversity) score that each farm can obtain per CHF 1000 of AES payments. To compare the CERs among farm types, the amount of payments paid out from AES is divided by EFA. Thus, the CER (unit: FBD score/CHF 1000) is computed per hectare as follows:

$$\text{CER} = (\text{FBD score}/(\text{payout from AES}/\text{EFA})) * 1000. \quad (3)$$

The producer rent quantifies how much the implementation of AES forgoes the income of farms [44]. It indicates the opportunity costs associated with impending AES. The producer rent (unit: CHF ha^{-1}) is calculated as follows:

$$\text{producer rent} = (\text{GM with AES} - \text{GM without AES})/\text{total farm size}. \quad (4)$$

Both CER and the producer rent were calculated per farm type and were also aggregated for the regional scale with the weights.

2.7. Map Regional Change of Orchard Meadows

To map the farm-level modeling results to each of the fields, we first identified which farm type was located in which field. Then, we assigned either of the land-use options in shares obtained with the model (grassland, orchard meadows or arable land) per farm type to each of the fields. Second, to determine which fields were most likely to belong to which land-use option, we assumed that fields with lower slope degrees would be covered by arable land for lower production costs, while on fields with higher slope degrees, trees would be planted on meadows to prevent erosion. This is consistent with a finding by Huber et al. [16] that fields with steeper slopes are more likely to enter the agglomeration scheme, which is a part of Swiss AES. The remaining fields were assigned as grassland.

3. Results

3.1. Farm Typology

We identified the following five representative farm types in the study region with the k-means clustering and the elbow methods in R (Version 1.2.1335) (Figure 5): 1. orchard farm without livestock (high-value trees and commercial cherry production, mainly cherries); 2. small-scale dairy farm; 3. large-scale dairy farm; 4. suckler farm; and 5. small-scale farms without livestock. Table 6 outlines the characteristics of these five farm types and their management. Given the number of farms and their farm size, large dairy and suckler farms are found to be the most prevalent farm types in the region. However, their management is contrastingly different. Large dairy farms tend to adopt intensive farming, as they own more arable land and less extensive grassland, whereas suckler farms utilize more extensive grassland. These results were confirmed by regional stakeholders.

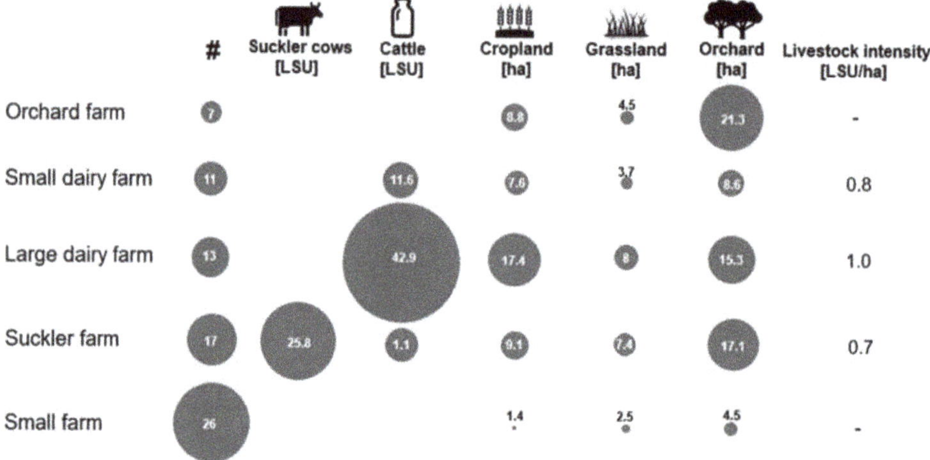

Figure 5. Identified representative farm types (left column) with k-means clustering in the study region and the average value of each explanatory variable. LSU: Livestock units. # indicates the number of farms found in each farm type.

Table 6. Modelled farm types and their characteristics.

	Orchard	Small Dairy	Large Dairy	Suckler Farm	Small Farm
Farm size (ha)	34.8	19.9	40.7	33.6	8.5
Weight for aggregation	14%	12%	30%	32%	12%
Initial grassland share	72%	60%	56%	70%	81%
Of which extensive grassland	41%	29%	19%	31%	46%
Permanent grassland	9.4%	10.5%	8.1%	5.9%	24.8%
Livestock	no	yes	yes	yes	No
Livestock intensity (LSU ha^{-1})	-	0.6	1.1	0.8	-
Capacity of livestock (LSU)	-	12	43	26	-
Labor availability in AWU	0.5	1.34	1.9	1.34	0.2
Grassland-based milk and meat program [1]	-	no	yes	yes	-

AWU stands for the annual working unit (1 AWU = 1800 h). We assumed that the orchard and small farms were part-time farms due to the small-scale farming. Grassland-based milk and meat program [1] provides farmers with an extra subsidy if they keep more than 75% of the share of fodder produced from grassland and less than 10% of the share of concentration (in weight of dry matter).

3.2. Orchard Meadows with the Baseline Scenario

Under this baseline scenario, the EFA of all farm types exceeds the obligatory level (7% of the total farmland) (Table 7). All farm types, except for the large dairy farm, chose orchard meadows for more than 50% of the total farmland. In particular, farm types without livestock (orchard and small farms) resulted in a higher share of orchard meadows. This led to a higher share of subsidy to the GM, which was more than 100%. By contrast, the large dairy farm resulted in the lowest share of orchard meadows. Regarding the FBD scores, orchard and small farms obtained higher values than the other farm types because of a relatively large share of orchard meadows and a lower share of arable land. Accordingly, their CERs were relatively high.

Table 7. Optimal baseline results per farm type with the current payments given the assumption that the area of grassland cannot be converted into arable land.

	GM	Subsidy to GM	EFA (%)	Trees	FBD Score	CER	Producer-Rent	Grass Land	Orchard	Arable Land
Orchard	73,322	113%	72%	755	12.2	6.0	541	0%	72%	28%
Small dairy	83,414	51%	52%	313	10.6	5.2	589	8%	52%	40%
Large dairy	236,446	31%	29%	350	9.4	4.6	256	28%	29%	44%
Suckler farm	101,440	77%	65%	659	11.8	5.8	702	5%	65%	30%
Small farm	17,048	122%	81%	206	13.2	6.4	605	0%	81%	19%

GM is gross margin and subsidy to GM indicates the share of the total amount of subsidies to the GM. EFA/FBD/CER indicate, ecological focused areas, farmland biodiversity (score), and cost-effectiveness ratio.

3.3. Policy Scenario

3.3.1. Role of AES in Land-Use and Sustaining Orchard Meadows

Regional land-use result: The regional result revealed that the share of grassland and orchard meadows at the current payment level was just under 20%, while arable land covered 80% of the land (Figure 6). Given the flexible land assumption, the arable land expanded considerably for all farm types at the current premium level, compared to the baseline result. As a result, the share of EFA dropped to 14%. However, as the payment level increased, the share of arable land decreased to 20% at 150% of the current AES premium, while EFA increased. This increase in EFA is mostly attributed to the expansion of the area of orchard meadow Type B.

Land-use differences across farm types: The EFA of the small and large dairy farms dropped just to the obligatory level, whereas the EFA of the suckler and small farms remained relatively high. Nonetheless, for all farm types, the arable land expanded considerably. Given the flexible land assumption, the difference in how much the arable land expands depends on the share of the permanent grassland. At the current payment level, all farm types except for large dairy farms expanded the arable land to the maximum possible area. Therefore, the share of land use at the current level would not change, even if the payment level was lowered from the current level.

Change in orchard meadows: The regional change in the number of trees is shown in Figure 7. The number of trees at the baseline is shown by a dot at 100%. In the policy scenario, the number of trees fell to 7627 from 29,847 at the current premium level. In the baseline scenario, where the conversion of land was not permitted, there was enough incentive to maintain orchards as shown in Table 7. However, if allowed, the arable land took over a large area of orchard meadows as they became less profitable. Increasing the payment level to 150%, however, restored the profitability of orchard meadows enough, allowing them to expand the area comparable to the baseline. The payment for orchard meadows at this level is around 1000 CHF ha^{-1}, higher than the current AES payments.

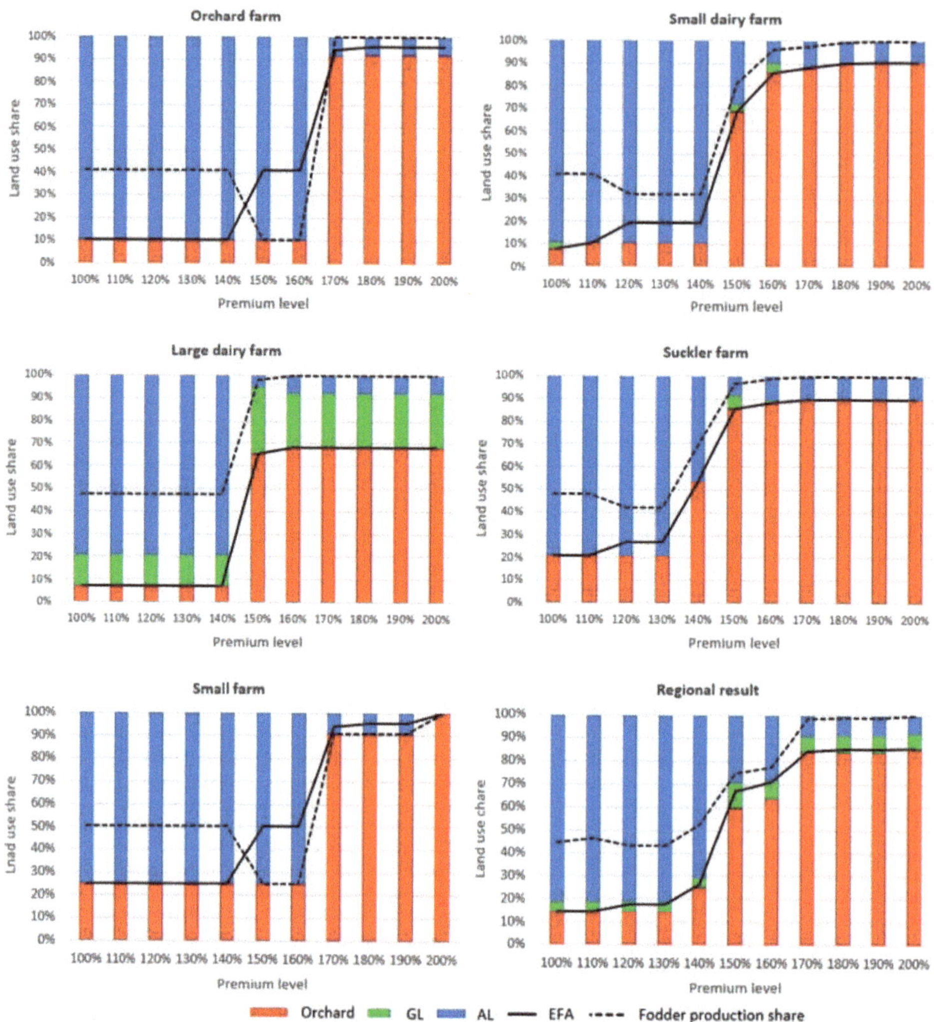

Figure 6. Policy scenario results with the increments of the payments from the current level up to 200%. The *X*-axis indicates the payment level compared to the current level. The *Y*-axis indicates land-use share. GL/AL/EFA designate grassland, arable land and ecologically focused area.

3.3.2. Difference in the Adoption of AES and the CER among Farm Types

The difference in the producer rents over farm types indicates the difference in the adoption costs of AES (Figure 8). The producer rent of large dairy farms stayed negative at lower payment levels, unlike other farm types. This reveals that for large dairy farms, the current AES payments cannot compensate for the cost of the mandatory implementation of the AES. The opportunity cost of adopting AES is the highest among all farm types due to the larger number of profitable dairy cows. Contrary to this result, all the other farm types had a positive producer rent with the current payment level. Among these farm types, the producer rent of the suckler farms was the highest: the opportunity cost of the suckler farm was the lowest. Nonetheless, the producer rents in the policy scenario are much less than the baseline producer rents.

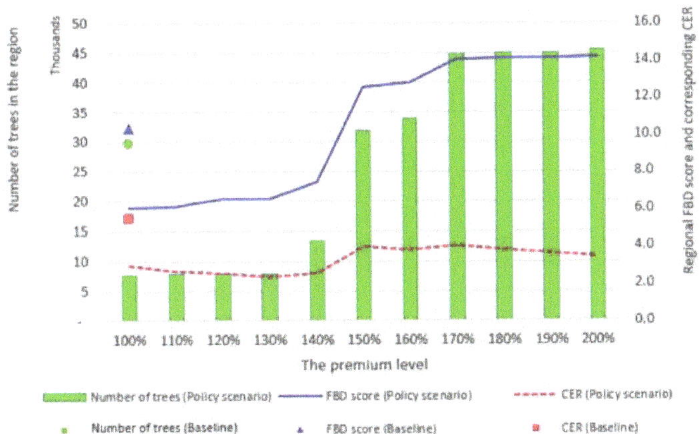

Figure 7. The regional change in the number of trees, regional farmland biodiversity score and the CER depending on the payment level (100% = the current level). FBD is farmland biodiversity.

All FBD scores at the initial level were approximately halved, despite the same level of payment, compared to the baseline scenario. The FBD scores of the orchard, large dairy and small farms remained the same until 140% of the current level. At a premium level of 150% premium, FBD scores increased sharply, as the area of orchard meadows tended to expand considerably.

The change of CER showed a similar trend. Up to a level of 140% premium, CERs tended to slightly decrease as the FBD score remained the same. However, due to the sharp increase in FBD scores above the 140% premium, CERs increased accordingly. For the regional level, it reached a maximum at 150% premium. Yet, they decreased eventually because there was little improvement in the FBD scores at higher payment levels. The highest CER in the policy scenario was even lower than the baseline CER.

3.3.3. Change in the Regional Land Use and Individual Species Suitability with AES

On the map of the case study region (Figure 9) with the baseline result, orchard meadows appeared more in the east and south, where the suckler, orchard and small farms tend to be located. On the other hand, more arable land and grassland appeared in the north and east, where the small and large dairy farms are more prevalent. The grassland mapped here is mostly with intensive pasture. The fields in the north and east area are relatively large and the slopes are flatter than the other areas. Thus, these fields are more suited to crop production and intensive grass production.

In the policy scenario at 100% premium (current level), most of the fields covered by orchard meadows disappeared, as Figure 7 shows that the number of trees is about one-quarter of the baseline number. When the premium was increased to 150%, the fields with orchard meadows appeared almost evenly on the map. However, 150% of the premium level is insufficient to sustain orchard meadows for orchard and small farms. Therefore, the fields belonging to these farm types remained as arable land, despite the increase in the payments. Figure 10 presents the variations in species suitability as a result of these regional-level results. Birds, butterflies, wild bees and grasshoppers are projected to be the most harmed by the expansion of arable land. All of these species groups are strongly linked to extensive grasslands.

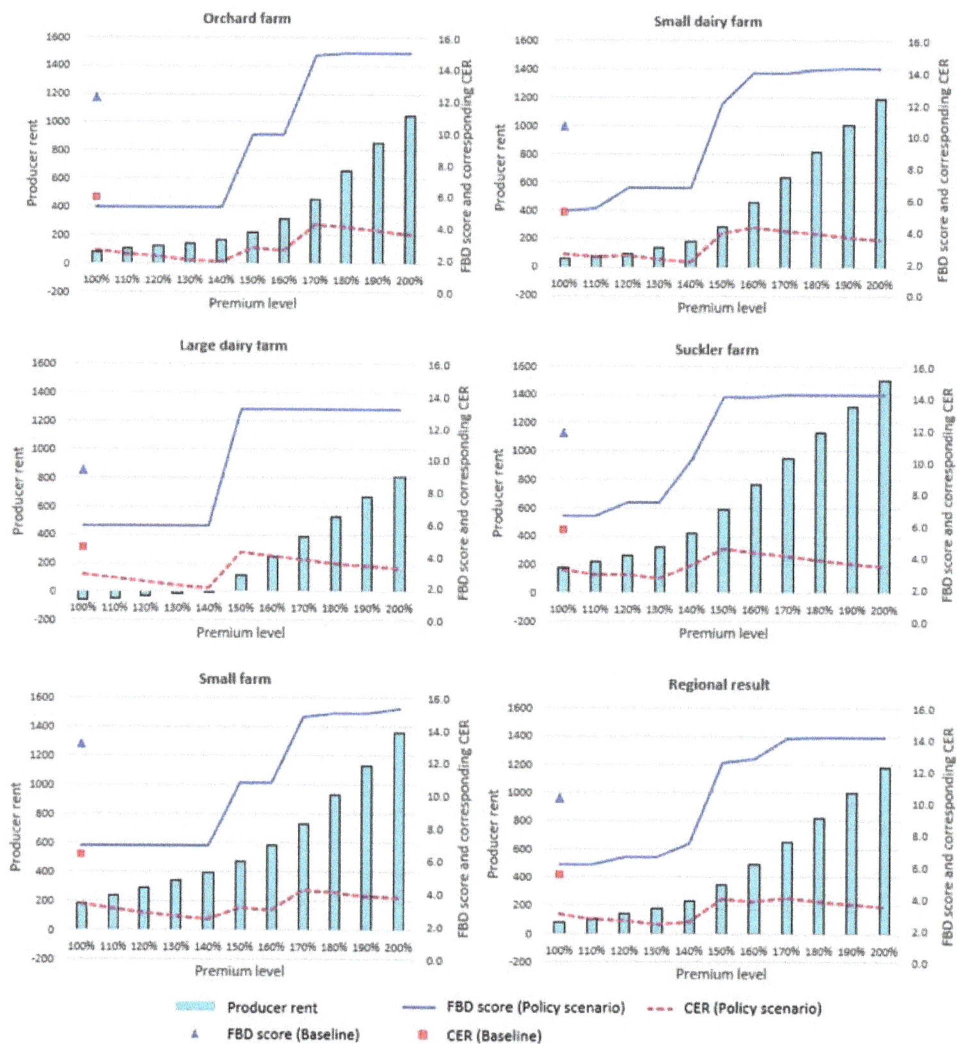

Figure 8. Policy scenario results with the increments of the payments from the current level up to 200%. The X−axis indicates the payment level compared to the current level. The Y−axis indicates producer rent in CHF per hectare. The Z−axis indicates the FBD (farmland biodiversity) scores and the CER (cost-effectiveness ratio). The triangle symbol in the graph indicates the level of FBD scores at the baseline, while the square symbol indicates the corresponding CER at the baseline.

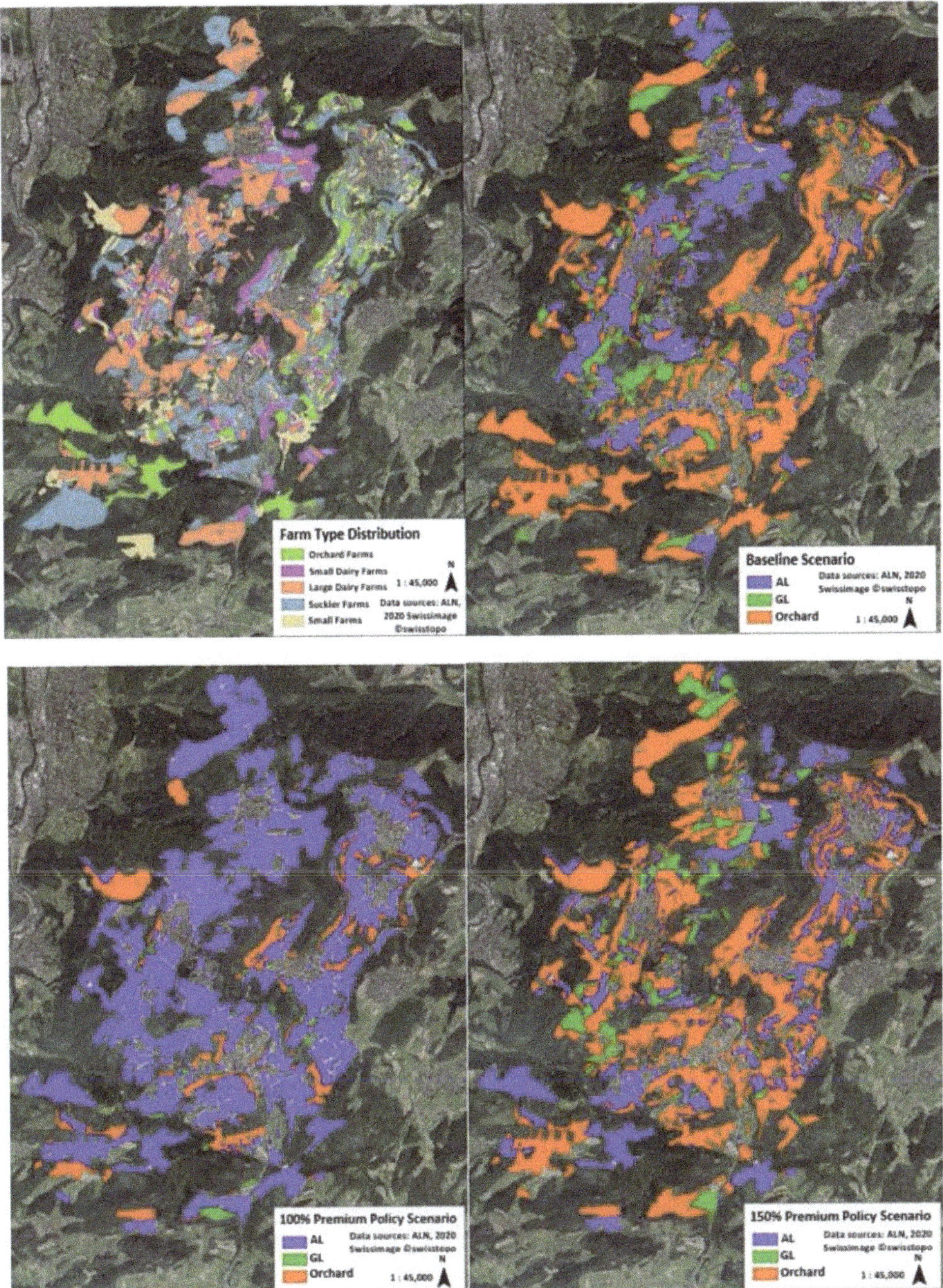

Figure 9. Maps of farm type distribution (above left) and the regional land use under the baseline scenario (above right) and policy scenarios—100% and 150% of the current premium level (bottom). AL/GL designate arable land and grassland.

	Flora of crops	Flora of grasslands	Birds	Small mammals	Amphibia	Molluscs
Baseline	4.2	9.1	17.3	12.5	5.0	4.9
100%	10.5	3.0	10.6	6.4	2.6	3.0
150%	5.3	9.0	18.5	13.5	5.1	5.1
	Spiders	**Carabids**	**Butterflies**	**Wild bees**	**Grasshoppers**	
Baseline	16.0	15.1	13.3	16.2	13.6	
100%	9.2	10.5	4.3	8.1	4.4	
150%	17.5	16.5	14.6	17.4	14.4	

Figure 10. Biodiversity scores of each species indicator group in the regional level with the baseline scenario and the policy scenario (100% and 150% premium levels).

4. Discussion

4.1. Orchard Meadows and Land-Use Change

We discovered that all farm types benefit from existing AES [34], and orchard meadows were well maintained in the baseline. Particularly, orchard and small farms favored orchard meadows to a large extent because the total gross margin for orchard meadows of Type B, including tree payments, was higher than the gross margins of pure meadows, due to the expected low selling price of hay (6 CHF dt^{-1}) [45]. Therefore, orchard and small farms rely on AES for a higher share of their income than the other farm types. This is also the case for suckler farms, which require less intensive meadows or pastures, compared to dairy [5]. Having dairy cows with high livestock intensity brings farms relatively high profits. Therefore, the share of orchard meadows on large dairy farms was the lowest of all farm types, as orchard meadows cannot provide the protein-rich fodder needed for dairy production.

In comparison, orchard meadow Type A was not chosen at any payment level. The sensitivity analysis with increasing cherry prices showed that it becomes only economically viable when the price of cherries exceeds 7 CHF kg^{-1}. The high production costs of cherries are mostly due to the high labor cost for harvesting. In reality, however, some orchard farms continue to produce cherries for profits. The discrepancy between our model's predictions and reality can be explained by traditional, family labor-based cherry production in the region: opportunity costs of labor may be low, and the local marketing of homemade products can be attractive. Nonetheless, the ecological benefits of trees alone justify public financial support.

A validity check of the modeling results with reality shows that the model tends toward orchard meadows, where, in reality, we find intensive meadows for dairy. As our model is a static comparative, it includes investments such as a dairy herd and its related infrastructure or planting of trees only as an annual average gross margin. Switching the production system between trees and cows is almost a once-in-a-lifetime decision, which does not depend on actual gross margins, as in our model. So, farmers, in general, have high resistance against such changes and can overcome smaller periods of lower gross margins in part of their production systems by compensating with income from other parts. Only if it becomes obvious that in the long run, the system will have low or even negative returns, do farmers change their production system. Often, this decision goes along with a generational change of ownership of the farm. Additionally, the timing of orchard-related labor peaks may play a role. Tree pruning can be done in winter, when labor pressure is low and the cherry harvest is in early summer, mostly after the labor peak of first hay making and before the start of crop harvesting. Nevertheless, our model shows this tendency under current circumstances. Should the performance relations between orchard meadows and dairy production remain the same for a longer period, we expect to see production shifts as projected in our model runs.

The policy scenario demonstrated that at the current payment level, regional biodiversity was considerably degraded as grassland and in particular, orchard meadows were often replaced by crop production. This implies that crop production in Switzerland is highly financially attractive if subsidies are considered [42]. Farmers receive a guaranteed payment of 1400 CHF ha^{-1} for crop production in addition to 120 CHF ha^{-1} as price support for supplying cereals [46], while a guaranteed premium for cultivating grassland is 1000 CHF ha^{-1} in the hilly regions [3].

4.2. Cost-Effectiveness and Its Difference over Farm Types

While the baseline maintains the current ratio of grassland and arable land, the policy scenarios allow flexible use of more than 75% of the land. With the current payment levels, this leads to lower biodiversity scores and also lower cost effectiveness. However, with increasing payments, the policy scenarios lead to high biodiversity impacts (see Figure 8). This is only possible as farmers are allowed to convert arable land into grassland, which goes along with decreasing cost effectiveness. This trade-off occurs as a result of higher payments, which reduce the cost effectiveness of AES.

Additionally, our results indicated that the producer rents over different farm types largely varied due to the different compliance costs of AES. Among the livestock farms, livestock intensity and type determine the producer rent. Large dairy farms still need to keep sufficient high-yield grassland to sustain high livestock intensity and fulfil the conditions of the grassland-based milk and meat program, which results in lower implementation rates from AES. In contrast, suckler farms gained relatively high implementation rates of AES. Mack et al. [5] verified these findings: the adoption of action-based EFA, which this study examined, is substantially influenced by farm type. Dairy farms are negatively correlated and suckler farms are positively correlated to the adoption rate. Our study demonstrated that farms with a higher implementation rate of AES tended to gain higher CER.

4.3. Methodological Limitations

We assumed that farmers would maintain or abandon orchard meadows depending on the economic profits in relation to the profitability of the other activities. However, we did not consider their non-market benefits, i.e., externalities, such as reduced soil erosion risk, carbon sequestration or regional identity, in the calculation of the economic profit. Although capturing the real value of orchard meadows is a core challenge in the economic assessment [11], accounting for such non-market benefits of orchard meadows in the decision process will improve the validity of results and help to determine a more appropriate level of financial support [42,47].

Another possible limitation of this study is that the evaluation of farmland biodiversity was neither contingent on the complexity of landscapes, such as the spatial configuration of semi-natural habitats [48], nor connectivity at different scales [49]. AES can be ineffective unless the ecological effects are observable at the landscape scale [50–52]. Spatial planning of biodiversity measures can enhance their benefits and reduce the opportunity cost for food production [53]. Understanding species dynamics and their relationships to landscape complexity, using a broader spatial scale and landscape indicators, could help improve biodiversity conservation in agricultural landscapes [43,54].

4.4. Policy Implications

Under the current situation, where it is possible to convert grassland into arable land, expanding arable land will increase the profitability of farms, especially for competitive farms, such as larger-scale dairy farms. The fact that these farms have a negative product rent at the present premium level implies that they will lose income and have no incentive to adopt AES beyond the obligatory level [55]. In contrast, extensively managed farms, such as suckler farms, are likely to profit from AES due to the lower compliance costs. They have more incentives to adopt biodiversity measures [56,57]. When the adoption rate of AES is high, the cost effectiveness tends to be higher. It can be more cost efficient to provide farm-type-specific payments rather than providing all farm types with the same payment level. This way of payments is in line with the claim of Armsworth et al. [18] that the inefficiency of the simplification of AES derives from their inability to address variation within and between farms in terms of private costs associated with providing biodiversity. Additionally, our study recommends a regulatory framework that incentivizes farmers to preserve the existing area of grassland. Under the current direct payments, crop production is far more financially attractive.

5. Conclusions

The purpose of this study is to provide policymakers with insight into the design of cost-effective AES for maintaining agroforestry systems and promoting farmland biodiversity by considering different farm types, using orchard meadows as an example. Based upon our results, the following can be concluded: 1. Higher AES payments increase orchard meadows and biodiversity scores. However, excessive payments would impede the improvement and lower the cost-effectiveness. 2. Farmers would maintain orchard meadows only with higher payments as compared to the current level, under the assumption that they can convert any grassland to arable land for maximizing their profit. However, if the conversion from grassland into arable land was not permitted, all farm types would maintain current orchard meadows. 3. Compliance costs and the adoption of AES vary considerably among farm types. Suckler farms and farms without livestock largely economically benefit from AES, while large dairy farms lose income under the flexible land-use assumption.

These findings can carry the following policy implications. First, AES can be more cost effective in targeting specific farm types and offer them the payments reflecting the compliance costs rather than paying all farm types with the same payments. Second, whether the current AES can contribute to the maintenance of orchard meadows is contingent on how far the conversion of land can be prevented. Under the current direct payments, crop production is significantly more profitable, which may encourage farmers to expand arable land. Therefore, this study recommends establishing a regulatory framework that incentivizes farmers to preserve existing grassland.

Supplementary Materials: The following supporting information can be downloaded at: https://www.mdpi.com/article/10.3390/su14095615/s1, Supplementary Materials S1: Documentation of the cluster analysis, Supplementary Materials S2: List of the parameters in the modeling.

Author Contributions: All authors contributed to the study's conception and design. Material preparation, data collection and analysis were performed by S.K., J.S. and T.N. The first draft of the manuscript was written by T.N. and all authors commented on previous versions of the manuscript. All authors have read and agreed to the published version of the manuscript.

Funding: This study was funded through the 2017–2018 Joint BiodivERsA-Belmont Forum Call on "Scenarios of Biodiversity and Ecosystem Services" (Grant ID: BiodivScen-157) and through the national funding agencies: BMBF (The German Federal Ministry of Education and Research) (Grant ID: 16LC1809A) and FNSNF (Swiss National Science Foundation) (Grant ID: 20BD21_184133/1).

Institutional Review Board Statement: Not applicable.

Informed Consent Statement: Informed consent to participate in the study was obtained for all participants in this study.

Data Availability Statement: 1: ALW Agrarinformationen Kanton Solothurn-Tierhaltung, Flächennutzung, Ressourceneffizienz, BFF, LQB. Amt Für Landwirtschaft. Canton Solothurn. Switzerland; Solothurn, 2020. 2: Erdin, D. AGRISTAT Erträge 2003–2020. 2021. 3: AGRIDEA (Swiss Association for the Development of Agriculture and Rural Areas) Deckungsbeiträge DBKAT; 2020.

Acknowledgments: The study presented in this paper was conducted under Project SALBES (Scenarios for Agricultural Landscapes' Biodiversity and Ecosystem Services). We would like to thank all those who participated in the interviews conducted by this project. We would also like to thank Seyed Ali Hosseini Yekani for his excellent support.

Conflicts of Interest: The authors declare no conflict of interest.

Appendix A. SALCA-BD Results

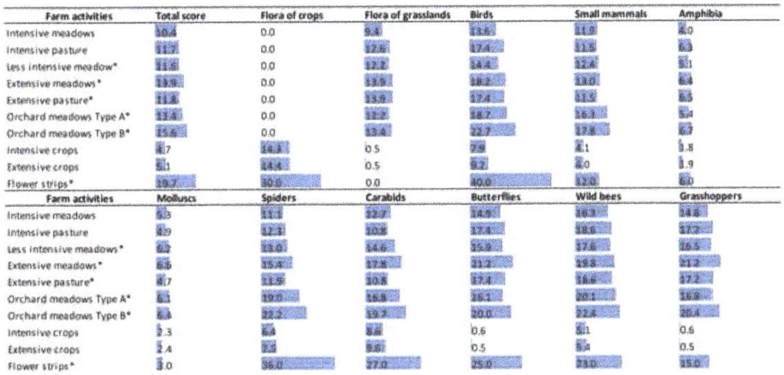

Farm activities	Total score	Flora of crops	Flora of grasslands	Birds	Small mammals	Amphibia
Intensive meadows	10.4	0.0	9.4	13.6	11.0	4.0
Intensive pasture	11.7	0.0	12.6	17.4	11.5	6.3
Less intensive meadow*	11.6	0.0	12.2	14.4	12.4	5.1
Extensive meadows*	13.9	0.0	13.9	18.2	13.0	6.4
Extensive pasture*	11.8	0.0	13.9	17.4	11.5	6.5
Orchard meadows Type A*	13.4	0.0	12.2	18.7	16.3	5.4
Orchard meadows Type B*	15.6	0.0	13.4	22.7	17.8	6.7
Intensive crops	4.7	14.3	0.5	7.9	4.1	1.8
Extensive crops	5.1	14.4	0.5	9.2	4.0	1.9
Flower strips*	19.7	30.0	0.0	40.0	12.0	6.0

Farm activities	Molluscs	Spiders	Carabids	Butterflies	Wild bees	Grasshoppers
Intensive meadows	5.3	11.1	12.7	14.9	16.3	14.8
Intensive pasture	4.9	12.3	10.8	17.4	18.6	17.2
Less intensive meadows*	6.2	13.0	14.6	15.9	17.6	16.5
Extensive meadows*	6.6	15.4	17.8	21.2	19.8	21.2
Extensive pasture*	4.7	11.5	10.8	17.4	18.6	17.2
Orchard meadows Type A*	6.1	19.0	16.8	18.1	20.1	16.8
Orchard meadows Type B*	6.4	22.2	19.7	20.0	22.4	20.4
Intensive crops	2.3	6.4	8.8	0.6	5.1	0.6
Extensive crops	2.4	7.5	9.6	0.5	5.4	0.5
Flower strips*	3.0	36.0	27.0	25.0	23.0	15.0

Figure A1. Biodiversity scores (0–50) of modelled farm activities per hectare estimated with SALCA-BD. The total score is the average of the scores of each ISG. * indicates the biodiversity measures under AES (ecological focused areas). Orchard meadow Type A corresponds to orchard meadows with commercial cherry production on less intensively managed meadows, while orchard meadows Type B orchard meadows without commercial cherry production on extensively managed meadows. The scores of intensive and extensive crops are aggregated over individual crops with the same weight. Online Resource 2 (Table S2, 6) provides the biodiversity scores of all of the modeled farm activities.

References

1. BAFU. *Biodiversität in der Schweiz: Zustand und Entwicklung*; Bundesamt Für Umwelt: Bern, Switzerland, 2017.
2. OECD. *Reforming Agricultural Subsidies to Support Biodiversity in Switzerland*; OECD: Paris, France, 2017.
3. Bundesrat Verordnung Über Die Direktzahlungen an Die Landwirtschaft. Available online: https://www.fedlex.admin.ch/eli/cc/2013/765/de (accessed on 26 February 2022).
4. OECD. *Agricultural Policy Monitoring and Evaluation 2021: Addressing the Challenges Facing Food Systems*; OECD: Paris, France, 2021. [CrossRef]
5. Mack, G.; Ritzel, C.; Jan, P. Determinants for the implementation of action, result and multi-actor-oriented agri-environment schemes in Switzerland. *Ecol. Econ.* **2020**, *176*, 106715. [CrossRef]
6. Herzog, F.; Jacot, K.; Tschumi, M.; Walter, T. The Role of Pest Management in Driving Agri-Environment Schemes in Switzerland. In *Environmental Pest Management*; John Wiley & Sons, Ltd.: Chichester, UK, 2017; pp. 385–403.
7. Van Der Meer, M.; Kay, S.; Lüscher, G.; Jeanneret, P. What evidence exists on the impact of agricultural practices in fruit orchards on biodiversity? A systematic map. *Environ. Evid.* **2020**, *9*, 2. [CrossRef]
8. Kay, S.; Crous-Duran, J.; García de Jalón, S.; Graves, A.; Palma, J.H.N.; Roces-Díaz, J.V.; Szerencsits, E.; Weibel, R.; Herzog, F. Landscape-scale modelling of agroforestry ecosystems services in Swiss orchards: A methodological approach. *Landsc. Ecol.* **2018**, *33*, 1633–1644. [CrossRef]
9. Bailey, D.; Schmidt-Entling, M.H.; Eberhart, P.; Herrmann, J.D.; Hofer, G.; Kormann, U.; Herzog, F. Effects of habitat amount and isolation on biodiversity in fragmented traditional orchards. *J. Appl. Ecol.* **2010**, *47*, 1003–1013. [CrossRef]
10. Plieninger, T.; Levers, C.; Mantel, M.; Costa, A.; Schaich, H.; Kuemmerle, T. Patterns and drivers of scattered tree loss in agricultural landscapes: Orchard meadows in Germany (1968–2009). *PLoS ONE* **2015**, *10*, e0126178. [CrossRef] [PubMed]
11. Schönhart, M.; Schauppenlehner, T.; Schmid, E.; Muhar, A. Analysing the maintenance and establishment of orchard meadows at farm and landscape levels applying a spatially explicit integrated modelling approach. *J. Environ. Plan. Manag.* **2011**, *54*, 115–143. [CrossRef]
12. Baroffio, C.A.; Richoz, P.; Fischer, S.; Kuske, S.; Linder, C.; Kehrli, P. Monitoring drosophila suzukii in Switzerland in 2012[1]. *J. Berry Res.* **2014**, *4*, 47–52. [CrossRef]
13. Knoll, V.; Ellenbroek, T.; Romeis, J.; Collatz, J. Seasonal and regional presence of hymenopteran parasitoids of drosophila in Switzerland and their ability to parasitize the invasive drosophila suzukii. *Sci. Rep.* **2017**, *7*, 40697. [CrossRef]
14. ALW. *Auswirkungen der AP 14 17 auf die Landwirtschaft im Kanton Solothurn*; Amt für Landwirtschaft: Kanton Solothurn, Switzerland, 2019.
15. Sereke, F.; Graves, A.R.; Dux, D.; Palma, J.H.N.; Herzog, F. Innovative agroecosystem goods and services: Key profitability drivers in Swiss agroforestry. *Agron. Sustain. Dev.* **2015**, *35*, 759–770. [CrossRef]
16. Huber, R.; Zabel, A.; Schleiffer, M.; Vroege, W.; Brändle, J.M.; Finger, R. Conservation costs drive enrolment in agglomeration bonus scheme. *Ecol. Econ.* **2021**, *186*, 107064. [CrossRef]

17. Hanley, N.; Acs, S.; Dallimer, M.; Gaston, K.J.; Graves, A.; Morris, J.; Armsworth, P.R. Farm-scale ecological and economic impacts of agricultural change in the uplands. *Land Use Policy* **2012**, *29*, 587–597. [CrossRef]
18. Armsworth, P.R.; Acs, S.; Dallimer, M.; Gaston, K.J.; Hanley, N.; Wilson, P. The cost of policy simplification in conservation incentive programs. *Ecol. Lett.* **2012**, *15*, 406–414. [CrossRef] [PubMed]
19. Bamière, L.; Havlík, P.; Jacquet, F.; Lherm, M.; Millet, G.; Bretagnolle, V. Farming system modelling for agri-environmental policy design: The case of a spatially non-aggregated allocation of conservation measures. *Ecol. Econ.* **2011**, *70*, 891–899. [CrossRef]
20. Wunder, S.; Brouwer, R.; Engel, S.; Ezzine-De-Blas, D.; Muradian, R.; Pascual, U.; Pinto, R. From principles to practice in paying for nature's services. *Nat. Sustain.* **2018**, *1*, 145–150. [CrossRef]
21. Ansell, D.; Freudenberger, D.; Munro, N.; Gibbons, P. The cost-effectiveness of agri-environment schemes for biodiversity conservation: A quantitative review. *Agric. Ecosyst. Environ.* **2016**, *225*, 184–191. [CrossRef]
22. Wuepper, D.; Huber, R. Comparing effectiveness and return on investment of action and results-based agri-environmental payments in Switzerland. *Am. J. Agric. Econ.* **2021**. [CrossRef]
23. Wunder, S.; Brouwer, R.; Engel, S.; Ezzine-de-Blas, D.; Muradian, R.; Pascual, U.; Pinto, R. Reply to: In defence of simplified PES designs. *Nat. Sustain.* **2020**, *3*, 428–429. [CrossRef]
24. ALW. *Agrarinformationen Kanton Solothurn—Tierhaltung, Flächennutzung, Ressourceneffizienz, BFF, LQB*; Amt für Landwirtschaft: Kanton Solothurn, Switzerland, 2020.
25. FSO. Farm Structure Survey. *Federal Statistical Office*. Available online: https://www.bfs.admin.ch/bfs/en/home/statistics/agriculture-forestry/farming.html/ (accessed on 26 February 2022).
26. Jeanneret, P.; Baumgartner, D.U.; Freiermuth Knuchel, R.; Koch, B.; Gaillard, G. An expert system for integrating biodiversity into agricultural life-cycle assessment. *Ecol. Indic.* **2014**, *46*, 224–231. [CrossRef]
27. Lüscher, G.; Nemecek, T.; Arndorfer, M.; Balázs, K.; Dennis, P.; Fjellstad, W.; Friedel, J.K.; Gaillard, G.; Herzog, F.; Sarthou, J.P.; et al. Biodiversity assessment in LCA: A validation at field and farm scale in eight European regions. *Int. J. Life Cycle Assess.* **2017**, *22*, 1483–1492. [CrossRef]
28. Bholowalia, P.; Kumar, A. EBK-means: A clustering technique based on elbow method and k-means in WSN. *Int. J. Comput. Appl.* **2014**, *105*, 975–8887.
29. Andersen, E.; Elbersen, B.; Godeschalk, F.; Verhoog, D. Farm management indicators and farm typologies as a basis for assessments in a changing policy environment. *J. Environ. Manag.* **2007**, *82*, 353–362. [CrossRef] [PubMed]
30. Schuler, J.; Toorop, R.A.; Willaume, M.; Vermue, A.; Schläfke, N.; Uthes, S.; Zander, P.; Rossing, W. Assessing climate change impacts and adaptation options for farm performance using bio-economic models in Southwestern France. *Sustainability* **2020**, *12*, 7528. [CrossRef]
31. Uthes, S.; Piorr, A.; Zander, P.; Bieńkowski, J.; Ungaro, F.; Dalgaard, T.; Stolze, M.; Moschitz, H.; Schader, C.; Happe, K.; et al. Regional impacts of abolishing direct payments: An integrated analysis in four European regions. *Agric. Syst.* **2011**, *104*, 110–121. [CrossRef]
32. Kaiser, H.M.; Messer, K.D. *Mathematical Programming for Agricultural, Environmental, and Resource Economics*; Wiley: Hoboken, NJ, USA, 2011; ISBN 9780470599365.
33. Mason, A.J. *OpenSolver—An Open Source Add-in to Solve Linear and Integer Programmes in Excel*; Springer: Berlin/Heidelberg, Germany, 2012; pp. 401–406. [CrossRef]
34. Böcker, T.; Möhring, N.; Finger, R. Herbicide free agriculture? A bio-economic modelling application to Swiss wheat production. *Agric. Syst.* **2019**, *173*, 378–392. [CrossRef]
35. Agristat, 2003-2020. In *Statistische Erhebungen und Schätzungen über Landwirtschaft und Ernährung (SES)*; Schweizer Bauernverband: Brugg, Switzerland, 2021.
36. AGRIDEA (Swiss Association for the Development of Agriculture and Rural Areas). *Deckungsbeiträge DBKAT*; AGRIDEA: Lausanne, Switzerland, 2020.
37. Agroscope. 8/Düngung von Ackerkulturen: Grundlagen für Die Düngung Landwirtschaftlicher Kulturen in der Schweiz. In *Grundlagen für die Düngung Landwirtschaftlicher Kulturen in der Schweiz (GRUD)*; Agroscope: Bern, Switzerland, 2017.
38. Reidsma, P.; Janssen, S.; Jansen, J.; van Ittersum, M.K. On the development and use of farm models for policy impact assessment in the European union—A review. *Agric. Syst.* **2018**, *159*, 111–125. [CrossRef]
39. Janssen, S.; Louhichi, K.; Kanellopoulos, A.; Zander, P.; Flichman, G.; Hengsdijk, H.; Meuter, E.; Andersen, E.; Belhouchette, H.; Blanco, M.; et al. A generic bio-economic farm model for environmental and economic assessment of agricultural systems. *Environ. Manag.* **2010**, *46*, 862–877. [CrossRef]
40. FEEDBASE (The Swiss Feed Database). Available online: https://www.feedbase.ch/ (accessed on 20 February 2022).
41. DLG-Akademie. *DLG Futterwerttabellen Wiederkäuer*; DLG-Verlag Frankfurt: Hessen, Germany, 1997.
42. Giannitsopoulos, M.L.; Graves, A.R.; Burgess, P.J.; Crous-Duran, J.; Moreno, G.; Herzog, F.; Palma, J.H.N.; Kay, S.; García de Jalón, S. Whole system valuation of arable, agroforestry and tree-only systems at three case study sites in Europe. *J. Clean. Prod.* **2020**, *269*, 122283. [CrossRef]
43. Schönhart, M.; Schauppenlehner, T.; Schmid, E. Integrated Bio-Economic Farm Modeling for Biodiversity Assessment at Landscape Level. In *Bio-Economic Models Applied to Agricultural Systems*; Flichman, G., Ed.; Springer: Dordrecht, The Netherlands, 2011; ISBN 978-94-007-1902-6.

44. Drechsler, M. *Ecological-Economic Modelling for Biodiversity Conservation*; Cambridge University Press: Cambridge, UK, 2020. [CrossRef]
45. Schweizer Bauernverband Preise Pflanzenbau. Available online: https://www.sbv-usp.ch/de/preise/pflanzenbau/ (accessed on 22 May 2021).
46. Bundesrat Verordnung Über Einzelkulturbeiträge Im Pflanzenbau Und Die Zulage Für Getreide. Available online: https://www.fedlex.admin.ch/eli/cc/2013/873/de (accessed on 26 February 2022).
47. Bethwell, C.; Burkhard, B.; Daedlow, K.; Sattler, C.; Reckling, M.; Zander, P. Towards an enhanced indication of provisioning ecosystem services in agro-ecosystems. *Environ. Monit. Assess.* **2021**, *193*, 269. [CrossRef]
48. Wrbka, T.; Erb, K.H.; Schulz, N.B.; Peterseil, J.; Hahn, C.; Haberl, H. Linking pattern and process in cultural landscapes. An empirical study based on spatially explicit indicators. *Land Use Policy* **2004**, *21*, 289–306. [CrossRef]
49. Wang, X.; Dietrich, J.P.; Lotze-Campen, H.; Biewald, A.; Stevanović, M.; Bodirsky, B.L.; Brümmer, B.; Popp, A. Beyond land-use intensity: Assessing future global crop productivity growth under different socioeconomic pathways. *Technol. Forecast. Soc. Chang.* **2020**, *160*, 120208. [CrossRef]
50. Uthes, S.; Matzdorf, B. Studies on agri-environmental measures: A survey of the literature. *Environ. Manag.* **2012**, *51*, 251–266. [CrossRef] [PubMed]
51. Kuhfuss, L.; Begg, G.; Flanigan, S.; Hawes, C.; Piras, S. Should Agri-Environmental Schemes Aim at Coordinating Farmers' Pro-Environmental Practices? A Review of the Literature. In Proceedings of the 172nd EAAE Seminar, Brussels, Belgium, 28–29 May 2019.
52. Manning, P.; Van Der Plas, F.; Soliveres, S.; Allan, E.; Maestre, F.T.; Mace, G.; Whittingham, M.J.; Fischer, M. Redefining ecosystem multifunctionality. *Nat. Ecol. Evol.* **2018**, *2*, 427–436. [CrossRef]
53. Fastré, C.; van Zeist, W.J.; Watson, J.E.M.; Visconti, P. Integrated spatial planning for biodiversity conservation and food production. *One Earth* **2021**, *4*, 1635–1644. [CrossRef]
54. Chopin, P.; Bergkvist, G.; Hossard, L. Modelling biodiversity change in agricultural landscape scenarios—A review and prospects for future research. *Biol. Conserv.* **2019**, *235*, 1–17. [CrossRef]
55. Mennig, P.; Sauer, J. The impact of agri-environment schemes on farm productivity: A DID-matching approach. *Eur. Rev. Agric. Econ.* **2020**, *47*, 1045–1093. [CrossRef]
56. Overmars, K.P.; Helming, J.; van Zeijts, H.; Jansson, T.; Terluin, I. A modelling approach for the assessment of the effects of common agricultural policy measures on farmland biodiversity in the EU27. *J. Environ. Manag.* **2013**, *126*, 132–141. [CrossRef]
57. Sinaga, K.P.; Yang, M.S. Unsupervised K-Means Clustering Algorithm. *IEEE Access* **2020**, *8*, 80716–80727. [CrossRef]

MDPI
St. Alban-Anlage 66
4052 Basel
Switzerland
Tel. +41 61 683 77 34
Fax +41 61 302 89 18
www.mdpi.com

Sustainability Editorial Office
E-mail: sustainability@mdpi.com
www.mdpi.com/journal/sustainability

www.ingramcontent.com/pod-product-compliance
Lightning Source LLC
LaVergne TN
LVHW071248170726
843515LV00006BA/1356

* 9 7 8 3 0 3 6 5 5 5 7 7 5 *